International Series in Operations Research & Management Science

Founding Editor
Frederick S. Hillier

Volume 367

The book series **International Series in Operations Research and Management Science** encompasses the various areas of operations research and management science. Both theoretical and applied books are included. It describes current advances anywhere in the world that are at the cutting edge of the field. The series is aimed especially at researchers, advanced graduate students, and sophisticated practitioners.

The series features three types of books:

- Advanced expository books that extend and unify our understanding of particular areas.
- Research monographs that make substantial contributions to knowledge.
- Handbooks that define the new state of the art in particular areas. Each handbook will be edited by a leading authority in the area who will organize a team of experts on various aspects of the topic to write individual chapters. A handbook may emphasize expository surveys or completely new advances (either research or applications) or a combination of both.

The series emphasizes the following four areas:

Mathematical Programming: Including linear programming, integer programming, nonlinear programming, interior point methods, game theory, network optimization models, combinatorics, equilibrium programming, complementarity theory, multiobjective optimization, dynamic programming, stochastic programming, complexity theory, etc.

Applied Probability: Including queuing theory, simulation, renewal theory, Brownian motion and diffusion processes, decision analysis, Markov decision processes, reliability theory, forecasting, other stochastic processes motivated by applications, etc.

Production and Operations Management: Including inventory theory, production scheduling, capacity planning, facility location, supply chain management, distribution systems, materials requirements planning, just-in-time systems, flexible manufacturing systems, design of production lines, logistical planning, strategic issues, etc.

Applications of Operations Research and Management Science: Including telecommunications, health care, capital budgeting and finance, economics, marketing, public policy, military operations research, humanitarian relief and disaster mitigation, service operations, transportation systems, etc.

This book series is indexed in Scopus.

Mohammad Zoynul Abedin • Wang Yong
Editors

Machine Learning Technologies on Energy Economics and Finance

Energy and Sustainable Analytics, Volume 1

Springer

Editors
Mohammad Zoynul Abedin
School of Management
Swansea University
Swansea, UK

Wang Yong
Dongbei University of Finance
and Economics
Dalian, China

ISSN 0884-8289 ISSN 2214-7934 (electronic)
International Series in Operations Research & Management Science
ISBN 978-3-031-94861-9 ISBN 978-3-031-94862-6 (eBook)
https://doi.org/10.1007/978-3-031-94862-6

This Springer imprint is published by the registered company Springer Nature Switzerland AG
The registered company address is: Gewerbestrasse 11, 6330 Cham, Switzerland

Preface

The global energy landscape is undergoing a transformative shift, driven by the twin imperatives of economic growth and sustainability. The rapidly evolving energy sector needs appropriate implementation of Machine Learning (ML) along with data analytics and energy economics at present. Price predictions together with demand-side management and rational decision-making processes for complex energy systems result from these data-driven approaches because of market volatility and weather pattern modifications and technology advancements. This book "Machine Learning Technologies on Energy Economics and Finance—Energy and Sustainable Analytics" aims to solve this essential need through combined advanced ML applications with energy economics and finance which establish a whole framework to enhance global energy systems.

This book demonstrates its purpose to address vital issues in the energy field because of unstable fossil fuel costs along with funding issues in renewable endeavors and essential Sustainable Development Goals (SDGs) requirements. Energy markets exhibit complex non-linear patterns which traditional economic and financial models usually fail to understand properly. The text enables the connection between machine learning and deep learning technologies to establish predictive models which support better energy domain decision-making. The integration of explainable AI (XAI) in predictive models guarantees transparent and interpretable energy-related forecasts which strengthens belief in artificial intelligence solutions for making decisions.

The primary audience of this book consists of researchers, academicians, business professionals, policymakers, data scientists, engineers, and students who want to investigate innovative ML approaches in energy economics and finance. The book connects theoretical expertise with operational expertise to deliver meaningful knowledge for those who want to use AI and ML technologies to develop energy economics and finance.

This book distinguishes itself from others by devoting its focus to real-time investigation outcomes with both industry expertise and state-of-the-art ML implementations for energy systems. This work exceeds traditional energy economics books through its incorporation of hands-on coding experiences alongside real-world case studies and predictive analytics for practical insights. Advanced ML techniques represent a substantial part of this book's content alongside explainable systems, practical applications, sustainable energy analytics, and the integration of machine learning with various disciplines which include energy economics, finance, and sustainability.

The book comprises thirteen chapters that group important thematic content areas. The initial part introduces ML for Global Energy Analysis and Forecasting through studies which analyze SDGs through clustering and trend prediction and examine explainable AI for natural gas consumption and develop methods for forecasting energy prices and predicting efficient gasoline spot prices. The second portion investigates Energy Economics and Financial Modeling through its exploration of energy finance and analysis of crude oil price forecasting together with sustainable energy applications of ML. Renewable Energy and Sustainability Analytics comprises three components which include energy transition assessment of emerging economies and CO2 emissions and economic growth as well as blending ensemble learning for energy consumption and biogas production analysis and ML strategies for renewable energy and energy transition case studies.

Section 1: Machine Learning for Energy Forecasting and Market Analysis

Energy forecasting and market analysis are crucial in making informed decisions in the energy sector. In the energy market context, ML transforms the forecasting skill of people when compared to traditional statistics in terms of optimization and modeling. The category consists of chapters that implement ML techniques to anticipate energy consumption along with crude oil market prices and natural gas utilization through explainable approaches for transparent modeling.

Chapter 1 develops a machine learning system which studies worldwide energy behavior and performs country SDG achievement clustering in addition to projecting essential energy measurements. Chapter 3 demonstrates how Categorical Boosting enables better natural gas consumption prediction accuracy through advanced ML modeling while demonstrating clear model decision logic. Chapter 4 describes the time-series modeling of crude oil price forecasting through analysis of ARIMA, SARIMA, and VAR alongside statistical approaches. Chapter 6 compares ML algorithms against traditional forecasting techniques by demonstrating superior performance in predicting crude oil and solar prices and electricity and natural gas values. Chapter 4 details an ensemble learning system optimized via hyperparameter optimization for gasoline spot price prediction as it demonstrates the value of ML in energy price modeling.

Collectively, these chapters showcase the role of ML in enhancing predictive capabilities, optimizing resource allocation, and improving decision-making in energy markets. By leveraging advanced algorithms and XAI tools, these studies offer more interpretable, efficient, and scalable forecasting solutions.

Section 2: Renewable Energy Transition, Sustainability, and Economic Impact

The world faces an extreme challenge to move toward renewable energy systems instead of maintaining dependence on fossil fuels for sustainable operations. Energy technologies built from renewable sources serve three essential functions: they decrease emissions of greenhouses gases, maintain stable energy costs, and protect future energy availability. The adoption of renewable energy depends on economic elements together with policy frameworks, regulatory conditions, and technology options and their economic and environmental impacts.

Financial limitations for developing renewable energy in Bangladesh receive detailed analysis in Chap. 2 to identify three alternative funding approaches through green bonds together with public–private partnerships and crowdfunding for breaking through investment obstacles. Chapter 5 examines developing sunbelt countries through an energy transition comparison which reveals policy strategies needed to reach sustainability goals. Chapter 11 explores ML and deep learning (DL) techniques used in renewable energy applications through a detailed analysis of advantages and drawbacks while introducing effective solutions to optimize renewable energy forecasting efficiency. Chapter 13 explores how China and India are handling their energy transition into sustainable models by analyzing both positive and negative aspects on their macroeconomies. The chapters demonstrate how implementing policies with financial strategies and ML solutions speeds up the transition toward improved energy systems that are clean and efficient.

The section demonstrates how policy frameworks intersect with financial innovations and AI-driven solutions to resolve obstacles in renewable energy implementation. The implementation of ML and XAI systems within sustainability analytics gives researchers data-based knowledge about energy transition methods as well as their sustained economic effects and environmental impact.

Section 3: Environmental and Financial Impact of Energy Consumption

The consumption of energy manifests effective outcomes for environment sustainability alongside financial market systems. The development of sustainable policies demands proper knowledge about the relationships between energy consumption and CO2 emissions together with economic stability and financial choices. The combination of ML and economic modeling works to study CO2 emissions together with financial stability along with primary energy production impacts.

The analysis in Chap. 7 uses an Explainable AI-driven model to study how CO2 emissions relate to economic growth through macroeconomic indicator assessments of SDGs. The conventional econometric models receive new insights through deep learning frameworks that include GRU, LSTM, and Bi-LSTM which analyze economic-environmental interactions differently. The authors introduce a primary energy consumption forecasting solution in Chap. 8 using ML techniques while showing its impact on national security alongside environmental sustainability and economic development. The analysis of biogas production through explainable ML models appears in Chap. 9 that delivers sustainable energy alternatives to waste-to-energy investment barriers. The application of ML within sustainable energy finance becomes the subject of Chap. 10 which displays how AI-powered financial forecasting tools boost investment choices for petroleum resources as well as natural gas and renewable energy.

This segment presents examples of how energy consumption optimization and environmental reduction combined with financial decision enhancement are made possible by ML and AI technologies in the energy industry. These studies turn sustainability analytics into an AI-driven comprehensive approach to sustainable development in decision-making toward securing a financial and environmental balance.

One of the primary challenges in developing this book was the integration of diverse ML methodologies while ensuring their applicability in energy economics and finance. Energy systems are influenced by dynamic factors such as global markets, geopolitical conflicts, climate policies, and technological disruptions. We have proven by utilizing sophisticated ML techniques that predictions concerning energy markets can be scientifically based yet both exact and simple to comprehend. The book highlights the significance of XAI for energy decision-making and solves AI interpretability issues by utilizing SHAP and ELI5 interpretability tools and others. The authors of this research book collected extensive information through collaborations with professionals from three fields: ML, Energy Economics, and Data Analytics. We profoundly thank the authors together with universities and research institutions which provide us with their endless backing and enriching knowledge. We extend our highest admiration to our families for staying supportive during this entire journey thus enabling us to complete our work. AI and ML together with Energy Economics will gain more importance for developing sustainable energy market strategies and policies since the energy sector continues its development. The publication provides researchers alongside professionals and policymakers with directions to implement AI-driven choice-based systems in resolving worldwide energy problems. The studied area requires sustained academic effort to create an operational system which ensures both economic sustainability and energy infrastructure effectiveness. Through this invitation we guide readers to investigate ML Technologies in Energy Economics and Finance to discover revolutionary opportunities in Energy and Sustainable Analytics.

Swansea, UK Mohammad Zoynul Abedin
Dalian, China Wang Yong

Contents

Analyzing Global Energy Patterns: Clustering Countries and Predicting Trends Toward Achieving Sustainable Development Goals

Mahmudul Hasan, Nusrat Afrin Shilpa, Ashrafuzzaman Sohag, Md. Mahedi Hassan, and Md. Jahangir Alam Siddikee

1 Introduction

Sustainability has become a prominent planning concept since its inception in the realm of economics and ecological thought (Nguyen et al., 2023). Described as the endeavor to satisfy current needs without comprising the ability of future generations to do the same, sustainability is multifaceted. The notion of energy sustainability essentially applies the fundamental principles of sustainability to the realm of energy (Khan et al., 2022). However, the concept of energy sustainability is intricate and multifaceted. It encompasses ensuring the delivery of energy services in a sustainable manner, thereby necessitating the provision of energy services that are adequate, affordable, environmentally friendly, and socially acceptable

M. Hasan (✉)
Department of Computer Science and Engineering, Hajee Mohammad Danesh Science and Technology University, Dinajpur, Bangladesh

School of Information Technology, Deakin University, Geelong, VIC, Australia

N. A. Shilpa
Department of Business Administration, Ishakha International University, Kishoreganj, Bangladesh

A. Sohag
Master's in International Management and Information Systems, South Westphalia University of Applied Sciences, Iserlohn, Germany

M. M. Hassan
Department of Computer Science and Engineering, World University of Bangladesh, Dhaka, Bangladesh

M. J. A. Siddikee
Department of Finance and Banking, Hajee Mohammad Danesh Science and Technology University, Dinajpur, Bangladesh

M. Z. Abedin, W. Yong (eds.), *Machine Learning Technologies on Energy Economics and Finance*, International Series in Operations Research & Management Science 367, https://doi.org/10.1007/978-3-031-94862-6_1

for all individuals both presently and in the future (Achuo et al., 2022). Energy sustainability mandates the sustainable utilization of energy within energy systems, encompasses various processes from sourcing energy to its conversion into usable forms, transportation, storage, and eventual consumption (Jiang et al., 2020). Energy is primarily utilized to deliver energy services like heating, transportation, lighting, and communication. Consequently, the promotion of energy sustainability hinges on the establishment and utilization of sustainable energy systems and practices (Zaharia et al., 2019). Energy sustainability is a vital component of the broader concept of sustainability. Numerous nations, regions, and urban areas are striving toward sustainability, promoting a reassessment of their current unsustainable energy practices (Qudrat-Ullah & Nevo, 2021). Simply put, sustainability is characterized by environmental, social, and economic dimensions, all of which are intricately linked to energy. Energy, being essential for various activities, plays a crucial role in achieving sustainability. The pursuit of energy sustainability is impeded by significant environmental, social, and economic obstacles, including climate change, escalating emissions, rapid resource depletion, affordability concerns, and social disparities (Bibi et al., 2021). Addressing these challenges effectively is imperative for the attainment of energy sustainability, yet it remains a daunting and intricate task. Factors like artificial energy prices influenced by taxes, incentives, economic, and political fluctuations further complicate the landscape of energy sustainability and consumption (Ozcan et al., 2019). The sustainable development goals (SDGs), which were endorsed by the United Nations General Assembly (UNGA) in 2015, present a robust framework for fostering international collaboration aimed at realizing a sustainable future for the globe. In 17 goals of SDGs, along with their 169 targets encapsulated in “Agenda 2030,” lay out a trajectory toward eradicating extreme poverty, combating inequality and injustice, and safeguarding the environment. The success of agenda 2030 hinges significantly on sustainable energy. SDG 7, the global energy objective, comprises three pivotal targets: ensuring accessible, dependable, and widespread modern energy services, substantially boosting the proportion of renewable energy in the global energy mix, and doubling the worldwide rate of enhancement in energy efficiency. The diverse targets within SDG 7 play a role in advancing other SDG objectives, a subject that has gained heightened research attention recently (Allen et al., 2016). Previous examinations of prospective energy trajectories demonstrate the technical feasibility of achieving enhanced energy accessibility, air quality, and energy reliability concurrently, while averting hazardous climate alterations. Indeed, various alternative technologies and strategies have been identified as capable of attaining these aims (Asmelash et al., 2020). During the twentieth century, mankind extended its endeavors by harnessing the power of scientific advancements and technological innovations. The outcomes are evident in the swift escalation of the global populace and the attainment of elevated living standards in developed nations. As societies embracing mass production and consumption patterns become customary in these regions, the excessive utilization of resources and energy has evolved into a persistent issue. The repercussions of these patterns are not confined solely to these nations; they have also exerted notable impacts on developing countries. The worldwide population has grown reliant on a plentiful energy supply. The necessity for a substantial energy reservoir has consequently led to a heavy reliance on fossil fuels, notorious for their emission of carbon dioxide. The looming threat of global warming triggered by the accumulation of carbon dioxide in the atmosphere has hence become a pressing subject of contention (Helten, 2013). The primary trigger of the fluctuation in supply and demand can be directly ascribed to the system of segregation. In the year 2019, China represented 24% of the total global energy consumption, positioning itself as the foremost importer of crude oil and natural gas worldwide (Guo et al., 2021). Nevertheless, stringent measures implemented in early 2020 brought numerous sectors to a near halt. Furthermore, various European nations, the United States, and India jointly responsible for almost one-third of the global energy consumption have enforced a series of isolation protocols (Verhoef et al., 2023). Within this context, the global energy demand has encountered an unparalleled downturn. Consequently, the outlook of the market has turned increasingly delicate. While a gradual easing of restrictions and a slow resurgence from isolation appears inevitable in due course, the substantial setback in economic activities may endure permanently. All these elements contribute to a crucial argument: the ambiguity stemming from the crisis and the necessity for a seamless evolution of the energy sector (Biazzi, 2022). The proliferation of renewable energy sources has initiated a global energy revolution with significant geopolitical ramifications. The emergence of a novel era in energy will revolutionize the interactions among nations and societies, ushering in a new era of energy security, independence, and prosperity for humanity. In contrast to fossil fuels, which are predominantly found in specific geographical areas, renewable energy sources (RESs) can be harnessed in any country. Due to its ability to be generated in various locations, renewable energy has the capacity to reshape the dynamics of energy trading (Vagiona & Kamilakis, 2018). Earlier, Vagiona and Kamilakis (2018) propose an integrated approach for the assessment and prioritization of appropriate sites for the establishment of sustainable offshore wind farms. Through the utilization of a combination of geographic information systems and multi-criteria decision-making techniques, the generated outcomes guarantee the spatial sustainability of these wind farms. Some researchers investigate the idea of hydrogen cities through the suggestion of hydrogen generation within urban areas. By employing Geographic Information System (GIS) tools, the monthly capacity for solar hydrogen production in urban regions of Mexico is evaluated (Juárez-Casildo et al., 2022). The study's findings reveal that the total annual hydrogen demand of the country could be met by the production from specific urban areas for just 1 month at a relatively economical expense. Furthermore, additional findings support earlier assertions regarding the minimal water demands and infrastructure footprint associated with metropolitan production. The list of objectives of this research is below:

The technical contributions of this chapter are as follows.

- We design an ML-driven framework to find the pattern and predicting trends toward achieving sustainable development goals.

- We employ unsupervised ML method to cluster the countries based on electricity access and renewable consumption for SDG.
- We design an ensemble ML model to predict the SDGs that outperforms existing ML models.

The structure of the remaining sections of this chapter is outlined as follows. The related works are outlined in Sect. 2. Section 3 is dedicated to presenting our proposed methodology and the experimental setup. We detail the approach we have taken to address the research problem, including the methods, techniques, and tools employed in our study. Within Sect. 4, we present the outcomes of our experiments. The chapter concludes in Sect. 5 with a summary of our findings and their significance. Additionally, we outline avenues for future research and development in this domain, emphasizing the potential directions for further exploration and enhancement.

2 Literature Review

The set of SDGs outlined by the United Nations serves as a strategic plan which aimed at enhancing global sustainability by the target year of 2030. These goals encompass various objectives such as combatting climate change, attaining gender parity, ensuring universal access to quality education, and promoting quality healthcare, among others, as part of the 17 specified targets (Sachs et al., 2019). As the global community progresses toward achieving the aspirations of the 2030 agenda, there is a growing interest among governments and societies in exploring strategies for attaining sustainable development. The advancement of technology has brought about significant transformations in our lifestyles and business practices (Stafford-Smith et al., 2017). Numerous nations have been promoting the sharing of building energy data for the development of innovative models, such as building energy benchmarks, across various building typologies. These initiatives are designed to stimulate investments in energy efficiency and mitigate building energy usage. The methodologies of benchmarking can be categorized as white-box, black-box, or gray-box, based on the classification of models utilized to forecast building energy efficiency (Papadopoulos et al., 2018). Information on building energy consumption and its characteristics is crucial for conducting benchmarking procedures. Nevertheless, the current absence of data presents a significant obstacle in this context. In order to tackle this issue, Juárez-Casildo et al. (2022) focused on exploring the utilization of machine learning to predict the energy use intensities of bank branches situated at Brazil (Veiga et al., 2021). The methodology applied in this research encompassed the acquisition of data pertaining to the typology of bank branches and the archetype model along with its fixed and variable inputs were identified to produce 48,000 samples that underwent simulation using EnergyPlus software. The result of this study revealed that the lighting power density and the weather parameter emerged as the most impactful variables in the estimation of energy consumption in bank branches. Previously, Veiga et al. (2021) examined the progression of the global energy consumption framework through the utilization of an evolutionary tree model (Hu et al., 2018). Initially, a total of 144 countries and regions were segmented into four distinct categories utilizing the k-means clustering technique. Nations and regions falling within the same category typically exhibit comparable evolutionary trajectories. Furthermore, nations classified as type IV, predominantly encompassing developed nations, showcase the most varied energy consumption frameworks. Countries can be positioned within the evolutionary tree of the global energy consumption framework, and such placements can serve as a foundation for elevating a country's energy consumption framework based on analogous countries with greater diversity. The analysis of smart meter data contributes significantly to enhancing the planning and operations of power system. This research endeavor of Tang et al. (2022) made investigation on identifying the determinants of residential energy consumption behaviors through a socioeconomic lens, utilizing machine learning techniques on consumption and demographic data. The study delves into the examination of real-world smart meter data, extracting load patterns through robust clustering methods. The correlation between consumer's load patterns and specific socioeconomic indicators was delineated through the application of machine learning algorithms. The proposed analytical framework, integrating feature selection and machine learning techniques, demonstrated superior effectiveness compared to XGBoost and traditional neural network models in capturing the relationship between load patterns and socioeconomic indicators. It is also noted that with the rise in population, urbanization, and standards of living, unprocessed wet waste presents a notable obstacle and offers unexplored possibilities. So, Zhu et al. (2023) concerned on this issue and presented an innovative framework that mobilized advance machine learning methodologies such as deep neural networks, random forest (RF), and extreme gradient boosting) with dual-objective optimization. This strategy facilitates a comparative evaluation of the solid byproducts generated from HTC and pyrolysis, with a focus on their Carbon Stability Index (CSI) and Return on Energy Investment (REI) metric (Zhu et al., 2023). The evaluation allowed for customizing char production for specific uses, resulting in optimal conditions for both high energy efficiency and stable carbon storage. A case study involving wet food waste revealed a substantial improvement from 4.83 to 14.43 in REI and an elevation from 47.4 to 57.98 in CSI when compared to traditional HTC methods. Lawrence et al. (2013) conducted a study on the worldwide distribution of energy consumption per capita. Their research revealed a decline in the Gini coefficient, G, from 0.66 in 1980 to 0.55 in 2010, indicating a reduction in inequality. The distribution of energy consumption per capita globally in 2010 closely resembled an exponential distribution, with a G value of 0.5, suggesting that the top third of the global population utilizes two-thirds of the energy produced. Chen and Chen (2011) undertook an analysis of the global energy landscape through a systematic input-output simulation which identified the United States as the largest importer of embodied energy but faced a deficit in energy reception. Fujimori et al. (2016) executed a hindcasting a global energy model using an integrated assessment model.

Their findings indicated a high level of reproducibility in global aggregated primary energy, with wealthier nations displaying greater reproducibility compared to lower-income nations. Through an analysis of entropy Information, Zhang et al. (2011) assessed the transformation of China's energy consumption pattern, highlighting a gradual enhancement in Chinese energy utilization. Nevertheless, past studies have neglected to explore the interconnections among energy consumption patterns in various nations, indicating a necessity for further investigation into the evolution of the global energy consumption structure.

3 Methodology

3.1 Approach Overview

The goal of this project is to explore the dataset and derive interesting insights from it. Throughout the work on it, I decided to focus on (1) clustering the countries and (2) generating predictions for the time until 2030, as 2030 is the target year for completion of many of the SDG targets. We have used some ML algorithm and an ensemble algorithm for better prediction and finally evaluate the performance of the models by some performance measure techniques.

3.2 Machine Learning Algorithms

3.2.1 K-Means Clustering

K-Means clustering is an unsupervised ML algorithm utilized to partition a dataset into K distinct clusters based on feature similarity. It is conducted by iteratively assigning data points to the nearest cluster centroid, recalculating the centroids until a stable solution is reached (Yang et al., 2024). This algorithm clusters objects such that those within the same cluster share similar characteristics, while objects in different clusters exhibit distinct characteristics. The function is as follows: minimize $J == C\|x - \|^2$, where k is the number of clusters, C_i is the set of points in cluster i, x is a data point, and μ_i is the centroid of cluster i.

3.2.2 Linear Regression (LR)

LR is a statistical approach which attempts to determine the relationship between two variables by fitting a linear equation to observed data. In LR, one variable is an explanatory variable and the other is a dependent variable (Hasan et al., 2024a). The basic form of the linear regression equation is Y = +X+, where Y is a dependent variable (the variable being predicted or explained), X an independent variable (the predictor variable) : intercept (the value of Y when X is 0), : the slope (the change in Y for a one-unit change in X) : the error term (the difference between the observed and predicted values of Y). RF is a widely used algorithm for ensemble learning that utilizes multiple decision trees (DTs) for classification and regression tasks. The algorithm builds a forest of numerous decision trees, each of which is trained employing different samples of training data and the input attributes. The average predictions of all the trees are considered (Mamun et al., 2024).

3.2.3 Light Gradient Boosting (LGB)

The LGB machine regressor is a breakthrough tree-based ensemble learning method which helps to overcome the efficiency and scalability limitations of XGBoost in massive dataset and high-dimensional input feature (Sajid et al., 2023). LGB is an updated gradient boosting framework that utilizes the prediction results from several DTs to make the final prediction. The LGB algorithm is comprised of two main approaches: gradient-based one-side sampling (GOSS) and exclusive feature bundling (EFB).

3.2.4 Decision Tree

A DT algorithm is a classic supervised learning model used for both classification and regression. DT demonstrates a diagram depicts like as tree. It makes sequential decisions based on attribute tests to classify data into different classes. DT is commonly used for decision-making and classification tasks in data science (Hasan et al., 2024b).

3.2.5 AdaBoost

AdaBoost (Adaptive Boosting) combines several weaker prediction algorithms into a robust regression model (Hasan et al., 2023a). Initially, equal weights are assigned to all data points. The model then processes the data, identifying misclassified instances. The weights of these misclassified points are increased to improve the model's accuracy. The final prediction for binary classification is mathematically represented as follows:

3.2.6 CatBoost

CatBoost is one of the newest boosting ensemble machine learning models. CatBoost utilizes ordered boosting and is an efficient enhancement of gradient boosting in addressing the issue of target leakage (Prokhorenkova et al., 2018). This is even effective in handling small datasets and categorical features.

3.2.7 B_DRRC

The B_DRRC model is a thoughtfully crafted blending ensemble that brings together four well-known algorithms—DT, RF, Ridge, and CatBoost—each contributing its unique strengths to improve predictions. Think of it as a team where each member has different expertise: The DT offers a straightforward and easy-to-understand structure, while RF adds reliability by averaging multiple trees to avoid overfitting. Ridge acts as a stabilizer, handling tricky correlations in the data, and CatBoost shines by efficiently processing complex categorical features (Hasan et al., 2023c).

By blending these models, the B_DRRC approach ensures that no single model's weaknesses hold back performance. It is like getting different perspectives to make the best possible decision, as one model might spot patterns that another misses, leading to more balanced and accurate predictions. This combination helps the B_DRRC model handle challenging datasets where simpler models might struggle, making it a versatile and powerful tool for solving real-world problems with more confidence and less bias.

3.3 *Performance Measure Metrics*

3.3.1 Mean Absolute Error (MAE)

MAE characterizes the alteration among the original and predictable values and is mined as the dataset's total alteration mean (Hasan et al., 2023d).

3.3.2 Mean Square Error (MSE)

The MSE is calculated to ensure that the original and decrypted images are in variations or not.

3.3.3 Root Mean Square Error (RMSE)

RMSE is defined as the measure of the differences between values that are predicted by a model and values that are observed.

3.3.4 Mean Absolute Percentage Error (MAPE)

MAPE is a widely used metric for assessing forecast accuracy. It is calculated as the average of absolute percentage errors (APEs). MAPE represents the actual and forecasted values at a given data point, respectively.

3.3.5 R-Squared

R-Squared is a statistical measure that represents the proportion of the variance for a dependent variable that is explained by an independent variable or variables in a regression model (Hasan et al., 2023b). It is often used to assess the goodness of fit of a model (0,1), indicating how well the model's predictions match the actual data.

4 Result Analysis

4.1 Descriptive Analysis

Table 1 deals with the descriptive statistics of variables where the access to electricity is 78.93% in average and the Skewness shows −1.2058, suggesting a

Table 1 Statistical characteristics of the variables of the SDG indicators

Variables	Mean	Std.	Variance	Skewness	Kurtosis
Access to electricity (% of population)	78.9337	30.2755	916.605	−1.2058	−0.0358
Access to clean fuels for cooking	63.2553	39.0437	1524.41	−0.5081	−1.4192
Renewable energy share in the total final energy consumption (%)	32.6381	29.8949	893.705	0.6709	−0.9058
Energy intensity level of primary energy (MJ/$2017 PPP GDP)	5.3073	3.5320	12.4750	2.5890	9.5037
Financial flows to developing countries (US $)	9.42e+07	2.98e+08	8.8804	8.3882	102.3670
Renewable electricity generating capacity per capita	113.137	244.167	59617.52	5.3669	40.4502

left skew in the distribution. This means that most countries have electricity access percentages higher than the mean, with a few having significantly lower access. Table 1 also informs that about 63.25% of population has the access to clean fuels for cooking while the variance is 1524.41, indicating a wide variance in population of getting access. Notably, in average, only 32.64% of total population gets renewable energy. However, there exists a positive skewness of 0.6709 that indicates while most regions have renewable energy shares lower than the mean, there exist a few with exceptionally high contributions. In average 94.3 million USD is allocated in developing countries, while the variance is 8.88 which indicates a wide disparity in financial flows among developing countries. A positive skewness of 8.39 demonstrates the mismatch in the allocation and indicates that a few countries get the most funds. Lastly, in average, the renewable energy generating capacity per capita is 113.137 where the standard deviation is 244.167, indicating significant variation among the different regions or countries.

Figure 1 illustrates the correlation matrix of variables. The correlation matrix depicts that access to clean fuels for cooking has high positive correlation with

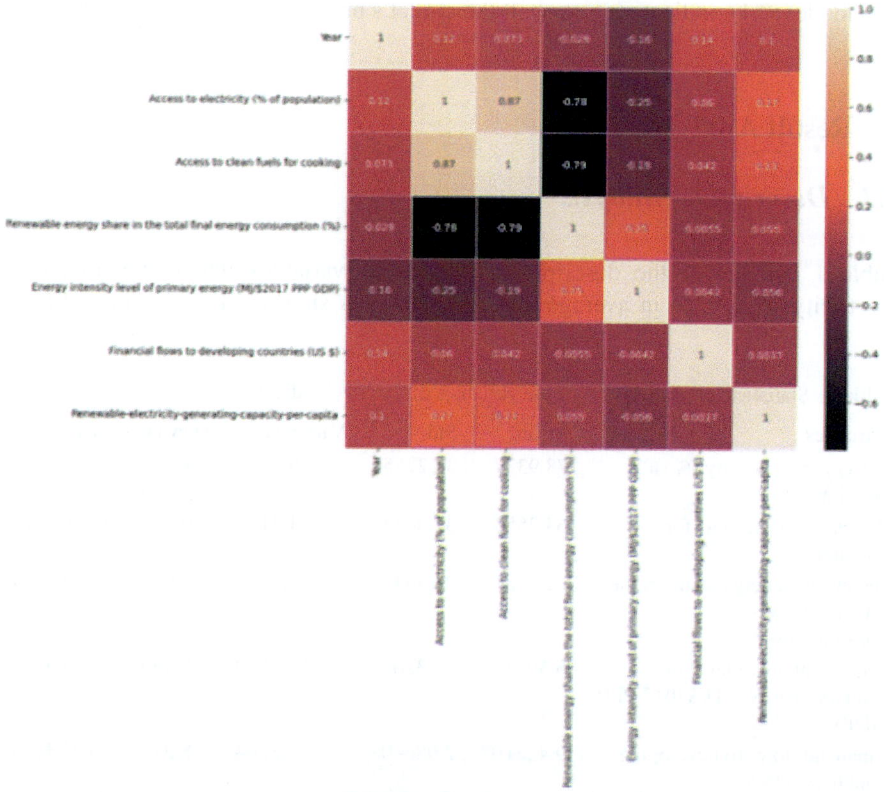

Fig. 1 Heatmap to represent the correlation among the variables

access to electricity (% of population) by 87%. That means the higher the portion of getting access to electricity, the higher to access to clean fuels for cooking. On the contrary, renewable energy share in the total final energy consumption (%) has a significantly negative correlation with access to clean fuels for cooking and access to electricity (% of population) by (−79%) and (−78%), respectively. The correlation reveals that countries with high share with electricity access and having clean fuel to cook are usually the lowest share of renewable energy as share of final energy consumption and vice versa. Notably, renewable energy share in the total final energy consumption (%) has a minimal positive (0.0037%) with financial flows to developing countries (US $). That indicates that financial allocation has a minor impact on renewable energy consumption percentage.

4.2 Results of the Clustering

Based on energy consumption, this study has clustered the dataset using K-means clustering. Analyzing the dataset, the elbow method suggests the number of countries that suite for the study. Figure 2 depicts that elbow three (3) is mostly curved. Therefore, this study considers three clusters (countries). The clusters are less, medium, and high energy consumed countries.

Figure 3 displays the scatter diagram of data points of three clusters (low, medium, and high) energy consumed nations derived from the elbow method K-

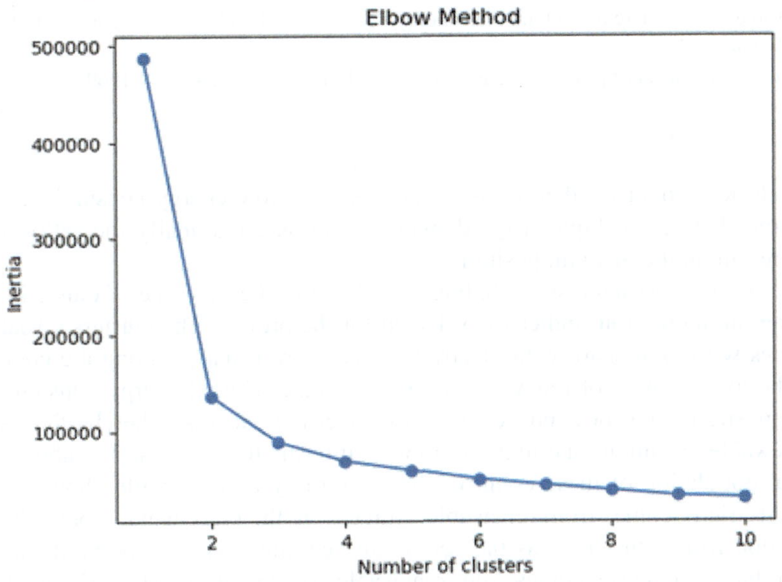

Fig. 2 Elbow method to determine the number of clusters

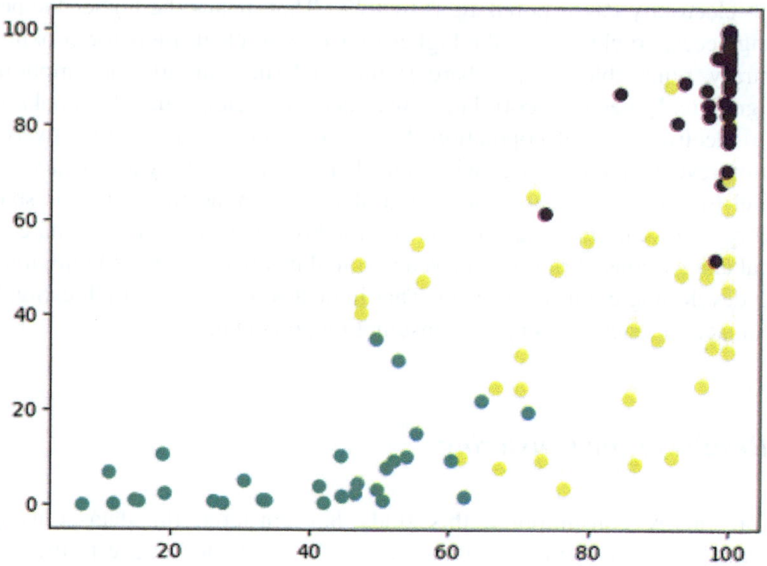

Fig. 3 Results of the clustered countries

Table 2 The value SDG indicators of three different clusters

Variables	Cluster 1	Cluster 2	Cluster 3
Access to electricity (% of population)	99.2806	39.6755	81.4477
Access to clean fuels for cooking	94.4446	7.2666	39.7324
Renewable energy share in the total final energy consumption (%)	15.1447	79.0230	50.0296
Energy intensity level of primary energy (MJ/$2017 PPP GDP)	4.8234	7.1969	4.4819

mean cluster. Countries that are in color green are low energy consuming, while the purple denotes the high energy demanding nations, and finally, the yellow color countries are in the medium position.

One of the most interesting findings is related to the outcome of clustering the countries along the four indicators selected for the project. The clusters reveal that countries with high share with electricity access are usually among the countries with the lowest share of renewable energy as share of final energy consumption. The opposite is true for countries with low electricity access, whereby their share of renewables is among the highest. Table 2 informs that in cluster 1, above 99% of total population of these countries has electricity access, while about 15% of these electricity comes from renewable sources. On the other hand, around 40% of population from Cluster 2 has the access of electricity. Most importantly, almost 80% of the total energy comes from renewable sources of this 40%. From Cluster 3, it is crystal that 81.46% has energy access, and from them 50.03% comes from renewable energy sources.

Table 3 Performance of the ML model to predict energy intensity level of primary energy

Model	MAE	MSE	RMSE	SMAPE	R-Squared
LR	0.0674	0.0119	0.1093	39.1958	0.33
RF	0.0439	0.0074	0.0862	25.1477	0.40
LGB	0.0401	0.0045	0.0670	24.2634	0.64
DT	0.0345	0.0037	0.0606	21.2577	0.70
AdaBoost	0.0787	0.0097	0.0989	45.0591	0.21
CatBoost	0.0361	0.0031	0.0560	22.6648	0.75
B_DRRC	0.0301	0.0028	0.0456	18.6472	0.84

4.3 Prediction of the SDG Indicators Using ML

This study has predicted the SDG based on different indicators such as energy intensity level of primary energy, access to electricity (% of population), and access to clean fuels for cooking.

Table 3 displays the performance matrix of prediction result of ML algorithms on energy intensity level of primary energy. The table shows that the novel algorithm blending decision tree, random forest, and ridge regression (B_DRRC) performs far better than the other ML algorithms. While the MAE, MSE, RMSE, and SMAPE are the lowest for B_DRRC than the others by 0.0301, 0.0028, 0.0456, and 18.6472 respectively, the value of R2 is considerably better than other by 84%. That means B_DRRC fits 84% with the dataset in predicting energy intensity level of primary energy which is impressive for an ML algorithm. MAE is high of AdaBoost (0.0787) followed by LR (0.0674). MSE and RMSE are comparably higher of LR by 0.0119 and 0.1093, respectively. After B_DRRC, CatBoost fits most with the prediction model by 75% followed by DT (70%). Notably, the R-Square of AdaBoost is 21% only which is too bad for this prediction model on this dataset.

The result of the prediction of energy intensity level of primary energy is demonstrated in Fig. 4 as well. In this graphical representation, it is crystal that all types of errors (MSE, MAE, and RMSE) are considerably lower than all other algorithms, while the closest one is CatBoost. Notably, LR has the highest MSE and RMSE, while AdaBoost has the highest MAE.

Table 4 displays the performance matrix of ML algorithms in predicting access to electricity (% of population). The table informs those errors (MAE, MSE, RMSE, and SMAPE) are minimal for B_DRRC in predicting electricity access populations by 3.65%, 0.68%, 0.642%, and 10.6731, while the model fits the most as the R2 is 95%. In terms of MAE, MSE, and RMSE, Linear Regression results out the highest error rate by 10.66%, 2.10%, and 21.4753 followed by AdaBoost 7.76%, 1.52%, and 18.1394, respectively. Considering RMSE, AdaBoost has the highest error rate than others by 12.33%. Though LR has the lowest R-square (77%) out of all algorithms, the model fits well as an ML algorithm. After B_DRRC, CatBoost performs well by having minimal errors (MAE: 4.64%, MSE: 0.71%, RMSE: 8.42%, and SMAPE: 11.3367) and also the second highest R2: 92%.

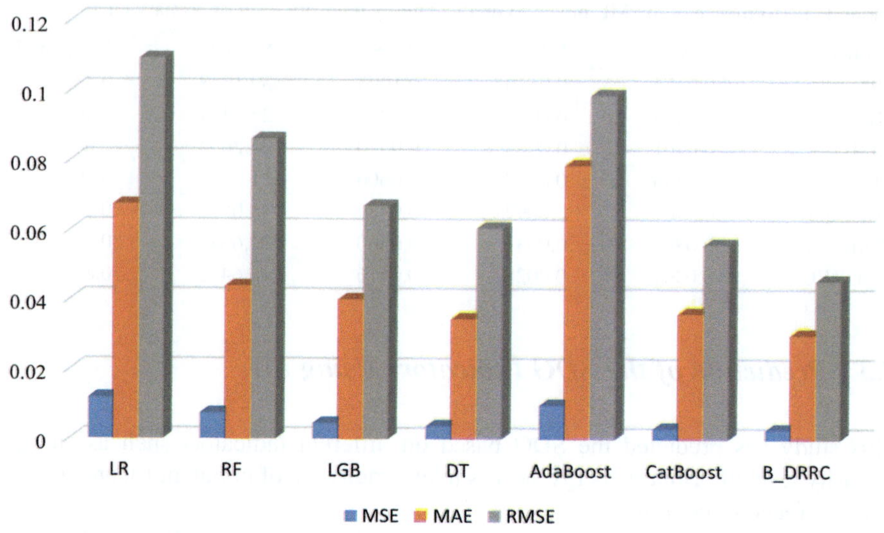

Fig. 4 Errors of the models to predict energy intensity level of primary energy

Table 4 Performance of the ML model to predict access to electricity (% of population)

Model	MAE	MSE	RMSE	SMAPE	R-Squared
LR	0.1066	0.0210	0.1149	21.4753	0.77
RF	0.0519	0.0138	0.1177	12.1570	0.85
LGB	0.0472	0.0072	0.0847	11.7098	0.90
DT	0.0421	0.0104	0.1019	10.5901	0.89
AdaBoost	0.0776	0.0152	0.1233	18.1394	0.84
CatBoost	0.0464	0.0071	0.0842	11.3367	0.92
B_DRRC	0.0365	0.0068	0.0642	10.6731	0.95

Figure 5 illustrates the errors of the models to predict access to electricity. The figure shows that LR has the most MSE and MAE followed by AdaBoost, while B_DRRC has the lowest. However, AdaBoost is containing the highest RMSE by above 12%, and B_DRRC has the lowest by around 6%. Overall, B_DRRC has the minimal error rate than other algorithms.

The performance of the ML model to predict "access to clean fuels for cooking" is tabulated in Table 5. Table shows that LR experienced most errors such as MAE 14.51%, MSE 3.21%, RMSE 17.91%, and SMAPE 48.4913 followed by AdaBoost (MAE:13.52%, MSE: 2.47%, RMSE: 15.73%, and SMAPE: 42.7043). Consequently, LR's goodness of fit for this model is the lowest by 78%. The table also depicts that B_DRRC has the minimal error rate by 5.10% of MAE, 1.01% of MSE, 9.85% of RMSE, and 25.6521 of SMAPE, while the R2 is 95%. That means the model 95% reads the dataset. Overall, this proposed model performs far better than other models.

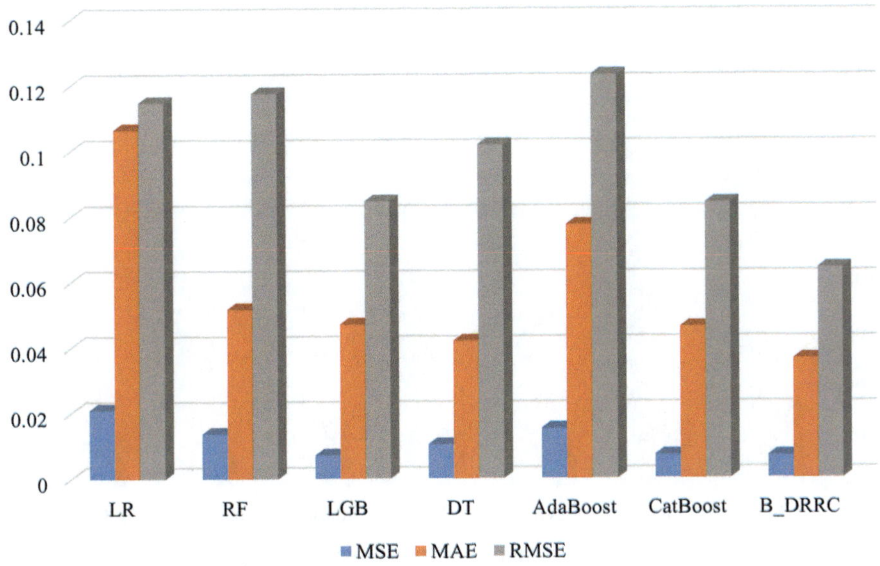

Fig. 5 Errors of the models to predict access to electricity

Table 5 Performance of the ML model to predict access to clean fuels for cooking

Model	MAE	MSE	RMSE	SMAPE	R-Squared
LR	0.1451	0.0321	0.1791	48.4913	0.78
RF	0.0758	0.0236	0.1536	27.2433	0.84
LGB	0.0711	0.0114	0.1067	30.2276	0.92
DT	0.0674	0.0196	0.1399	23.8521	0.87
AdaBoost	0.1352	0.0247	0.1573	42.7043	0.83
CatBoost	0.0707	0.0111	0.1054	29.5615	0.92
B_DRRC	0.0510	0.0101	0.0985	25.6521	0.95

To support Fig. 6 and visually illustrate the performance of selected models in predicting access to clean fuel, Fig. 7 is presented. From the figure, it is visualized that Linear Regression has experienced most error in all three performance measures (MSE, MAE, and RMSE). Following LR, AdaBoost and Random Forest are positioning second and third, respectively. Most importantly, B_DRRC, the proposed model, outperforms other algorithms that this has the lowest errors in all three parameters.

Figure 8 represents the R2 values of all ML algorithms in three different conditions such as energy intensity level, access to electricity, and access to fuels. In all three categories, B_DRRC outperforms all other algorithms significantly. In terms of energy intensity level, B_DRRC can read the dataset more than 80%, while AdaBoost can read the dataset only by around 20%. The second-best model is CatBoost which fits for the dataset by around 75%. Considering access to electricity,

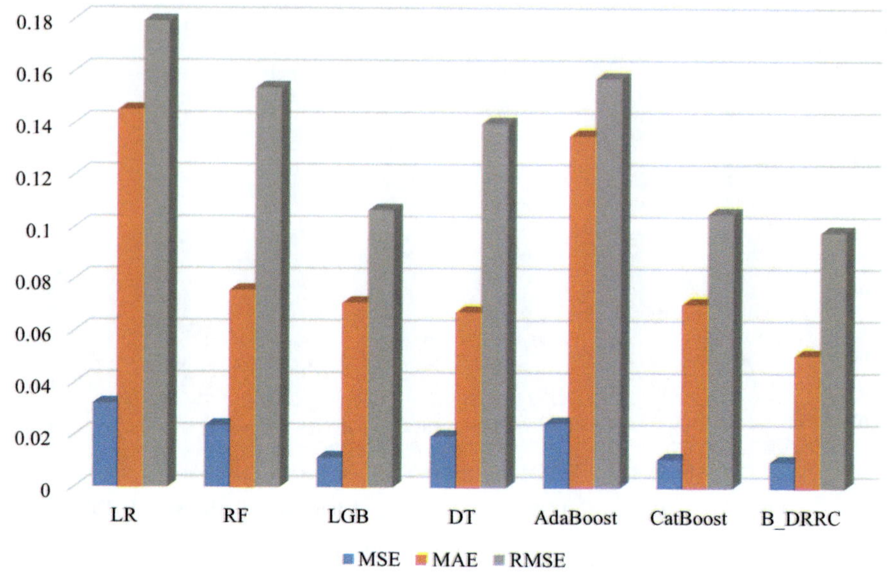

Fig. 6 Errors of the models to predict access to clean fuel

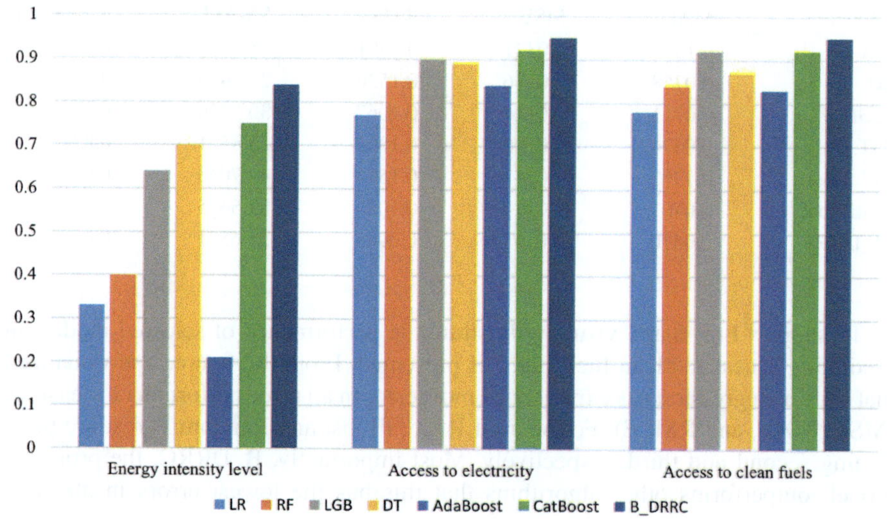

Fig. 7 R-Squared values of all

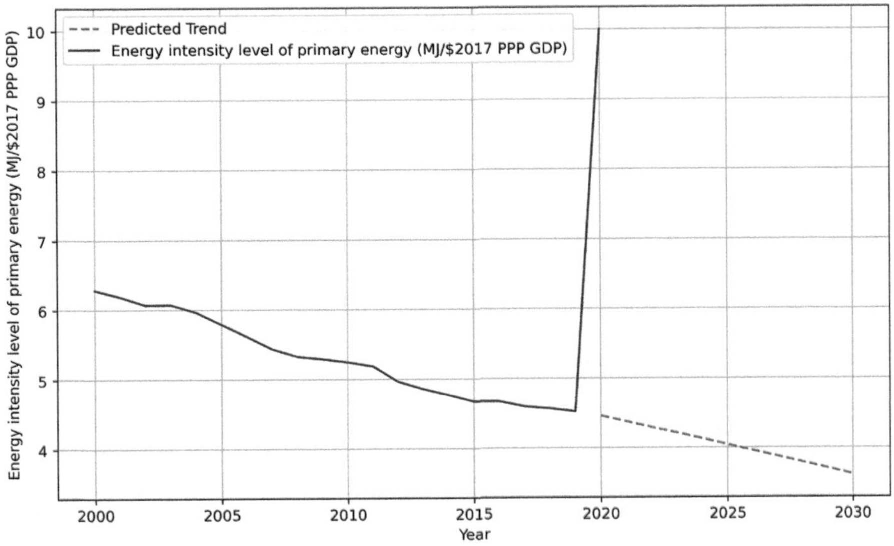

Fig. 8 Predicted trend in energy intensity level of primary energy

all the algorithms performed well, but the B_DRRC is better here again by more than 93%. The same happens for the access to clean fuel data that all the models experienced satisfactory performance but B_DRRC outperformed all.

4.4 Predictive Trend up to 2030 of Different variables

Figure 8 shows the prediction result of primary energy intensity level using B_DRRC, proposed model. In the Y axis, energy intensity level of primary energy is given, while X axis deals with consecutive years from 2000 to 2030. The figure informs that there remained a gradual decrease in energy intensity level from above 6 in 2000 to below 5 in 2019. However, the trend experienced a steep increase in 2020 to about ten levels. Most importantly, the prediction suggests that though there was a sudden raise in intensity level in the early 2020, it will fall to around the same level at the end of 2020. Then, the intensity level will continue decreasing gradually in the preceding years and expected to reach below four (4) by 2030.

Figure 9 displays the prediction result of B_DRRC model on access to electricity. Utilizing dataset from 2000 to 2020, this novel algorithm predicts the probable average access to electricity till 2030. Average access to electricity experienced a continuous increase over the entire span of 2000–2020. The average access to electricity was around 73% of total population at the beginning which touched 85% by 2020. Moreover, from 2012 to 2013, the access to electricity remained constant by around 80.1%. Notably, it is expected to increase in the same pace in the next 10

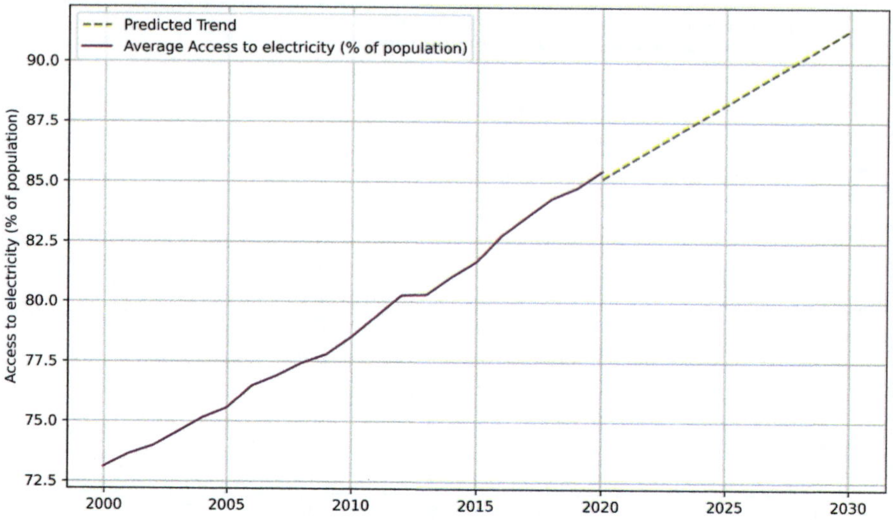

Fig. 9 Predicted trend in access to electricity

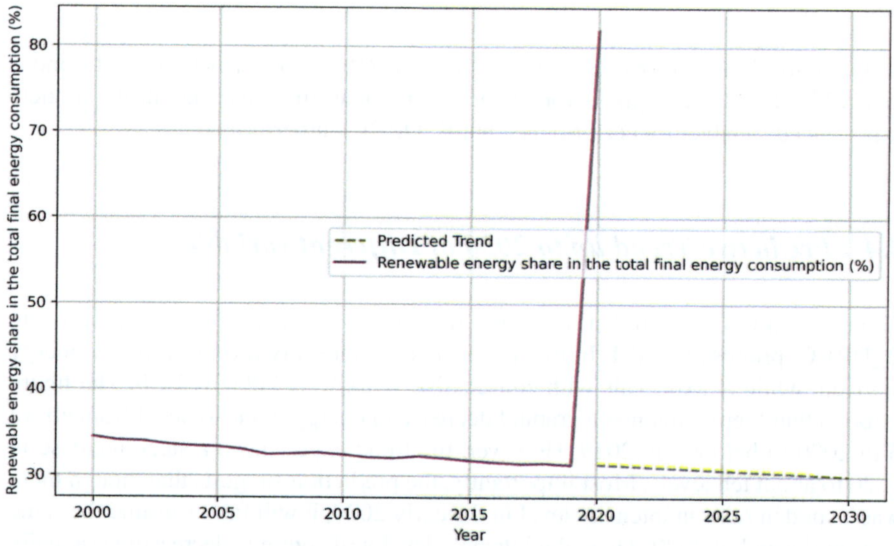

Fig. 10 Predicted trend in renewable energy share

years as well. The prediction result of this novel algorithm informs that the average can be up to 92% within 2030.

Figure 10 shows the prediction result of renewable energy share in the total energy consumption utilizing B_DRRC, a novel model. The figure illustrates that the renewable energy usages experienced a gradual decrease in against total energy

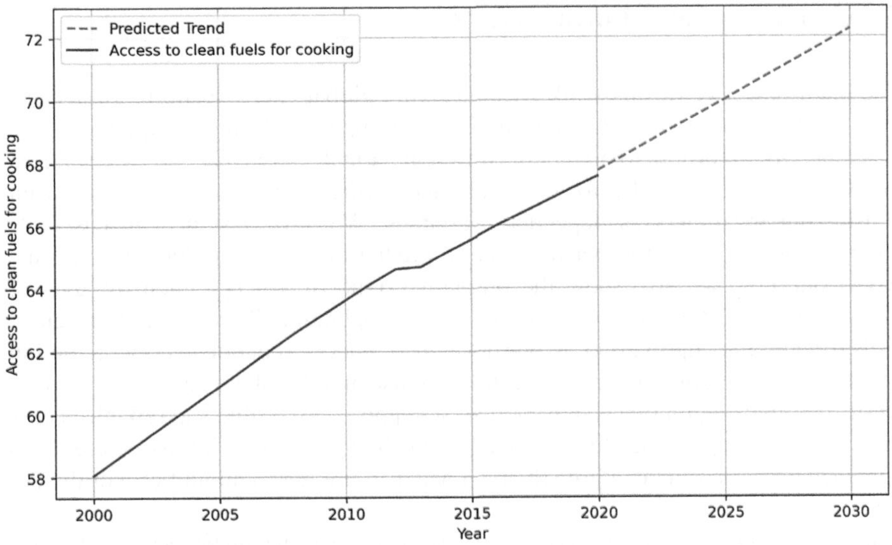

Fig. 11 Predicted trend in clean fuels for cooking access

consumption percentage. While the proportion was around 35% in 2020, the actual condition after 19 years was more severe by around 31% in 2019. However, the trend experienced a steep increase in 2020 to around 83%. Most importantly, the prediction informs that though there was a sudden raise in renewable energy use in the early 2020, it will fall to around the same level at the end of 2020. Then, the consumption of renewable energy will experience a gradual downfall in the preceding years and expected to reach around 30% in 2030 and expected to be lowest of these three decades.

Figure 11 demonstrates the prediction result of B_DRRC model on access to clean fuel for cooking. Employing dataset from 2000 to 2020, this novel algorithm predicts the probable access to clean fuels for cooking till 2030. Average access to electricity experienced a considerable increase over the years from 2000 to 2020. The average access to clean fuel for cooking was about 58% of total population in 2000 which reached around 68% by 2020. Notably, the trend remained constant for 1 year (2012–2013) by around 65%. The proposed algorithm suggests that there will be an increase in next 10 years as the same pace. The prediction of this novel algorithm results out that the average will reach above 72% in 2030.

5 Conclusion and Future Work

The main objective of this work was analyzing global energy pattern. To achieve this, we have employed ML algorithms both supervised and unsupervised and ensemble ML algorithm to predict the different variables of SDGs. We have selected the SDGs-related variables for our study and predicted each if the variables with considering the others as dependent variables. We have designed an ensemble algorithm to get the better performance to predict the SDG variables. Besides the model building, we also show the predictive trend of "energy intensity level of primary energy," "access to electricity (% of population)," and "access to clean fuels for cooking" up to 2030 by our developed model. Accurate energy prediction helps us make smarter choices about how we use and distribute energy. By cutting waste and focusing on renewable sources, it supports affordable and reliable access for everyone, helps fight climate change, and builds stronger infrastructure. This approach drives sustainable growth and a better future for communities worldwide.

Acknowledgments We extend our deepest gratitude to the Center for Multidisciplinary Research and Development (CeMRD) for their resources, guidance, and support. The expertise and encouragement from CeMRD members have been invaluable. This research was made possible by the collaborative environment and cutting-edge facilities at CeMRD.

References

Achuo, E. D., Miamo, C. W. & Nchofoung, T. N. (2022). Energy consumption and environmental sustainability: What lessons for posterity? *Energy Reports, 8*, 12491–12502.

Allen, C., Metternicht, G. & Wiedmann, T. (2016). National pathways to the sustainable development goals (SDGs): A comparative review of scenario modelling tools. *Environmental Science & Policy, 66*, 199–207.

Asmelash, E., Prakash, G., Gorini, R. & Gielen, D. (2020). Role of Irena for global transition to 100% renewable energy. *Accelerating the Transition to a 100% Renewable Energy Era*, 51–71.

Biazzi, R. (2022). Saudi energy sector developments within the energy transition process: Strategies and geopolitical impact. International Master in Security, Intelligence and Strategic Studies, University of Trento.

Bibi, A., Zhang, X. & Umar, M. (2021). The imperativeness of biomass energy consumption to the environmental sustainability of the united states revisited. *Environmental and Ecological Statistics, 28*, 821–841.

Chen, Z.-M. & Chen, G.-Q. (2011). An overview of energy consumption of the globalized world economy. *Energy Policy, 39*, 5920–5928.

Fujimori, S., Dai, H., Masui, T. & Matsuoka, Y. (2016). Global energy model hindcasting. *Energy, 114*, 293–301.

Guo, S., Yan, D., Hu, S. & Zhang, Y. (2021). Modelling building energy consumption in china under different future scenarios. *Energy, 214*, 119063.

Hasan, M., Das, U., Datta, R. K. & Abedin, M. Z. (2023a). *Model development for predicting the crude oil price: Comparative evaluation of ensemble and machine learning methods* (pp. 167–179). Springer.

Hasan, M., Sajid, S. W. & Al Amin, M. (2023b). *An effective three-layer network security to prevent distributed denial of service (DDoS) attacks in early stages* (pp. 174–183). Routledge.

Hasan, M. et al. (2023c). Ensemble machine learning-based recommendation system for effective prediction of suitable agricultural crop cultivation. *Frontiers in Plant Science, 14*, 1234555.

Hasan, M., Hassan, M. M., Faisal-E-Alam, M. & Akter, N. (2023d). *Empirical analysis of regression techniques to predict the cybersecurity salary* (pp. 65–84). Routledge.

Hasan, M. et al. (2024a). A blending ensemble learning model for crude oil price forecasting. *Annals of Operations Research,* 1–31.

Hasan, M., Ahmed, T., Islam, M. R. & Uddin, M. P. (2024b). Leveraging textual information for social media news categorization and sentiment analysis. *PLoS one, 19*, e0307027.

Helten, R. (2013). *Environment in the 21st century and new development patterns,* vol. 2. Springer Science & Business Media.

Hu, Y. et al. (2018). A novel evolution tree for analyzing the global energy consumption structure. *Energy, 147*, 1177–1187.

Jiang, Z., Lyu, P., Ye, L. & wenqian Zhou, Y. (2020). Green innovation transformation, economic sustainability and energy consumption during china's new normal stage. *Journal of Cleaner Production, 273*, 123044.

Juárez-Casildo, V., Cervantes, I. & González-Huerta, R. d. G. (2022). Solar hydrogen production in urban areas of Mexico: towards hydrogen cities. *International Journal of Hydrogen Energy, 47*, 30012–30026.

Khan, I., Zakari, A., Ahmad, M., Irfan, M. & Hou, F. (2022). Linking energy transitions, energy consumption, and environmental sustainability in OECD countries. *Gondwana Research, 103*, 445–457.

Lawrence, S., Liu, Q. & Yakovenko, V. M. (2013). Global inequality in energy consumption from 1980 to 2010. *Entropy, 15*, 5565–5579.

Mamun, M., Hasan, M. & An, K. G. (2024). Advancing reservoirs water quality parameters estimation using Sentinel-2 and Landsat-8 satellite data with machine learning approaches. *Ecological Informatics, 81*, 102608.

Nguyen, T. T. H. et al. (2023). Impact of natural resources extraction and energy consumption on the environmental sustainability in Asian countries. *Resources Policy, 85*, 103713.

Ozcan, B., Tzeremes, P. & Dogan, E. (2019). Re-estimating the interconnectedness between the demand of energy consumption, income, and sustainability indices. *Environmental Science and Pollution Research, 26*, 26500–26516.

Papadopoulos, S., Bonczak, B. & Kontokosta, C. E. (2018). Pattern recognition in building energy performance over time using energy benchmarking data. *Applied Energy, 221*, 576–586.

Prokhorenkova, L., Gusev, G., Vorobev, A., Dorogush, A. V. & Gulin, A. (2018). CatBoost: unbiased boosting with categorical features. *Advances in Neural Information Processing Systems, 31*.

Qudrat-Ullah, H. & Nevo, C. M. (2021). The impact of renewable energy consumption and environmental sustainability on economic growth in Africa. *Energy Reports, 7*, 3877–3886.

Sachs, J. D. et al. (2019). Six transformations to achieve the sustainable development goals. *Nature Sustainability, 2*, 805–814.

Sajid, S. W., Hasan, M., Rabbi, M. F. & Abedin, M. Z. (2023). *An ensemble LGBM (light gradient boosting machine) approach for crude oil price prediction* (pp. 153–165). Springer.

Stafford-Smith, M. et al. (2017). Integration: the key to implementing the sustainable development goals. *Sustainability Science, 12*, 911–919.

Tang, W., Wang, H., Lee, X.-L. & Yang, H.-T. (2022). Machine learning approach to uncovering residential energy consumption patterns based on socioeconomic and smart meter data. *Energy, 240*, 122500.

Vagiona, D. G. & Kamilakis, M. (2018). Sustainable site selection for offshore wind farms in the south Aegean–Greece. *Sustainability, 10*, 749.

Veiga, R., Veloso, A., Melo, A. & Lamberts, R. (2021). Application of machine learning to estimate building energy use intensities. *Energy and Buildings, 249*, 111219.

Verhoef, P. C., Noordhoff, C. S. & Sloot, L. (2023). Reflections and predictions on effects of covid-19 pandemic on retailing. *Journal of Service Management, 34*, 274–293.

Yang, F., Qiao, Y., Hajek, P. & Abedin, M. Z. (2024). Enhancing cardiovascular risk assessment with advanced data balancing and domain knowledge-driven explainability. *Expert Systems with Applications, 255*, 124886.

Zaharia, A., Diaconeasa, M. C., Brad, L., Lădaru, G.-R. & Ioanăş, C. (2019). Factors influencing energy consumption in the context of sustainable development. *Sustainability, 11*, 4147.

Zhang, Z.-h., Chen, Y.-j. & Dai, L.-Y. (2011). *Evolution of china's energy consumption structure based on entropy* (pp. 1–4). IEEE.

Zhu, S., Preuss, N. & You, F. (2023). Advancing sustainable development goals with machine learning and optimization for wet waste biomass to renewable energy conversion. *Journal of Cleaner Production, 422*, 138606.

Mahmudul Hasan is currently pursuing a PhD in Information Technology (IT) at Deakin University, Melbourne, Australia. He earned his BSc (Eng.) and MSc (Eng.) degrees in Computer Science and Engineering (CSE) from Hajee Mohammad Danesh Science and Technology University, Dinajpur, Bangladesh, in 2021 and 2023, respectively. He previously served as a lecturer in the Department of CSE at the University of Creative Technology, Chittagong (UCTC), Bangladesh. He is the Founder and Director of the Center for Multidisciplinary Research and Development (CeMRD) and a moderator of "Be Researcher BD," the largest online research forum in Bangladesh. Additionally, he has taught online as a Data Science instructor to students in the USA, Italy, Denmark, South Korea, and Australia. He is also the founder of the online educational platform "MHM Academy." His research interests include federated learning, machine learning, deep learning, cybersecurity, health informatics, renewable energy, computational sociology, and business intelligence.

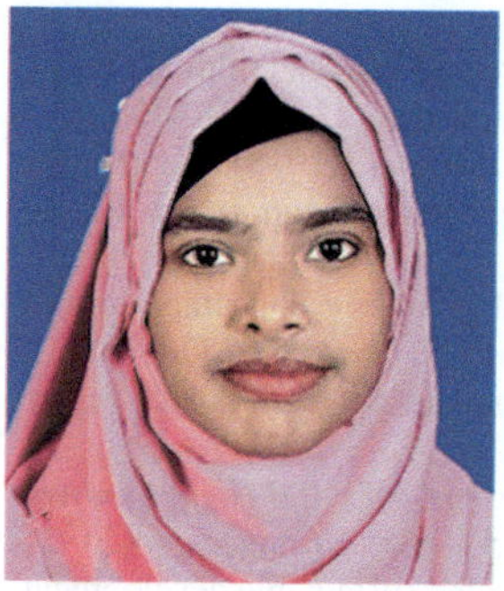

Nusrat Afrin Shilpa is an enthusiastic young lecturer at the Department of Business Administration in Ishakha International University, Bangladesh. She is currently teaching finance-related courses and several graduate courses focusing on the domain of finance. She is a passionate, ambitious, responsible, and student-oriented person. She completed her MBA in Finance from the Department of Finance & Banking at Hajee Mohammad Danesh Science and Technology University, Dinajpur, Bangladesh in 2020 and BBA in Finance from the same university in 2018. Her research interests include FinTech, Corporate Finance, Capital Market, Share Market, Financial Inclusion, Cybersecurity, Blockchain, and Machine Learning.

Ashrafuzzaman Sohag is currently pursuing a master's degree in international management and information systems (IMIS) at South Westphalia University of Applied Sciences, Germany. He holds both an MBA and a BBA in Finance from Hajee Mohammad Danesh Science and Technology University (HSTU), Dinajpur, Bangladesh. He has previously contributed as a Research Assistant in the Department of Finance and Banking at HSTU. His research interests encompass a wide range of areas, including Sustainable and Green Finance, FinTech, Decentralized Finance (DeFi), Financial Inclusion, Data Analytics, Supply Chain Management (SCM), Nonperforming Loans, VAT, Cybersecurity, Business Intelligence, Credit Default Prediction, and Stock Market Analysis.

Md. Mahedi Hassan is a lecturer at the CSE Department of the World University of Bangladesh (WUB). He completed his MSc in Computer Science and Engineering from Hajee Mohammad Danesh Science and Technology University, Dinajpur, in 2023. Before that, he also completed his BSc (Engineering) in Computer Science and Engineering from the same university in 2021. Besides being a teacher, he has also devoted himself to research activities. His focal interest is in Machine Learning (ML), Deep Learning (DL), Artificial Intelligence (AI), Cybersecurity, and Data Science.

Md. Jahangir Alam Siddikee finished his Bachelor of Business Administration (BBA) with distinction from Hajee Mohammad Danesh Science and Technology University (HSTU) in 2007. Following this, he earned a Master of Business Administration (MBA), specializing in Finance, from the same institution in 2010, also achieving outstanding academic results. In the same year, he began his career as a food inspector in the food department under the Ministry of Food of the Bangladesh government. In addition to this role, he served as a guest lecturer at KBM College in Dinajpur, Bangladesh, from 2008 to 2010. In 2012, Siddikee became a lecturer in the Department of Finance and Banking at HSTU. Two years later, in 2014, he was promoted to an assistant professor in the same department. His academic profession continued to progress, and in 2019, he was promoted to an associate professor. Since then, he has been serving in this role while also pursuing a PhD at the Institute of Bangladesh Studies (IBS) at Rajshahi University. Over his scholarly career, he has published more than 20 articles, and his work has been cited 40 times.

Access to Energy Finance: Development of Renewable Energy in Bangladesh

Mohammad Monzur Morshed Bhuiya and Aminul Haque Russel

1 Introduction

This is universally accepted that energy is central to development. It enables investments, innovations, and the emergence of new industries that create employment opportunities, vital to alleviating extreme poverty, foster inclusive growth and promote shared prosperity on a more sustainable world. However, it is disheartening that there are still 685 million people live without electricity globally, and approximately 2.1 billion people depend on traditional polluting fuels and technologies for cooking their meals. Thus, keep energy access affordable, reliable, and sustainable (The World Bank, 2018). On the other side, global warming has become one of the most pressing challenges of our time, with human activities like burning fossil fuels and deforestation exacerbating climate change through increased greenhouse gas (GHG) emissions. Scaling up renewable energy and energy efficiency can help to mitigate the adverse effects of climate change and environmental pollution (Lam & Law, 2016). Renewable energy resources are natural resources which have a vital role to meet up the energy demand (Islam et al., 2006; Ahmed et al., 2014). Bangladesh's renewable energy journey began in 2008, when the Ministry of Power, Energy and Mineral Resources published their policy guidelines. Bangladesh has set a clean energy or renewable target of 40 percent by 2041. Since then, up until 2024, the sustainable energy niche in Bangladesh has been making sluggish progress compared to other countries. According to SREDA (2020), in Bangladesh, renewable energy sources make up only 3.1% of the national energy mix. Solar is

M. M. M. Bhuiya (✉)
Jagannath University, Dhaka, Bangladesh

A. H. Russel
Daffodil Institute of IT, Dhaka, Bangladesh
e-mail: aminul.bba@diit.info

M. Z. Abedin, W. Yong (eds.), *Machine Learning Technologies on Energy Economics and Finance*, International Series in Operations Research & Management Science 367, https://doi.org/10.1007/978-3-031-94862-6_2

responsible for the lion's share of current renewable energy capacity, with 1080.36 MW. The country generates less than 1% of its electricity from hydropower. These levels are far below the 13% global average (SREDA Homepage, n.d.; Tachev, 2024).

In contrast, developing countries have taken the lead in the global shift toward renewable energy, with China adding the top annual solar power capacity, and India and Brazil ranking among the top five. India, in particular, has made significant strides toward renewable energy, targeting 50% of its energy mix from renewable sources by 2030. By early 2023, India had already achieved an installed solar capacity of 64 GW. The world is headed toward renewables, e.g. Iceland and Norway generate nearly all their energy needs from renewable sources. Inspired by these countries' advancements, Denmark has set a goal to transition its entire energy supply to renewable energy by 2050 (Chowdhury, 2024).

However, existing studies evidence that the lethargic progression of renewable energy in Bangladesh may be due to various factors, ranging from government policies to household awareness and reactions. It has been identified that knowledge and information, regulatory frameworks, financial-economic conditions, market dynamics, lack of adequate financing, technological issues, institutional challenges, and behavioral aspects are barriers to the expansion of renewable energy such as solar and wind energy (Mahmud & Roy, 2021a). On the other hand, the energy sector's dependence on large-scale projects, advanced technologies, and complex infrastructure underscores its dependency on project finance and significant investments. Globally, the average annual investment in energy is approximately $413 billion, a figure that is growing, particularly in the developing world. Developing countries will need an estimated $165 billion annually in electricity investments through 2010, with this figure expected to rise by around 3% per year through 2030. Because of the magnitude of their investments in the energy sector, international financial institutions (IFIs) have the potential to profoundly affect future energy paths (World Bank Group, 2007).

Furthermore, renewable energy sector are still relatively new, the research and development efforts aimed at their further exploitation require significant investments (Gielen et al., 2019; Strielkowski et al., 2021). However, due to high upfront costs and the risk of commercializing renewable energy initiatives compared with conventional energy like fossil fuel, a barrier exists in securing financing of renewable energy projects (Warren, 2013). Also, the cost of solar cells in Bangladesh is decreasing every day. This offers an opportunity for the Bangladesh to invest in renewable energy sectors, thereby reducing overall infrastructure costs and enhancing energy security systems (Bhuiyan et al., 2021).

Moreover, previous studies mainly focus on prospects, challenges of renewable energy in Bangladesh and policy related to renewable energy (Bhuiyan et al., 2021; Hossain et al., 2023; Uddin & Park, 2021; Abdullah-Al-Mahbub, & Islam, A. R. M. T., 2023). But, few researches are conducted on energy finance. On the other hand, existing research recommended that investment in renewable energy is immediately needed to address the rising energy demand, mitigate climate change, and foster sustainable development. Such investments bring about substantial socioeconomic, environmental, and health benefits. Although falling renewable energy technology costs have significantly lowered the upfront capital needed, financing renewable energy projects remains difficult (Zhang & Wang, 2019; Michaelowa et al., 2020). Therefore, the current study emphasizes on access to energy finance for the development of renewable energy sector in Bangladesh because access to finance is a critical component in the global effort to ensure that all individuals and communities have reliable and sustainable energy sources.

This study is organized in the following manner. In the introduction section, the background, problem, and aim of the study are presented; the following section provides a comprehensive literature review, covering prior relevant studies and key concepts that are essential for this research. This review establishes the foundation for understanding the theoretical and empirical insights related to the study's focus. Afterward, we discuss the methodological framework, including the strategies for data collection and an outline of the data analysis process. The subsequent section is structured into two parts: the first part addresses the orientation on current scenario of renewable energy, rationality of renewable energy development, barriers to renewable financing as well as others obstacle for the development of renewable energy sector in Bangladesh and the second section shows policy related to renewable energy, green energy finance mechanism, and data related to financing in renewable energy sector including refinancing scheme by IDCOL, Bangladesh Banks, and loan from different international development partners. The concluding section offers recommendations for the government and key stakeholders, highlighting the need for strategic initiatives to improvement in renewable energy finance. Furthermore, it highlights areas of improvement for fostering investments in renewable energy projects.

2 Literature Review

2.1 Concept of Renewable Energy

Renewable energy refers to energy sources that are naturally replenished and can be sustainably recovered from the environment. These include solar, wind, hydropower, biomass, waves, tidal, and geothermal energy, all of which offer cleaner alternatives to traditional fossil fuels. With the characteristics of sustainability and low environmental pollution, the issue of renewable energy has received huge attention (Lai et al., 2020). Renewable energy technologies like solar, mini/micro hydro, wind, and biomass systems offer modern, sustainable solutions for rural electrification. These systems are cost-effective, environmentally friendly, and can be easily operated and managed by local communities, making them ideal for expanding access to clean energy in remote areas. The development of rural renewable energy is an effective way of reducing poverty and promoting sustainable development (Sapkota et al., 2013). Islam et al. (2008) conducted a study on sustainable energy resources and technologies for development activities in Bangladesh, focusing on the electricity challenges faced in rural areas. The authors concluded that renewable energy could serve as the primary energy source to address the electricity issues in these regions (Islam et al., 2008). Islam et al. (2011) discussed renewable energy technologies that can reduce energy shortage, environmental degradation, and climate change effects in Bangladesh (Islam et al., 2011). Therefore, as previous studies highlighted, renewable energy, branded by sustainability and low environmental impact, is important for modern, sustainable electrification and development. Technologies like solar, wind, and biomass are effective in addressing electricity shortages and promoting sustainable development particularly in Bangladesh's rural areas.

2.2 *Key Renewable Energy Sources in Bangladesh*

The key renewable energy sources include solar energy from the sun, biomass, wind, tidal, geothermal, and hydro. The availability of these resources determines the extent to which each type of renewable energy can be utilized in a country. But, Bangladesh lacks geothermal potential and has limited hydro potential, particularly those reliant on elevation. Tidal energy is still in its early stages and has not yet been commercialized. As a result, solar, wind, and biomass remain the only viable alternatives. However, a significant challenge with biomass is the high demand for agricultural and animal waste as fuel for cooking in rural areas. Solar energy is the most reliable renewable energy resource that can be utilized on a large scale. In contrast, wind has consistently been a challenging resource to assess within the context of Bangladesh. From Bangladesh perspectives, the core barriers with biomass are: the price is high, and accumulating huge quantities is tough, hence costly. However, there is considerable potential through the bio gasification process, although it would necessitate careful planning and effective management of bio-resources. Therefore, this expectation is turning out to be true with local and foreign investments occurring in grid-tied utility-scale solar parks and industrial rooftop projects (Hossain & Chisti, 2022). A block diagram in Fig. 1 shows the list of renewable energy sources in Bangladesh.

2.3 *Concept of Energy Finance*

Energy finance is an emerging interdisciplinary area that primarily focuses on the connections between energy markets and financial markets. However, it is also exploring energy products and markets through a financial lens. Zhang (2018) explain that energy finance is the combination of six broad themes such as energy and financial markets, pricing mechanisms, energy corporate finance, green finance and investment, energy derivative markets, and energy risk management

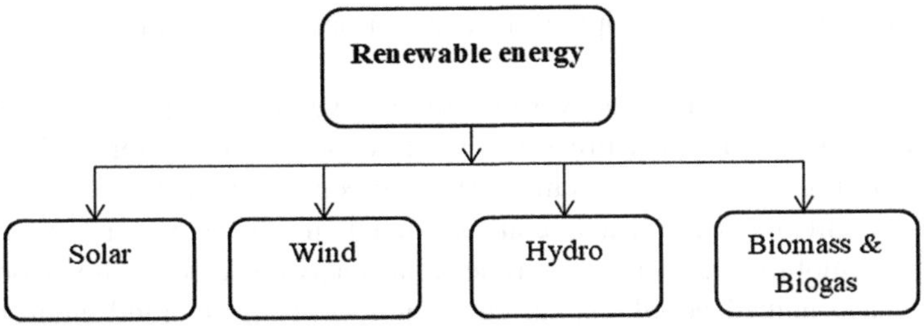

Fig. 1 Major renewable energy sources in Bangladesh

(Zhang, 2018). According to Friebe et al. (2013), sustainable energy finance is the structuring of financial instruments and the mobilization of capital specifically for the development and expansion of renewable energy sources and energy efficient technologies, considering both environmental and economic objectives. In this study energy finance is interchangeably used to mainly focus on financing in renewable energy or the sustainable finance (Friebe et al., 2013).

Therefore, access to renewable energy finance refers to the ability of individuals, businesses, and government to obtain financial resources necessary for the development, deployment, and maintenance of renewable energy technologies such as solar, wind, hydro, and biomass etc. Since supply of finance is crucial for the development, deployment, and scaling of energy solutions that are both affordable and sustainable, this study highlights the various ways of financing in renewable energy sector in Bangladesh such as green finance by commercial banks and non-bank financial institutions, IDCOL and Bangladesh Bank refinancing scheme, loan and grants international development partners, green bond and different subsidies, tailored to meet the specific needs of clean energy projects.

2.4 Financing in Renewable Energy Sector

Financing plays a vital role in the development of renewable energy projects. As global concerns over climate change and environmental sustainability intensify, the transition from fossil fuels to renewable energy sources has become more urgent. However, this transition is heavily dependent on the availability and accessibility of finance, which remains a significant barrier in many parts of the world. In the early 2000s, renewable energy investments were seen as high risk due to the nascent state of the technologies and the uncertainty of returns (International Energy Agency (IEA), 2003). Conventional, financial institutions were often unwilling to invest in renewable energy projects because of these perceived risks, comprising technology performance risks, market risks, and policy-related risks (Beck & Martinot, 2004). These barriers were compounded by the higher upfront capital costs related to renewable energy projects compared to conventional energy sources (Cochran et al., 2014).

In spite of these challenges, various innovative financing mechanisms have emerged over the years to address the specific needs of the renewable energy sector. According to the World Bank (2018), green bonds, feed-in tariffs, power purchase agreements (PPAs), and concessional loans have become increasingly significant in mobilizing capital for renewable energy projects. Green bonds, in particular, have gained popularity as they allow investors to fund environmental projects while making returns on their investments (The World Bank, 2018). The issuance of green bonds reached a record $269.5 billion in 2020 and built investors' confidence in the renewable energy sector (Climate Bonds Initiative, 2021). Public sector participation has also been critical in sinking the perceived risks and attracting private investment in renewable energy. Government-backed financial instruments, such as guarantees and subsidies, have been instrumental in creating a conducive environment for renewable energy financing (Polzin et al., 2015). Feed-in tariffs have been successfully implemented in countries like Germany and China to provide long-term price guarantees for renewable energy producers, thus ensuring stable revenues and attracting investments (Zhang et al., 2013). Similarly, public–private partnerships (PPPs) have emerged as a viable model for financing large-scale renewable energy projects, with governments sharing the financial risks with private entities (Reiche & Bechberger, 2004). On the other side, in developing nations, access to financing for renewable energy remains a significant challenge due to fragile financial systems and lower investor confidence (Bhattacharya et al., 2019). Microfinance institutions and development banks have started to play a more prominent role in financing small-scale renewable energy projects, particularly in rural areas where traditional banking services are limited. These institutions often provide loans at concessional rates, enabling households and small businesses to invest in renewable energy technologies like solar home systems (Dib et al., 2013). However, the existing studies also highlight ongoing challenges in financing the renewable energy sector. One major issue is the lack of identical metrics and benchmarks for measuring the financial performance of renewable energy projects, which disguises the investment decision-making (Inderst et al., 2012). Likewise, the regulatory environment in many countries is still not fully supportive of renewable energy financing, with inconsistent policies and bureaucratic difficulties preventing potential investors (REN21, 2020). In the context of Bangladesh's ambitious renewable energy goals, achieving a 40 percent renewable energy capacity by 2041 presents significant financial challenges. Estimates suggest that the country would need to invest between $1.53 billion and $1.71 billion annually from 2024 to 2041 to meet this target.

However, this amount does not account for the extra costs related to grid modernization and the development of storage facilities, both of which are crucial for the incorporation of renewable energy into the national grid. Current funding available for sustainable energy projects is substantially lower than what is required, indicating a significant financing gap. This shortfall underscores the need for innovative financing mechanisms and greater international financial support to bridge the gap and ensure the successful transition to a sustainable energy future in Bangladesh (Hossain, 2024). While significant progress has been made in developing innovative financing mechanisms for the renewable energy sector, challenges remain. The literature suggests that overcoming these challenges will require continued public sector support, further development of financial instruments tailored to the unique needs of renewable energy projects, and stronger regulatory frameworks to ensure investor confidence. As the world transitions toward a low-carbon future, effective financing strategies will be critical in scaling up renewable energy deployment globally.

3 Methodology

The current study is going to investigate the access to energy finance for the development of the renewable energy sector in Bangladesh, following the following analytical framework along with the method and tools for data collection. Both primary and secondary data were collected to conduct this research. For secondary information, particularly on the policy and legal regime, a comprehensive review of policies, articles, and reports linked to renewable energy and its finance was explored, e.g. Power System Master Plan, Renewable Energy Policy, SREDA-produced analysis, NDC, ADB reports, Bangladesh Bank's Sustainable Finance Policy, Private Sector Power Generation Policy of Bangladesh, Perspective Plan of Bangladesh, 2021–2041, etc. Qualitative data are obtained from unstructured discussion with three banks personnel and one government official. We discussed with them the barriers of renewable energy financing, existing financing mechanisms, new financing methods and then interpreted the discussion. At the beginning of the study, we discussed the renewable energy concept, classification of major renewable energy products, existing renewable energy finance mechanism as well as reviewing the why renewable energy from a theoretical perspective through a literature review. Also, in order to understand the barriers of access to energy finance implementation toward development of renewable energy sector in Bangladesh, we took the views of key personnel of banks through unstructured discussion and attach with the findings of previous studies. For measuring the trend of renewable financing in Bangladesh, we used quantitative data and followed Johnson's et al. (2011)'s research lineup, i.e., Bangladesh Bank's sustainable finance report, policy documents, and renewable energy-related organization's reports were reviewed for the investigation (Johnson et al., 2011).

4 Current Scenario of Renewable Energy in Bangladesh

Bangladesh has great prospects for accelerating renewable energy deployment, current targets remain weak. In Bangladesh, renewable energy sources make up

Table 1 Renewable energy scenario in Bangladesh

Technology Off-grid (MW)	On-grid (MW)	Total (MW)	Technology Off-grid (MW)
Solar	373.84	706.52	1080.36
Wind	2	60.9	62.9
Hydro	0	230	230
Biogas electricity	0.69	0	0.69
Biomass electricity	0.40	0	0.40
Total	376.93	997.42	1374.35

Source: SREDA

only 3.1% of the national energy mix. Within this percentage, solar energy accounts for 63.7%, followed by hydro at 35.7%, wind at 0.4%, and biogas at 1.4%, of the installed capacity. In this situation, small-scale renewables, particularly Solar Home Systems (SHS), offer greater promise. In 2018, the number of green energy users reached 18 million (SREDA, 2024). For example, up to year 2018, Grameen Shakti alone had installed over 4.13 million SHS, making it a leading player in the sector (Mahmud & Roy, 2021b). Overall, the company experienced a 40% increase (1.6 million). But, in recent years, this sector has faced challenges for causing sluggish growth. This decline signals a worrying trend for the clean energy industry as a whole (Masukujjaman et al., 2021).

Table 1 presents the current scenario of renewable energy in Bangladesh which shows both on-grid and off-grid renewable energy application. Solar energy has the most substantial contribution, with 373.84 MW from off-grid and 706.52 MW from on-grid installations, totaling 1,080.36 MW. It's essential to note that due to the expansion of rural electrification through grid extension, a significant number of Solar PV Home Systems, which once carried great recognition to Bangladesh, are now unused. Wind energy follows, predominantly in on-grid systems, contributing 60.9 MW on-grid and only 2 MW off-grid. Hydropower is solely on-grid, providing 230 MW. Biogas and biomass electricity contribute minimally, both in off-grid setups. Overall, the total installed capacity for renewable energy in Bangladesh amounts to 1,374.35 MW, with on-grid systems making up 997.42 MW and off-grid systems contributing 376.93 MW.

5 Rationality to Development of Renewable Energy Sector

Renewable energy brings numerous benefits that extend beyond environmental sustainability, positively impacting public health, agriculture, women's empowerment, and employment generation. In addition, renewable energy sources (RES) offer several advantages, including a reduction in energy dependence on foreign countries, and the potential for cost savings (Gielen et al., 2019; Benti et al., 2023). These benefits contribute to the holistic development of societies, environmental and economies, particularly in developing countries like Bangladesh.

Environmental benefits—Transitioning to green energy sources, such as solar, wind, hydroelectric, and geothermal power, is a crucial component of climate mitigation strategies. Unlike fossil fuels, renewable energy technologies offer clean, abundant, and sustainable alternatives that can significantly reduce carbon emissions and mitigate the impacts of climate change (Tiruye et al., 2021). In response to the questions of—what is rationality to behind development renewable energy in Bangladesh, Government officials who said

> In think in Bangladesh, where air pollution is a critical problem, transitioning to renewable energy can significantly improve air quality. The Renewable Energy Policy of Bangladesh aims to generate 40% of total electricity from renewable sources by 2041, a goal that has pushed significant investment in solar and wind energy projects. Achieving this target is expected to reduce the carbon footprint and help mitigate the adverse effects of climate change.

Health benefits—The adoption of renewable energy significantly improves public health by reducing air pollution and associated health problems. Conventional energy sources, such as coal and oil, emit pollutants that contribute to respiratory diseases, cardiovascular conditions, and other health issues. According to a study, air pollution from fossil fuels is responsible for an estimated 8.7 million premature deaths annually worldwide (Vohra et al., 2021). Government officials and bankers who said that

> In Bangladesh, a transition to cleaner energy sources can substantially reduce the health burden caused by air pollution. For example, the widespread use of solar energy can cut down the reliance on biomass and kerosene, which are significant sources of indoor air pollution and related health issues.

Agricultural benefits—Renewable energy can enhance agricultural productivity and sustainability. For instance, solar-powered irrigation systems provide a reliable and cost-effective water supply for farming, reducing dependency on erratic electricity supply and expensive diesel pumps. A study shows that solar irrigation can increase crop yields by up to 20% and reduce water usage by 30% (Burney & Naylor, 2012). Government officials given opinion

> The adoption of solar irrigation has the potential to significantly improve agricultural outcomes, given the country's reliance on agriculture for livelihood. The government has already installed over 1,500 solar irrigation pumps, benefiting thousands of farmers by providing a sustainable and cost-effective water supply.

Women empowerment—Renewable energy projects can empower women by providing them with new opportunities for economic participation and reducing the time and labor burden associated with traditional energy collection methods. For instance, access to clean and efficient energy sources can free up time spent on collecting firewood, allowing women to engage in educational and entrepreneurial activities (Clancy & Skutsch, 2013). The entire four experts given the opinion

> Renewable energy particularly solar home systems program helps to women's bring empowered, if it connect with micro-credit program.

Also existing study evidence that in Bangladesh, renewable energy initiatives such as the Solar Home System (SHS) program have empowered over 4 million households, many of which are led by women, by providing access to clean and reliable energy (Khandker et al., 2014).

Employment generation—The renewable energy sector is a significant source of job creation. It generates employment opportunities in various stages of the value chain, including manufacturing, installation, maintenance, and operations. According to the International Renewable Energy Agency (IRENA), renewable energy jobs worldwide reached 11.5 million in 2019, with solar photovoltaic being the largest employer (IRENA (International Renewable Energy Agency), 2020). Bankers and govt. officials opined that

> The renewable energy sector has created thousands of jobs, particularly in the solar energy industry. The SHS program alone has generated employment for over 100,000 people in manufacturing, sales, installation, and maintenance roles.

Therefore, the transition to renewable energy presents numerous benefits across different sectors. By improving public health, enhancing agricultural productivity, empowering women, and generating employment, renewable energy can play a pivotal role in fostering sustainable and inclusive development. Addressing the barriers to renewable energy adoption and leveraging these benefits is essential for achieving a greener and more equitable future.

6 Barriers to Renewable Energy Development in Bangladesh

There are several barriers that have been underlined as the causes of slow development of renewable energy in Bangladesh including policy and legal factors, financial, technological, infrastructural, social, and environmental. These constraints prevent the expansion and successful execution of renewable energy projects even though the country has massive prospects for solar, wind, and biomass resources, i.e. overall renewable energy sector. The following sections outline the significant barriers to renewable energy development in Bangladesh. These barriers are found from the literature review and discussion with government officials and bankers who are working with project financing and renewable energy project.

Predominantly, financial challenges for renewable energy investments include weak local financial markets and unfavorable project scales. Limited access to private sector equity funding exacerbates this issue, forcing projects to rely heavily on bank credit, which can restrict necessary financial resources. But in countries where there is a lack of bank credit, the high costs of debt and limited length of loan tenure can be issues (IRENA (International Renewable Energy Agency), 2018). Unfavorable project scale also impacts renewable energy finance, as the scale of investment in these projects is usually small and transaction costs are high, which makes these projects particularly undesirable for bankers (Rahman, 2021). Furthermore, there is a lack of innovative financial instruments and products tailored to the needs of renewable energy developers. Conventional financing mechanisms are not always suitable for the unique characteristics of renewable energy projects, which require different risk assessment and management approaches. One of the primary barriers to renewable energy development in Bangladesh is the lack of a comprehensive and consistent policy framework. While there are policies in place to promote renewable energy, they are often fragmented and not effectively enforced. Additionally, regulatory uncertainties and bureaucratic delays can impede the approval and implementation of renewable energy projects (Mahmud & Roy, 2021b). Regulatory challenges that are hindrances in renewable energy projects include unclear legal and regulatory frameworks including weak feed-in-tariff pricing and non-bankable public–private agreements are major barriers (IRENA (International Renewable Energy Agency), 2018; Moazzem & Hridoy, 2023).

The growth and deployment of renewable energy technologies in Bangladesh are hampered by a lack of technical expertise and infrastructure. Limited research and development (R&D) capabilities and the absence of local manufacturing facilities for renewable energy equipment further obstruct technological advancement (Chowdhuri et al., 2023). The dominance of conventional energy sources, such as natural gas and coal, which are often subsidized and thus more economically attractive. The lack of a competitive market structure for renewable energy, coupled with insufficient market incentives, also poses significant challenges. The development of renewable energy infrastructure, such as grid connections and storage facilities, is often lacking in Bangladesh. Many renewable energy projects, especially those in remote or rural areas, face significant challenges in connecting to the national grid. The absence of adequate energy storage solutions further complicates the integration of intermittent renewable energy sources like solar and wind. There is a shortage of skilled professionals and technicians required to design, install, operate, and maintain renewable energy systems (Hossain et al., 2023; Tarik-ul-Islam & Ferdousi, 2007). The economic sustainability of renewable energy projects can be hindered by the higher initial costs compared to conventional energy sources. Additionally, the absence of adequate incentives, such as feed-in tariffs or tax breaks for renewable energy investments, can reduce the attractiveness of renewable energy projects to investors. Renewable energy projects, particularly large-scale installations, can face environmental challenges such as land use conflicts, biodiversity impacts, and water resource management issues. These environmental considerations can slow down project approvals and lead to public opposition (Karim et al., 2023). We summarize the barriers in Fig. 2.

Therefore, addressing the barriers to renewable energy development in Bangladesh requires a multifaceted approach that involves policy reforms, financial innovations, technological advancements, and capacity building. By tackling these challenges, Bangladesh can unlock its renewable energy potential, promote sustainable development, and reduce its reliance on fossil fuels.

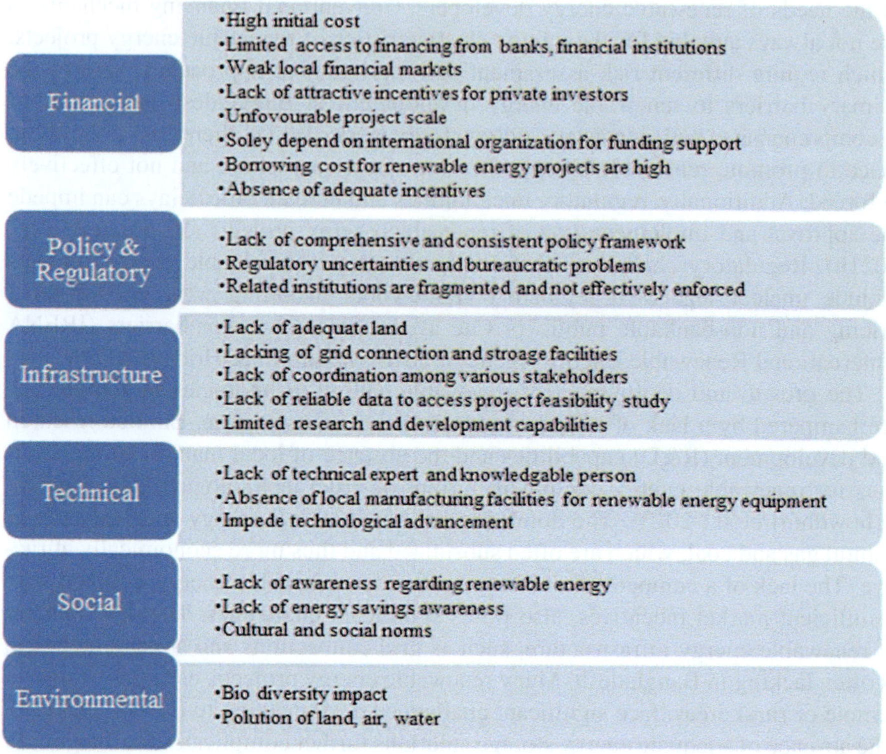

Fig. 2 Renewable energy development barriers in Bangladesh

7 Renewable Energy Policy, Legal Framework, and Financial Incentives for RE Development

In Bangladesh, both GDP and population have been increasing steadily. Subsequently, the demand for electricity is projected to reach 34,000 MW by 2030. To address this rising demand, which outpaces the electricity generation capacity, the government has executed many initiatives through policy-making, rigid regulation, and extensive investments in the sector. The government of Bangladesh has committed to investing USD 70 billion over the next 15 years to create a sustainable and green energy future for the country (Masud et al., 2019). Regulatory measures and government policies greatly influence renewable energy finance, i.e. it's also called indirect financing or public financing mechanism. Public financing mechanisms, including government grants, subsidies, and tax incentives, play a pivotal role in catalyzing renewable energy projects. These mechanisms reduce the financial burden on developers and investors, making renewable energy projects more attractive. In addition, policies that stabilize the market for RECs offer legal rights to the "renewable-ness" of electricity, making green energy projects more appealing financially. For example, in Bangladesh, the government has executed numerous financial incentives to support renewable energy development, such as the Sustainable and Renewable Energy Development Authority (SREDA) and the Infrastructure Development Company Limited (IDCOL), which provide financing and technical support for solar home systems and other renewable projects (Khan et al., 2014). However, public financing alone is often insufficient to meet the large-scale capital needs of renewable energy projects, especially in developing countries where government budgets are limited (Zhang & Wang, 2019). Besides, policies aim to reduce the risk associated with renewable energy investments and guide the direction of finance flows. Here are number of distinguished policies and regulatory measures take on by the government for its renewable energy infrastructure:

7.1 *Policies and Legal Framework for Renewable Energy Development*

Government policies play a crucial role in shaping the pace and direction of economic development. By establishing an enabling environment, these policies can encourage private sector participation and attract private investments into various economic activities. In accordance with the vision of the Article 16 of "The Constitution of the People's Republic of Bangladesh," which is to eradicate discrepancies in the living condition of living between urban and rural areas through electrification and development, the government of Bangladesh has ratified numerous policies and legal framework over the past few decades (Masud et al., 2019). To achieve the aims of electrification through the development of both conventional and alternative energy sources, several policies and legislations have been established. These outlines are designed to facilitate the growth of energy infrastructure, promote the use of renewable energy, and ensure energy security and sustainability for the future. They provide guidelines for private sector involvement, incentives for clean energy projects, and regulatory mechanisms to support the development of a diversified energy portfolio. Table 2 shows the related policies.

In Bangladesh, the Sustainable and Renewable Energy Development Authority (SREDA) was established as the government's central body to promote renewable energy and energy efficiency initiatives in both the public and private sectors. To

Table 2 Renewable energy related policies

Name of plan/Regulations	Issued on
Renewable Energy Policy of Bangladesh (draft)	2022
Mujib Climate Prosperity Plan	2021
Nationally Determined Contributions	2020
Energy Efficiency and Conservation Rules	2015
The Sustainable and Renewable Energy Development Authority Act	2012
Renewable Energy Policy of Bangladesh	2008

Source: SREDA

foster the growth of renewable energy, Bangladesh introduced its Renewable Energy Policy in 2008, aiming to have 10% of total power generation come from renewable sources by 2020, which equates to at least 2000 MW.

7.2 *Investment and Fiscal Incentives*

To encourage the development of renewable energy projects, all stakeholders, including private sector participants and investors, are offered tax concessions and fiscal incentives.

- With approval from the Bangladesh Securities and Exchange Commission (BSEC), renewable energy companies in Bangladesh will be able to issue corporate bonds in both bearer and registered forms (Ministry of Power, Energy and Mineral Resources, 2011).
- In December 2020, Bangladesh Bank announced its sustainable finance policy, mandating that banks and non-bank financial institutions (NBFIs) allocate 2% of all loans to renewable energy facilities and green projects.
- The government of Bangladesh will not regulate the price of electricity produced from renewable energy sources. In its place, the price will be negotiated between the owners and consumers.
- According to the Ministry of Power, Energy and Mineral Resources (2016), companies and NGOs involved in renewable energy projects, whether semi-government, foreign, or locally private, will be granted a 15-year exemption from corporate income tax (Ministry of Power, Energy and Mineral Resources, Government of Bangladesh, 2016).
- The government will provide companies with up to 100% depreciation in the first year for solar thermal and solar photovoltaic projects. Furthermore, projects in biomass, geothermal, tidal, small hydro, and wind energy will be eligible for 100% depreciation over the first five years (Rasel, 2018).
- No restrictions will apply to issuing work permits for foreign personnel and employees involved in renewable energy projects.
- According to the Ministry of Power, Energy and Mineral Resources (2002), foreign employees working on a renewable energy project will receive up to 50% of their salary remitted and will be provided with retirement benefits throughout their tenure (Ministry of Power, Energy and Mineral Resources, Government of Bangladesh, 2002).
- The existing renewable energy financing facility will be expanded to include diverse funding sources, such as public and private investments, donor contributions, carbon emission trading (CDM), and carbon funds, enhancing financing options for renewable energy investments.
- To encourage renewable energy adoption in the power sector, the Ministry of Power, Energy, and Mineral Resources (2002) has implemented a policy that exempts all renewable energy equipment and related raw materials from a 15%

VAT charge (Ministry of Power, Energy and Mineral Resources, Government of Bangladesh, 2002).

- Beyond commercial lending, SEDA will establish a micro-credit support network specifically targeted at rural and remote areas, providing financial assistance for purchasing renewable energy equipment.
- The Power Division of MPEMR will lead initiatives to promote investments in renewable energy and energy efficiency projects. SEDA, in collaboration with local government offices, will implement an outreach program to support renewable energy development.
- SEDA is considering providing subsidies to utilities for installing renewable and clean energy projects, including solar, wind, and biomass technologies.
- Private sector participation, particularly through joint venture initiatives, will be actively encouraged and supported in the development of renewable energy. The Power Division of MPEMR/SEDA will provide assistance in identifying suitable projects and acquiring land for these renewable energy initiatives.
- Investors in renewable energy projects, whether they be from the public or private sectors, will not have to pay corporate income tax for a 5-year period starting from the date when this policy is officially announced in the gazette. The extension of this exemption will be determined based on a regular assessment of its impact on renewable energy.
- Consider establishing an incentive tariff for electricity generated from renewable energy sources, set at a rate 10% above the utility's highest purchase price for electricity from private generators.

7.3 Tariff Policies

Feed-in tariffs (FITs) have proven to be an effective policy tool for promoting renewable energy development in both developed and developing nations. FITs are designed to encourage investment in renewable energy technologies by offering a tariff above the retail electricity rate, thus making renewable energy projects more financially viable. In Bangladesh, implementing FITs could significantly boost electricity supply to the grid, especially if policy frameworks support it. For instance, solar parks could thrive with long-term contracts, such as a 25-year agreement. This is essential, as numerous small and medium-scale homeowners and real estate developers may be willing to invest in rooftop solar PV systems. To support this, compensation packages for rooftop solar development should align with FITs policies. Additionally, small-scale solar parks (1 to 5 MW) with long-term contracts, alongside solar-powered irrigation projects, could contribute significantly toward Bangladesh's goal of sourcing 10% of its energy from renewables. Similarly, medium-scale solar parks could further enhance the country's renewable energy growth (The Daily Star, 2016).

8 Financing Mechanism for Renewable Energy Development in Bangladesh

Key funding sources for advancing renewable energy in Bangladesh include the central bank's green refinancing scheme, IDCOL's refinancing program, green financing through banks and NBFIs, green bonds, and loans from international organizations. This section examines various financing avenues available for promoting renewable energy development.

8.1 Green Finance by Banks and Financial Institutions

8.1.1 Investment in Renewable Energy

Figure 3 describes the investment in renewable energy by banks and non-bank financial institutions from 2016 to 2023. In 2016, the investment was BDT 4599.13 million, which decreased significantly to BDT 3018.73 million in 2017. The following year, 2018, saw a moderate increase to BDT 3636.57 million, and this upward trend continued in 2019 with BDT 3712.76 million. However, 2020 experienced a slight decline to BDT 3669.83 million. The investment picked up again in 2021, rising to BDT 4339.55 million. This positive trend continued into 2022, with a substantial increase to BDT 6417.67 million, and finished in a significant surge to BDT 7421.78 million in 2023. This data highlights a fluctuating yet overall upward trend in investments, particularly with marked increases in the last two years, indicating a growing commitment to renewable energy by commercial banks and non-bank financial institutions.

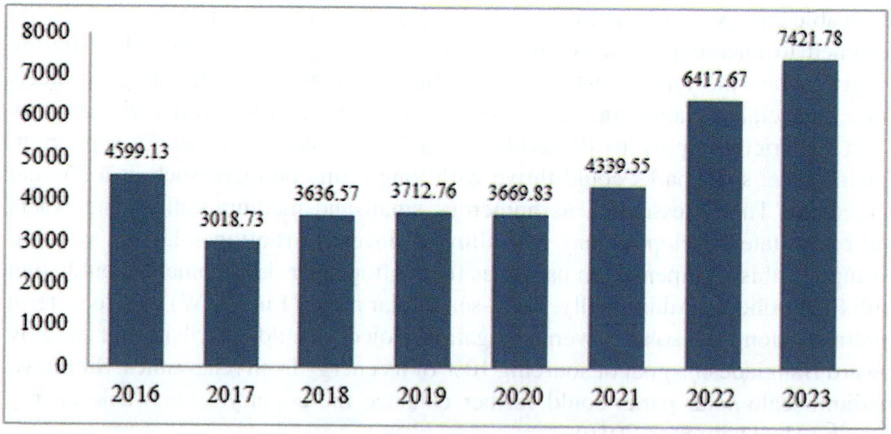

Fig. 3 Investment scenario of banks & financial institutions in renewable energy sector

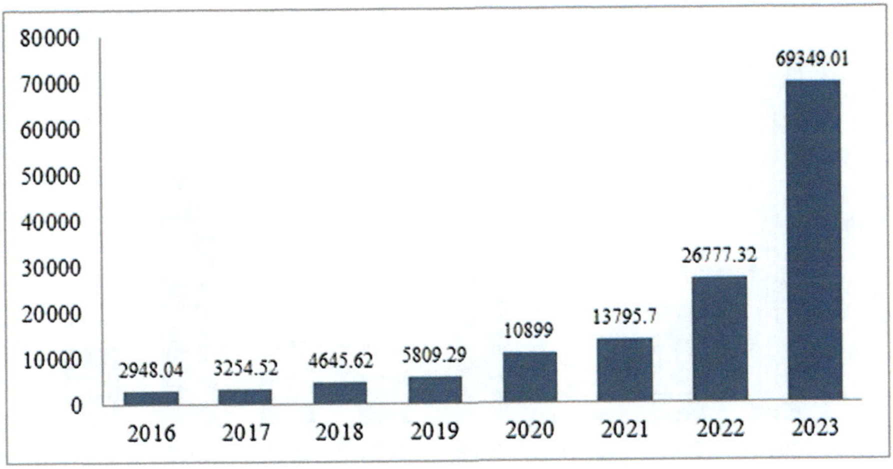

Fig. 4 Investment scenario of banks & financial institutions in energy efficiency

8.1.2 Investment in Energy Efficiency

Figure 4 illustrates the combined annual investment in energy efficiency by both bank and non-bank financial institutions from 2016 to 2023. In 2016, the investment was BDT 2948.04 million, which modestly increased to BDT 3254.52 million in 2017. A more significant rise occurred in 2018 with BDT 4645.62 million, followed by BDT 5809.29 million in 2019. The investment saw a notable jump to BDT 10899 million in 2020. This growth continued, with investments reaching BDT 13795.7 million in 2021 and surging to BDT 26777.32 million in 2022. The most dramatic increase happened in 2023, with investments soaring to BDT 69349.01 million. This data indicates a strong and accelerating trend of investment for energy efficiency by financial institutions, particularly in the last three years, highlighting an increasing commitment to renewable energy development.

8.1.3 Investment in Alternative Energy

Figure 5 depicts the annual investment in alternative energy by bank and non-bank financial institutions from 2016 to 2023, showing significant fluctuations. In 2016, the investment was at a high of BDT 281.36 million, but it dropped sharply to BDT 91.67 million in 2017 and further plummeted to BDT 7.72 million in 2018. A recovery occurred in 2019 with investments rising to BDT 94.84 million, but this was followed by another decline to BDT 38.37 million in 2020. In 2021, investments slightly increased to BDT 43.84 million. A substantial surge was seen in 2022, with investments reaching BDT 206.85 million, before dropping again to BDT 61.36 million in 2023. This data indicates significant volatility in investment levels, highlighting periods of both sharp declines and strong recoveries, suggesting

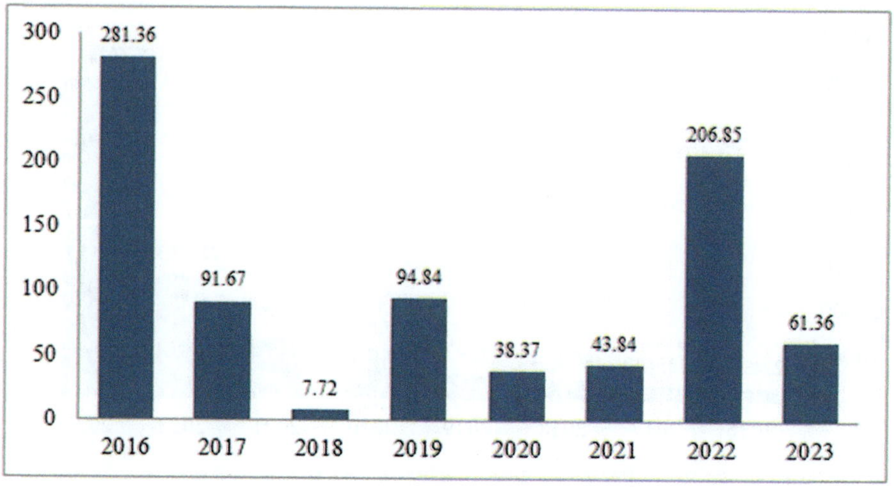

Fig. 5 Investment scenario of banks and financial institutions in alternative energy

varying degrees of commitment and external factors influencing the investment trends in the alternative energy sector.

8.2 *Refinancing Scheme*

8.2.1 IDCOL Refinancing Scheme

IDCOL's most successful Solar Home System (SHS) Program, this reputed government financial institutions so far has introduced many refinancing schemes and concerted programs to diversify the RE installations in areas like Biogas and Biomass based power and energy generation, solar micro and mini-grid, solar irrigation, and other types of commercial-scale RE projects (SREDA Homepage, n.d.).

Under the IDCOL Solar Home System Program and Domestic Biogas Program, loans are not issued directly to end users; instead, they are distributed through Participating Organizations (POs). This lending model also applies to other renewable energy projects, including solar-diesel hybrid systems for telecom base stations, solar-powered transport, rooftop solar installations, solar cold storage and dryers, battery charging stations, and community biogas initiatives. In contrast, larger grid-tied renewable energy Independent Power Producer (IPP) projects will be financed on commercial terms and may qualify for loans denominated in USD.

8.2.2 Bangladesh Bank Refinancing Scheme

In August 2009, the central bank of Bangladesh introduced a BDT 2 billion green banking refinance scheme aimed at promoting solar panels, biogas plants, and effluent treatment plants (ETPs) to reduce industrial pollution and boost power supply. The scheme offers loans to commercial banks at interest rates between 5% and 12%, enabling them to provide loans to entrepreneurs at a maximum interest rate of 12%. The initiative allows for 100% refinance facilities for rural and urban solar panel installations, biogas power plants, and other green products. The scheme, aligned with the government's targets of meeting 5% and 10% of electricity demand from green energy by 2015 and 2020 respectively, has expanded to include 47 green products, with a specific focus on household and business enterprises.

Financing from Re-finance Scheme in Solar Home System

Individuals or entities who install solar panels for personal, joint, business, or cooperative purposes in both urban and rural areas and obtain financing from banks will be eligible for refinancing under this scheme. The sub-sectors covered include solar home systems, solar mini-grids, solar irrigation pumping systems, and solar photovoltaic assembly plants.

Figure 6 depicts the investment scenario in solar home systems through the Central Bank refinancing scheme from 2016 to 2023. Initially, there was a substantial investment of 108.29 million BDT in 2016, representing the highest point within this period. This significant investment suggests a strong initial push toward promoting solar home systems. However, the following year, 2017, saw a dramatic

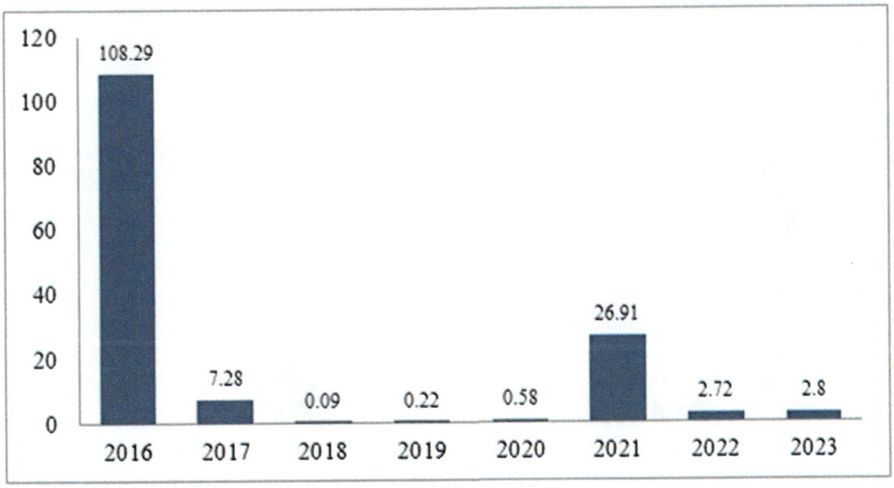

Fig. 6 Financing scenario in solar home system from Bangladesh Bank refinance scheme

decline to 7.28 million BDT, and the trend continued downward, reaching near negligible amounts in 2018 (0.09 million BDT) and 2019 (0.22 million BDT). A slight recovery occurred in 2020 with an investment of 0.58 million BDT, followed by a notable increase to 26.91 million BDT in 2021, indicating renewed interest or possibly new policy incentives. Despite this resurgence, investments dropped again to 2.72 million BDT in 2022 and slightly rose to 2.8 million BDT in 2023. This trend reflects significant volatility in investments, likely influenced by changing government policies, market dynamics, and financial conditions. The overall trend indicates an initial high investment followed by a sharp decline, with a brief resurgence in 2021, reflecting volatility and possibly the impacts of policy changes, market conditions, and financial accessibility on the solar home system sector.

Financing from Re-finance Scheme in Biogas

Likewise, those who take loans from banks for producing and using biogas in rural or urban areas will also be eligible for this refinancing scheme. Sub-sectors eligible for this support include setting up biogas plants in existing cattle or poultry farms, combined cattle rearing with biogas plants, producing organic fertilizer from slurry, and establishing medium-scale biogas plants.

The graph in Fig. 7 illustrates the investment scenario in biogas projects through the Central Bank refinancing scheme from 2016 to 2023. In 2016, there was a significant investment of 93.35 million BDT, the highest in the given period, indicating a strong initial commitment to biogas development. However, this investment sharply declined to 11.29 million BDT in 2017 and continued to decrease to 6.48 million BDT in 2018. The downward trend persisted in subsequent years, with investments

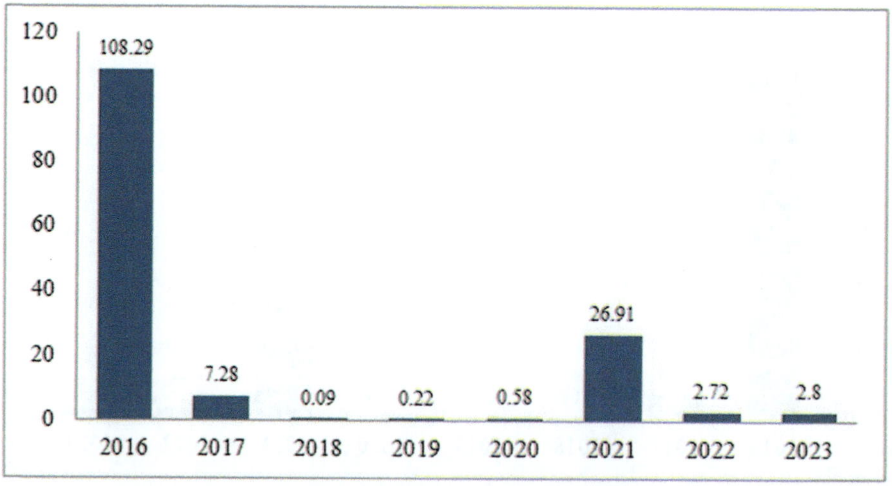

Fig. 7 Financing scenario in biogas system from Bangladesh Bank refinance scheme

dropping to 3.02 million BDT in 2019, 1.18 million BDT in 2020, and further to 0.69 million BDT in 2021. A slight increase was observed in 2022 with an investment of 1.47 million BDT, followed by another minor decline to 0.8 million BDT in 2023. This overall declining trend highlights the challenges and perhaps diminishing focus on biogas projects within the refinancing scheme, suggesting a need for renewed policy support and incentives to revitalize investments in this sector. Therefore, to boost the capacity of renewable energy in alignment with the government's ongoing emphasis, Bangladesh requires a significant increase in financing from diverse and additional sources.

9 International Financing and Multilateral Institutions

International financing, through multilateral institutions such as the World Bank, Asian Development Bank (ADB), and Green Climate Fund (GCF), provides substantial support for renewable energy projects, especially in developing countries. These institutions offer concessional loans, grants, and guarantees that lower the cost of capital and reduce investment risks (Sovacool, 2012). For instance, the World Bank has financed multiple renewable energy projects in Bangladesh, including solar power initiatives, through its IDA credits (World Bank, 2020). This bank recently signed a $515 million agreement with the government of Bangladesh to support the country in its clean energy transition by developing battery storage systems and distributed renewable energy (World Bank, 2022). Previously, in year 2019, government of Bangladesh receives $ 185 million from World Bank for financing in renewable energy (World Bank, 2019). In May 2022 the Asian Infrastructure Investment Bank extended a $200 million long-term credit line to Bangladesh under which IDCOL will on-lend to eligible projects renewable energy, energy efficiency, and related projects (Asian Infrastructure Investment Bank (AIIB), 2022).

The Asian Development Bank (ADB) has entered into a financing agreement worth $121.55 million with Dynamic Sun Energy Private Ltd. to construct and maintain a 100 MW grid-connected solar photovoltaic power plant in Pabna, Bangladesh. The plant is the country's first private sector utility-scale solar facility to secure support from global financiers (Asian Development Bank, 2024). Institutions like IDCOL (Infrastructure Development Company Limited) and the recently established Super ESCO have the potential to secure credit lines from multilateral agencies to facilitate renewable energy projects in Bangladesh. The International Finance Corporation (IFC) estimates that Bangladesh possesses a climate-smart investment potential totaling $172 billion from 2018 to 2030, spanning various sectors, including green buildings, transportation infrastructure, urban water, agriculture, waste management, and renewable energy. These investments are essential for achieving the country's Nationally Determined Contribution (NDC) goals. Out of $172 billion, $3.2 billion is invested in renewable energy projects (IFC, 2020). So, international financing is crucial, it often comes with stringent conditions and requires extensive documentation, which can be challenging for developing countries to meet (Buchner et al., 2019). Bangladesh secured € 400 million in funding from the European Investment Bank (EIB) and the European Union (EU) for renewable energy generation and capacity building (Khan & Sultana, 2024).

10 Innovative Financing Mechanism

Green bond—Green bonds are designed to raise funds specifically for green projects, including clean energy initiatives. Green sukuk (financial certificates) operate in the same manner, with the exception that instead of fixed interest, the income of investors follows Sharia principles (Islamic law) (World Bank, 2020). Both green bonds and green sukuk serve as financing mechanisms for large-scale clean energy projects. Notably, the inaugural issuance of green sukuk occurred in June 2017, and by 2019, the annual issuance of this financial instrument had escalated to $4 billion. From 2017 to September 2020, green sukuk worth $10 billion was issued in Indonesia, Saudi Arabia, the United Arab Emirates, and Malaysia (Asian Development Bank (ADB), Asian Development Outlook, 2021). In Bangladesh, IDCOL issued its first green bond in 2019 to finance renewable energy projects, marking a significant step toward diversifying financing sources (Uddin et al., 2019). Last year, a green sukuk amounting to 30 billion BDT ($300 million) was issued for the development of a 230 MW capacity solar project in Bangladesh (Babu, 2023). Investment in electricity is mostly public-funded and the private sector accommodates a minor share. Issuing green bonds to increase investment in reshaping domestic electricity production can be a step forward (Khan & Ali, 2021). Therefore, issuing green sukuk (Islamic bonds) for renewable energy projects would serve as a strong foundation for financing large-scale renewable energy initiatives in Bangladesh. In addition, leveraging green bonds could help overcome financial barriers and catalyze investments in renewable energy at a significant scale. *Crowdfunding*—Crowdfunding stands as a powerful tool to democratize access to financing in renewable energy projects. Individuals collectively contribute small amounts of capital to support large-scale projects. This model not only garners public interest but also fosters a sense of community involvement in the transition to green energy, as seen with platforms that directly connect investors to renewable projects (Lam & Law, 2016). *Pay-as-you-go (PAYG)* approach—The PAYG model, particularly in off-grid solar systems, allows consumers to pay for energy services in installments, making renewable energy more accessible in low-income regions (Rolffs et al., 2015). *Foreign private investment*—Foreign private investment also presents significant potential for the country's renewable energy sector. For instance, a U.S.-based company has shown interest in investing in solar power projects in Bangladesh, which could help reduce the nation's reliance on fossil fuels and lower its environmental impact. *Public–private partnership*—Bangladesh could rapidly add an estimated 7500 MW of solar power to its energy mix through effective public–private partnerships and adequate funding mechanisms like green bonds.

Developing capacity through these partnerships and focusing on green finance is vital. Both public and private financial institutions must prioritize their lending portfolios to support the shift toward renewable energy. Commitment and accountability from all stakeholders are crucial for making the necessary investments in renewable energy development. Therefore, to accelerate the transition to renewable energy, it is essential to enhance the capacity of financial institutions, improve policy frameworks, and develop innovative financing solutions that can effectively mobilize the required capital.

11 Discussion

Access to energy finance plays an essential role in the development of renewable energy in Bangladesh. Despite the country's significant impending for renewable energy, for the most part in solar and wind power, the expansion of these resources has been slow. A major factor contributing to this sluggish growth is the lack of adequate financing mechanisms tailored to the needs of renewable energy projects. Renewable energy projects, by nature, often require high upfront capital investments, which can be a barrier for many small and medium-sized enterprises (SMEs) and individual investors in Bangladesh. A rough estimate suggests that reaching the 40% renewable energy capacity target could incur cost for Bangladesh ranging from $1.53 billion to $1.71 billion each year from 2024 to 2041. This figure does not account for additional expenses like grid modernization and storage facilities, which are critical for integrating renewable energy into the national grid (Hossain, 2024).

Furthermore, banks and non-bank financial institutions in Bangladesh have been reluctant to lend to renewable energy projects due to the perceived risks associated with new and relatively untested technologies and the long payback periods associated with these investments. This has created a financing gap, limiting the ability of project developers to access the funds needed to scale up renewable energy initiatives. The limited availability of concessional financing, high interest rates, and stringent collateral requirements further exacerbate this challenge, making it difficult for many potential investors to pursue renewable energy projects. As of 2023, total investment in renewable energy was significantly lower than needed, with the central bank's refinancing scheme contributing only a fraction of the necessary funds. For instance, the total investment in solar home systems through the central bank's refinancing scheme plummeted from BDT 93.35 million in 2016 to just BDT 0.8 million in 2023 (Bangladesh Bank, 2023). This decline highlights the inadequacy of current financial support mechanisms in driving large-scale renewable energy adoption. Moreover, high interest rates, which range between 8% and 12% depending on whether the loan is provided directly or through microfinance institutions (MFIs), further, deter potential investors (Bangladesh Bank, 2022).

The central bank's refinancing schemes and other government-backed financial initiatives have been crucial in promoting renewable energy investments. However, these efforts are not sufficient to meet the growing demand for renewable energy. For example, investments in biogas projects fell drastically from BDT 108.29 million in 2016 to a mere BDT 2.8 million in 2023 (Bangladesh Bank, 2023). Moreover, the lack of awareness among financial institutions and investors about the profitability and long-term benefits of renewable energy investments has hindered the growth of this sector. There is also a need for more innovative financing solutions, such as green bonds, blended finance, and public–private partnerships, to attract more investment in renewable energy. This sharp decline underscores the need for more robust financial instruments and policies that can sustain and increase investment in the sector.

12 Recommendations

To ensure the successful development and expansion of renewable energy in Bangladesh, it is crucial to address the multifaceted challenges that hinder progress in this sector. The recommendations provided to create a more advantageous environment for renewable energy investments, enhance financial accessibility, and promote stakeholder collaboration. By focusing on strategic financial allocations, streamlining regulatory processes, and fostering partnerships between the public and private sectors, these recommendations seek to accelerate the transition to a sustainable energy future. Some effective recommendations are discussed below:

Expand Government Support and Incentives: Another study finds that distribute funds purposefully in order to maximize the investment opportunity of $10 billion in renewable energy production over the next 10 to 12 years. Simplify subsidies to align with the projected $2 billion needed to achieve the 40% renewable energy goal, thus decreasing the existing subsidy burden of $2.82 billion. Therefore, government of Bangladesh should consider expanding existing financial incentives, such as subsidies, tax breaks, implement carbon tax and low-interest loans, to encourage more investment in renewable energy. Additionally, the government could establish a dedicated renewable energy fund to provide concessional financing to SMEs and individual investors. Besides, develop financial products that are explicitly designed to address the risks and credit concerns that are typically associated with renewable energy projects. Likewise, to foster renewable energy development, it is crucial to implement motivating and efficient incentives for renewable energy entrepreneurs while leveraging international funding.

Furthermore, the current tax structures that favor fossil fuel investments create barriers for renewable energy adoption. Expanding tax holidays for renewable energy power plants from five to ten years, providing full duty exemptions for small-scale solar projects, and lowering the overall tax rates on solar-related equipment would create a more favorable environment for advancing clean energy initiatives in Bangladesh. These reforms would stimulate investment in renewable energy, making it more competitive and attractive.

Focus on Clean Energy Financing: Focusing on clean energy financing is indispensable for Bangladesh to meet its renewable energy targets. A study done by the Change Initiative, the country will need around $26.5 billion to attain its clean energy goals. Furthermore, 39% of the promised funding from development partners for the energy sector remains undistributed. In the coming years, the government must confirm the release of this committed support and enthusiastically seek additional funding sources. Moreover, intensified efforts in securing funds from bilateral, multilateral, and regional partners are necessary. Accessing global climate funds, as well as clean energy and green technology funds should also be prioritized to accomplish these goals.

Stakeholder's Collaboration: Foster greater collaboration among stakeholders to advance renewable energy by involving civil society, the private sector, academia, and the media in the development, implementation, and monitoring of policies. Additionally, seek new credit lines from international financial institutions to empower IDCOL and BIFFL to offer extended loan terms or more competitive financing options for renewable energy projects.

Improve Capacity Building for Financial Institutions: The results of previous research and bankers' opinion show that government authority and field-level bankers do not have sufficient awareness and also fail to understand the critical need for financing in green energy projects. Additionally, there is no set of guidelines for commercial banks to determine what qualifies as a green or renewable project. As a result, ground-level bankers consider green and renewable projects in the same way as other commercial projects when it comes to financing them (Rahman, 2021). So, banks and non-bank financial institutions (NBFs) need to develop better capabilities like knowledge, tools necessary to assess and manage the risks associated with renewable energy projects, and to build dedicated teams or units within banks and NBFIs that focus solely on renewable energy financing. Also, regular training programs and workshops should be conducted in collaboration with international experts to improve understanding of the financial dynamics of renewable energy projects. As well as increase the Bangladesh Bank's monitoring system, i.e. to verify whether banks and NBFIs have been obeying to the regulator's instructions or policy to secure the required financing for green and renewable energy sectors. Finally, to ensure the successful implementation of renewable energy projects and initiatives, the lending process will be streamlined and reinforced. Simplifying and strengthening these procedures will help facilitate easier access to financing and promote broader adoption of renewable energy solutions.

Promote Innovative Financing Mechanisms: The introduction of innovative financing mechanisms like green bond, crowdfunding, blended finance, and other financing tools could attract a wider range of investors including international development partners, private sector investors, and impact investors as well as help to bridge the financing gap in the renewable energy sector. Green bonds, for instance, could raise significant funds for renewable energy projects, leveraging the growing global demand for sustainable investment opportunities. The government could also explore the potential of blended finance, which mixed public and private funds to reduce investment risks.

Green Financing Framework: A robust green financing framework, along with enhanced capacity among stakeholders, is key to improving access to financing for the promotion of renewable energy and energy efficiency. This is fundamental for attaining Bangladesh's sectoral targets and ensuring the country's long-term energy security. Importantly, the advantages of such a strengthened green financing framework will not be confined solely to the sustainable energy sector. Instead, its positive effects will extend across a variety of green sectors, fostering a transition toward a more sustainable and environmentally friendly economy. This holistic impact underscores the significance of green financing in promoting broader economic and environmental sustainability in Bangladesh.

Strengthen Public–Private Partnerships (PPPs): Public–private partnerships (PPPs) have the potential to significantly accelerate the development of renewable energy projects in Bangladesh. By fostering collaboration between the government and private sector, PPPs can leverage shared resources and expertise to scale up clean energy initiatives. To encourage more PPPs, the government should provide clear guidelines and reduce bureaucratic red tape, which often hinders private investment. Moreover, offering co-financing opportunities can help reduce the financial burden on private investors, making renewable energy projects more attractive. This collaborative approach will play a crucial role in meeting the country's renewable energy targets while enhancing energy security and sustainability.

Expand the Competitive Private Sector for Renewable Energy: Privately owned power plants play a noteworthy role in the renewable energy sector. But, in Bangladesh, private sectors are more reluctant to invest in renewable energy projects due to the lengthy process, i.e., at least 30 numerous approvals are needed for green energy development projects (Rahman, 2021). These bureaucratic difficulties slow down the implementation of green energy projects in the country. Therefore, encourage and support the private sector in participating in incentives, confirming arrangement with national renewable energy goals and regulations while maintaining transparency. Provide training to develop bankable projects and offer guidance on directing regulations and securing funding.

Raise Awareness and Education: Rising awareness regarding the profitability and environmental benefits of renewable energy among potential investors, financial institutions, and the general public is crucial. Public education campaigns, coupled with targeted outreach to financial institutions and potential investors, could help shift perceptions and increase investment in the renewable energy sector and highlight their long-term profitability.

Develop a Robust Regulatory Framework: A clear, supportive, and strong regulatory framework is essential for attracting investment in clean energy projects. So, the government should work on creating clear, consistent, and supportive policies that provide long-term certainty for investors, including streamlined procedures for project approvals and clear guidelines on tariff structures.

Leverage International Financial Support: Bangladesh should actively seek international financial support through grants, low-interest loans, and technical assistance from global organizations like the World Bank, Asian Development Bank, and Green Climate Fund. These resources can help reduce the financing gap and take international preeminent practices to the native perspective.

Reduce Customs Duties on Solar Products: Reducing customs duties on solar products could significantly boost the adoption of solar energy in Bangladesh. Estimates from several global organizations indicate that rooftop solar systems could generate 5,000 MW of electricity, with 400 MW coming from the textile sector. This transition to solar energy has the potential to yield significant fiscal savings, allowing the government to save between Tk. 5,230 crore and Tk. 11,032 crore annually from a 2,000 MW rooftop solar capacity. However, the expansion of solar energy is hindered by high import duties and taxes on solar products, which inflate installation costs. While solar panels attract a low tax rate of just 1 percent, solar inverters are subject to a hefty 37% duty. Furthermore, total taxes and tariffs (TTI) on solar equipment range from 26.2% to 58.6%, further exacerbating costs. The current Net Energy Metering (NEM) policy, which imposes a 10 MW cap, also limits the growth of solar energy. By reducing import duties on solar products and providing subsidies, particularly for SMEs, installation costs could be significantly lowered. Additionally, simplifying and standardizing the NEM process would facilitate greater expansion opportunities.

Therefore, by addressing the financing challenges in the renewable energy sector, Bangladesh can unlock its full possibility for sustainable energy development. Through the implementation of targeted policies, innovative financing mechanism and other recommendations will not only enhance access to energy finance but also quicken the transition to a greener, more sustainable energy future for the country.

13 Conclusion

The progress of renewable energy in Bangladesh is critically dependent on overcoming significant financial barriers that currently hinder progress. Whereas the country has fixed ambitious goals to intensify its renewable energy capacity, the present investment amount fall far short of what is necessary. The study highlights the insufficiencies in present financial instruments, mainly the central bank's refinancing scheme, which has seen a sharp decline in investment in main areas such as solar home systems and biogas projects. This investigation underscores the need for a comprehensive and strategic approach to energy financing, which comprises scaling up government intervention, boosting the capabilities of financial institutions, and fostering innovative financing models like green bonds and combined finance. Furthermore, enhancing regulatory frameworks and fostering strong public–private partnerships are critical to unlocking the prospective for domestic and international investment in the sector. The incorporation of these components is essential to bridging the existing financial gap, thereby enabling Bangladesh to meet its renewable energy targets and ensure a sustainable, resilient, and inclusive energy transition.

References

Abdullah-Al-Mahbub, & Islam, A. R. M. T. (2023). Current status of running renewable energy in Bangladesh and future prospects: A global comparison. *Heliyon, 9*(e14308) (2021).

Ahmed, S., Islam, T., Karim, M. A., & Karim, N. M. (2014). Exploitation of renewable energy for sustainable development and overcoming power crisis in Bangladesh. *Renewable Energy, 72*, 223–235.

Asian Development Bank. (2024). *ADB Leads Financing of First Private Sector Solar Project in Bangladesh by International Lenders*, 29 April 2024, Available from. https://www.adb.org/news/adb-leads-financing-first-private-sector-solar-project-bangladesh-international-lenders.

Asian Development Bank (ADB), Asian Development Outlook. (2021, April). *2021: Financing a Green and Inclusive Recovery*. https://www.adb.org/publications/asian-development-outlook-2021

Asian Infrastructure Investment Bank (AIIB). (2022). *AIIB Signs USD200M On-lending Facility to Support Sustainable Infrastructure Development in Bangladesh*. Available from. https://www.aiib.org/en/news-events/news/2022/AIIB-Signs-USD200M-On-lending-Facility-to-Support-Sustainable-Infrastructure-Development-in-Bangladesh.html.

Babu, M. U. (2023, August 17). Beximco Green Sukuk: A fixed income opportunity overlooked. *The Business Standard*. Available from. https://www.tbsnews.net/economy/beximco-green-sukuk-fixed-income-opportunity-overlooked-683706

Bangladesh Bank. (2022). *Policy on Green Bond Financing for Banks and FIs*. Bangladesh Bank.

Bangladesh Bank. (2023). *Quarterly review report on sustainable finance*. Bangladesh Bank.

Beck, F., & Martinot, E. (2004). Renewable energy policies and barriers. *Encyclopedia of Energy, 5*, 365–383.

Benti, N. E., Mekonnen, Y. S., & Asfaw, A. A. (2023). Combining green energy technologies to electrify rural community of Wollega. *Western Ethiopia Scientific African, 19*, e01467.

Bhattacharya, A., Gallagher, K. P., Muñoz Cabré, M., Jeong, M., & Ma, X. (2019). *Aligning G20 infrastructure investment with climate goals and the 2030 Agenda. G20 Insights*. Available from https://www.g20-insights.org/policy_briefs/aligning-g20-infrastructure-investment-with-climate-goals-and-the-2030-agenda/

Bhuiyan, M. R. A., Mamur, H., & Begum, J. (2021). A brief review on renewable and sustainable energy resources in Bangladesh. *Cleaner Engineering and Technology, 4*, 100208.

Buchner, B., Clark, A., Falconer, A., Macquarie, R., Meattle, C., Tolentino, R., & Wetherbee, C. (2019). Global landscape of climate finance. *Climate Policy Initiative*. https://www.climatepolicyinitiative.org/wp-content/uploads/2019/11/2019-Global-Landscape-of-Climate-Finance.pdf

Burney, J. A., & Naylor, R. L. (2012). Smallholder irrigation as a poverty alleviation tool in Sub-Saharan Africa. *World Development, 40*(1), 110–123. https://doi.org/10.1016/j.worlddev.2011.05.007

Chowdhuri, M. Y., Khatun, E., Hossain, M. M., & Halim, M. A. (2023). Current challenges and future prospects of renewable energy: A case study in Bangladesh. *International Journal of Innovative Science and Research Technology, 8*, (4).

Chowdhury, S. A. (2024). *Where is Bangladesh in the global race towards renewable* energy? Available from, https://www.thedailystar.net/opinion/views/news/where-bangladesh-the-global-race-towards-renewable-energy-3297006. Accessed 2024/05/17

Clancy, J. S., & Skutsch, M. (2013). The gender-energy-poverty nexus: Finding the energy to address gender concerns in development. *Renewable and Sustainable Energy Reviews, 25*, 409–419. https://doi.org/10.1016/j.rser.2012.10.035

Climate Bonds Initiative. (2021). *2020 Green Bond Market Summary*. Climate Bonds Initiative. Available from https://www.climatebonds.net/resources/reports/2020-green-bond-market-summary

Cochran, I., Hubert, R., Marchal, V., & Youngman, R. (2014). *Public financial institutions and the low-carbon transition: Five case studies on low-carbon infrastruc-*

ture and project investment (OECD Environment Working Papers, No. 72). OECD Publishing, https://documents1.worldbank.org/curated/en/720991536072019085/pdf/129292-REVISED-Innovative-Financing-for-Development-FINAL.pdf. Accessed 2024/04/20

Dib, L., Falk, J., & Schneider, T. (2013). *Microfinance and renewable energy: Investing in a sustainable future*. UNEP Collaborating Centre for Climate & Sustainable Energy Finance. Available from https://www.eib.org/attachments/ev/ev_microfinance_and_renewable_energy_en.pdf

Friebe, C. A., von Flotow, P., & Täube, F. A. (2013). Exploring the link between products and services in low-income markets—Evidence from solar home systems. *Energy Policy, 52*, 760–769.

Gielen, D., Boshell, F., Saygin, D., Bazilian, M. D., Wagner, N., & Gorini, R. (2019). The role of renewable energy in the global energy transformation. *Energy Strategy Review, 24*, 38–50.

Hossain, I. (2024). Refinancing sustainable energy projects in Bangladesh. *The Daily Star*, Feb 20, 2024. Available from. https://www.thedailystar.net/anniversary-supplement-2024/innovation-key-the-future/news/lighting-the-future-renewables-3547416

Hossain, I., Chisti, I. A. (2022). Does renewable energy have a future in Bangladesh? *The Daily Star*, (Nov 28, 2022). Available from. https://www.thedailystar.net/opinion/views/news/does-renewable-energy-have-future-bangladesh-3181761. Accessed 2024/05/08

Hossain, M. S., Biswas, S., & Uddin, R. M. (2023). Sustainable energy transition in Bangladesh: Challenges and pathways for the future. *Engineering Reports, 6*(1), e12752.

IFC. (2020). *IFC supports Bangladesh Bank to promote domestic green bond market*. October 31, 2020. Available from. https://www.ifc.org/en/pressroom/2019/ifc-supports-bangladesh-bank-to-promote-domestic-green-bond-market

Inderst, G., Kaminker, C., & Stewart, F. (2012). *Defining and measuring green investments: Implications for institutional investors' asset allocations* (OECD Working Papers on Finance, Insurance and Private Pensions, No. 24). OECD Publishing.

International Energy Agency (IEA). (2003). *Renewables for power generation: Status and prospects* (p. 18). OECD.

IRENA (International Renewable Energy Agency). (2018). *Renewable Energy Market Analysis Southeast Asia*. IRENA. https://www.irena.org/-/media/Files/IRENA/Agency/Publication/2018/Jan/IRENA_Market_Southeast_Asia_2018.pdf

IRENA (International Renewable Energy Agency). (2020). *Renewable Energy and Jobs – Annual Review 2020*. IRENA. Available from IRENA website: https://www.irena.org/publications/2020/Sep/Renewable-Energy-and-Jobs-Annual-Review-2020

Islam, A. S., Islam, M., & Rahman, T. J. R. E. (2006). Effective renewable energy activities in Bangladesh. *Renewable Energy, 31*, 677–688.

Islam, M. R., Islam, M. R., & Beg, M. R. A. (2008). Renewable energy resources and technologies practice in Bangladesh. *Renewable and Sustainable Energy Reviews, 12*, 299–343.

Islam, M. S., Khan, A. M. H. R., Nasreen, S., Rabbi, F., & Islam, M. R. (2011). Renewable energy: the key to achieving sustainable development of rural Bangladesh. *Journal of Chemical Engineering, 26*, 9–15.

Johnson, B. R., Connolly, E., & Carter, T. S. (2011). Corporate social responsibility: The role of Fortune 100 companies in domestic and international natural disasters. *Corporate Social Responsibility and Environmental Management, 18*(6), 352–369.

Karim, A. T., Hossain, S., Jami, T. A., & Zaman, A. (2023). *Unlocking Bangladesh's renewable energy potential*, July 13, 2023. Available from: https://www.thedailystar.net/opinion/views/news/unlocking-bangladeshs-renewable-energy-potential-3367231

Khan, M., Ahsan, T., & Alam, M. (2014). Renewable energy in Bangladesh: Prospects and policy issues. *Renewable Energy, 62*, 1–9.

Khan, M. Z. H., & Sultana, S. (2024). Renewable energy: A nature-based sustainable solution for Smart Bangladesh. *TBS News*, 29 January, 2024, Available from. https://www.tbsnews.net/supplement/renewable-energy-nature-based-sustainable-solution-smart-bangladesh-783198#google_vignette

Khan, S. M., & Ali, S. (2021). *Green bonds in Bangladesh: A new incentive to save our environment.* Light Castle Partners. Available from. https://www.lightcastlebd.com/insights/2022/02/green-bonds-in-bangladesh-a-new-incentive-to-save-our-environment/

Khandker, S. R., Samad, H. A., Sadeque, Z. K. M., Asaduzzaman, M., Yunus, M., & Haque, A. K. E. (2014). *Surge in solar-powered homes: Experience in off-grid rural Bangladesh.* World Bank Publications. https://doi.org/10.1596/978-1-4648-0190-1

Lai, J.-P., Chang, Y.-M., Chen, C.-H., & Pai, P.-F. (2020). A survey of machine learning models in renewable energy predictions. *Applied Sciences, 10*(5975).

Lam, P. T. I., & Law, A. O. K. (2016). Crowdfunding for renewable and sustainable energy projects: An exploratory case study approach. *Renewable and Sustainable Energy Reviews, 60*, 11–20.

Mahmud, H., & Roy, J. (2021a). Barriers to overcome in accelerating renewable energy penetration in Bangladesh. *Sustainability, 13*(14), 1–27.

Mahmud, H., & Roy, J. (2021b). Barriers to overcome in accelerating renewable energy penetration in Bangladesh. *Sustainability, 13*, 7694.

Masud, M. H., Ananno, A. A., Arefin, A. M. E., Ahamed, R., Das, P., & Joardder, M. U. H. (2019). Perspective of biomass energy conversion in Bangladesh. *Clean Technologies and Environmental Policy, 21*(4), 719–731. https://doi.org/10.1007/s10098-019-01668-2

Masukujjaman, M., Alam, S. S., Siwar, C., & Halim, S. A. (2021). Purchase intention of renewable energy technology in rural areas in Bangladesh: Empirical evidence. *Renewable Energy, 170*, 639e651.

Michaelowa, A., Hoch, S., Weber, A. K., Kassaye, R., & Hailu, T. (2020). Mobilising private climate finance for sustainable energy access and climate change mitigation in Sub-Saharan Africa. *Climate Policy, 21*(1), 47–62. https://doi.org/10.1080/14693062.2020.1796568

Ministry of Power, Energy and Mineral Resources. (2011). Power System Master Plan 2010.

Ministry of Power, Energy and Mineral Resources, Government of Bangladesh. (2002). *Renewable Energy Policy of Table of Contents.*

Ministry of Power, Energy and Mineral Resources, Government of Bangladesh. (2016). *Power System Master Plan 2016 Summary*. September.

Moazzem, K. G., & Hridoy, M. A. (2023). *Renewable energy policy (draft) 2022 – A comprehensive assessment.* Centre For Policy Dialogue, Policy Brief 2.

Polzin, F., Migendt, M., Täube, F. A., & von Flotow, P. (2015). Public policy influence on renewable energy investments—A panel data study across OECD countries. *Energy Policy, 80*, 98–111.

Rahman, M. (2021). *Renewable energy development and sustainable project financing*. The Financial Express, Aug 23, 2021. Available from. https://thefinancialexpress.com.bd/views/views/bangladeshs-economic-outlook-for-2024-25#google_vignette

Rasel, A. R. (2018). *Guideline for net-metering for on-grid consumers finalized.* Dhaka Tribune.

Reiche, D., & Bechberger, M. (2004). Policy differences in the promotion of renewable energies in the EU member states. *Energy Policy, 32*, 843–849.

REN21. (2020). *Renewables 2020: Global Status Report.* REN21. Available from https://www.ren21.net/reports/global-status-report/

Rolffs, P., OckWell, D., & Byrne, R. (2015). Beyond technology and finance: Pay-as-you-go sustainable energy access and theories of social change. *Environment and Planning A, 47*, 2609–2627.

Sapkota, A., Yang, H., Wang, J., & Lu, Z. (2013). Role of renewable energy technologies for rural electrification in achieving the Millennium Development Goals (MDGs) in Nepal. *Environmental Science & Technology, 47*(3), 1184–1185.

Sovacool, B. K. (2012). Energy security: Challenges and needs. *WIREs Energy and Environment, 1*(1), 1–117. https://doi.org/10.1002/wene

SREDA. (2024). *National Database of Renewable Energy.* Available from: https://ndre.sreda.gov.bd/index.php?id=8

SREDA Homepage., https://sreda.gov.bd/, Accessed 2024/04/20.

Strielkowski, W., Civín, L., Tarkhanova, E., Tvaronaviciene, M., & Petrenko, Y. (2021). Renewable energy in the sustainable development of electrical power sector: a review. *Energies, 14*, 8240.

Tachev, V. (2024) *Renewable Energy in Bangladesh – Current Trends and Future Opportunities.* Available from, https://energytracker.asia/renewable-energy-in-bangladesh-current-trends-and-future-opportunities/. Accessed 2024/06/09

Tarik-ul-Islam, M., & Ferdousi, S. (2007). Renewable energy development – challenges for Bangladesh. *Energy & Environment, 18*(3–4), 421–430. https://doi.org/10.1260/095830507781076149

The Daily Star. (2016). *Feed-in-Tarrif (FITs): Rays of hope for renewable green energy - Feb 1, 2016*, Available from. https://www.thedailystar.net/25th-anniversary-special-part-1/feed-tarrif-fits-rays-hope-renewable-green-energy-210454

The World Bank. (2018). Innovative financing for development: Scaling up after the Addis Ababa financing for development conference. In *Innovative financing for development.* Available from. https://documents1.worldbank.org/curated/en/720991536072019085/pdf/129292-REVISED-Innovative-Financing-for-Development-FINAL.pdf. Accessed 2024/04/20

Tiruye, G. A., Besha, A. T., Mekonnen, Y. S., Benti, N. E., Gebreslase, G. A., & Tufa, R. A. (2021). Opportunities and challenges of renewable energy production in Ethiopia. *Sustainability, 13*, 10381.

Uddin, M. M., & Park, J. (2021). Renewable Energy Resources in Bangladesh: Current Status and Future Prospects. In: *IEEE 48th Photovoltaic Specialists Conference (PVSC) | 978-1-6654-1922-2/21/$31.00*. IEEE | https://doi.org/10.1109/PVSC43889.2021.9518861.

Uddin, M. N., Rahman, M. A., Mofijur, M., Taweekun, J., Techato, K., & Rasul, M. G. (2019). Renewable energy in Bangladesh: Status and prospects. In *2nd International Conference on Energy and Power*, ICEP2018, 13–15 December 2018, Sydney, Australia (pp. 655–661). Energy Procedia, 160. https://doi.org/10.1016/j.egypro.2019.02.206

Vohra, K., Vodonos, A., Schwartz, J., Marais, E. A., Sulprizio, M. P., & Mickley, L. J. (2021). Global mortality from outdoor fine particle pollution generated by fossil fuel combustion: Results from GEOS-Chem. *Environmental Research, 195*, 110754. https://doi.org/10.1016/j.envres.2021.110754

Warren, C. M. J. (2013). Business models for renewable energy in the built environment IEA-RETD. *Property Management, 31*(4), 359–360.

World Bank. (2019). *Bangladesh Receives $185 Million World Bank Financing for Renewable Energy*. Available from. https://www.worldbank.org/en/news/press-release/2019/08/29/bangladesh-receives-185-million-world-bank-financing-for-renewable-energy

World Bank. (2020). *Pioneering the Green Sukuk: Three years on*. World Bank. Available from. https://www.worldbank.org/en/country/malaysia/publication/pioneering-the-green-sukuk-three-years-on

World Bank. (2022). *Bangladesh: World Bank Supports Reliable Electricity Supply, Clean Energy for 9 Million People*. World Bank, June 29, 2022. Available from. https://www.worldbank.org/en/news/press-release/2022/06/29/bangladesh-world-bank-supports-reliable-access-to-electricity-supply-clean-energy-for-9-million-people.

World Bank Group. (2007). *Energy*. Available from, http://www.worldbank.org/energy, accessed 2024/06/20

Zhang, B., & Wang, Y. (2019). The effect of green finance on energy sustainable development: A case study in China. *Emerging Markets Finance and Trade, 57*(12), 3435–3454.

Zhang, D. (2018). Energy finance: Background, concept, and recent developments. *Emerging Markets Finance and Trade, 54*(8), 1687–1692.

Zhang, S., Andrews-Speed, P., & Zhao, X. (2013). Political and institutional analysis of the successes and failures of China's wind power policy. *Energy Policy, Elsevier, 56*(C), 331–340.

Mohammad Monzur Morshed Bhuiya is a Professor and Dean of Business Studies faculty at Jagannath University, Bangladesh, specializing in Finance and Banking, Fintech, Financial Inclusion, E-Commerce, and Microfinance. He has published extensively in renowned journals, authored book chapters, and presented at international conferences. Known for his expertise in research methodology and data analysis, he conducts training in STATA, Econometric and Financial Modeling, and SPSS. Currently, he guides to his MPhil & PhD fellow on the research projects titled—Determinants of Efficiency, Profitability, and Stability of Commercial Banks in Bangladesh; Simultaneous Relationship among Risk, Profitability, and Efficiency of Banks, the inception of Bangladesh's derivatives market; microcredit's role in poverty alleviation; and Role of Fintech Adoption on digital financial inclusion.

Aminul Haque Russel is currently working as an Assistant Professor, Department of Business Administration at the Daffodil Institute of Information Technology (DIIT), Bangladesh. Mr. Russel has published many research papers in renowned peer-reviewed journals, book chapters and presented his research works at international conferences. He is conducted different training and workshop on research methodology, literature review, and data analysis by using SMRTPLS 4.0, SPSS, and AMOS. In addition to his academic achievements, Mr. Russel has completed different funded research projects. In 2020, he achieved "Best Faculty Awards". His research work has focused on financial technology (fintech) & digital financial inclusion, and green banking & green finance.

Explainable AI in Energy Forecasting: Understanding Natural Gas Consumption Through Interpretable Machine Learning Models

Farhana Sultana Eshita, Tasnim Jahin Mowla, and Abu Bakar Siddique Mahi

1 Introduction

In 2024, it is forecasted that the global demand for natural gas might rise by 2.5%. Predicted chillier winters in 2024, compared to the mild ones in 2023, might lead to an increased demand for heating in domestic and commercial heating (2024). Compared to other fossil fuels, natural gas is preferred for its large supply, adaptability, low price, and ecologically friendly nature. This makes it an ideal option for many uses in households and businesses, such as power generation, heating, and vehicle fueling. As a steady supply of gas is so important, precise consumption prediction is a must for efficient energy management. This ensures a consistent and efficient utilization of energy resources. Figure 1, a visual depiction is presented of the statistics regarding United States' natural gas consumption from 1995–2023 (2024). The research shows that in 2021, the United States consumed a total of 32.51 trillion cubic feet of natural gas. This natural gas consumption has been steadily increasing over time.

From an analytical point of view, it is essential to develop a comprehensive prediction model which is capable of predicting natural gas consumption in all sectors, although explicit prediction models (2023) for distinct sectors—like industrial, commercial, or residential consumers—are currently available. Employing such models for each industry can be laborious and expensive. The utilization of a comprehensive model which includes every sector enables a more efficient and effective method of predicting natural gas consumption. A more precise representation of the total gas consumption scenario, improved resource allocation, and enhanced decision-making in energy management are all feasible through the

F. S. Eshita · T. J. Mowla · A. B. S. Mahi (✉)
Department of Computer Science and Engineering, University of Asia Pacific, Dhaka, Bangladesh
e-mail: 20101023@uap-bd.edu; 21101069@uap-bd.edu; 20101021@uap-bd.edu

M. Z. Abedin, W. Yong (eds.), *Machine Learning Technologies on Energy Economics and Finance*, International Series in Operations Research & Management Science 367, https://doi.org/10.1007/978-3-031-94862-6_3

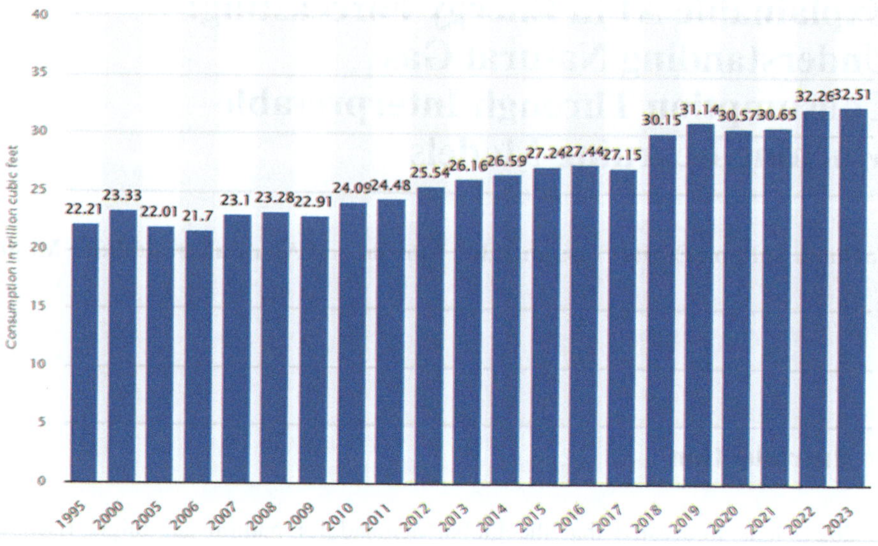

Fig. 1 Natural gas use in the United States from 1995 to 2023

implementation of a comprehensive model. In (2022), the authors developed an Android app to read gas meters. The app uses optical character recognition (OCR) which is expensive, complicated and requires several devices for support.

Song et al. (2020) present a method for monitoring substation instruments based on image recognition. The instrument displays are located and interpreted by the system which uses a Gaussian difference model and SIFT. The experiment's outcomes demonstrate how well the system recognizes graphical interfaces on substation equipment. Although the instrument identification system works well for substation equipment with graphical user interfaces, it is not suitable for monitoring natural gas meters because they are characterized by their numerical displays. An automated monitoring system for gas alarm devices in coal mines was created by Liu et al. (2013) using image recognition technology. By automatically identifying the images captured by methane detectors, their proposed digital recognition algorithm obtained an accuracy of over 99.9%. One downside of this method is that its algorithm has a relatively high computational complexity, which requires further optimization for real-time application. We present an innovative approach in this paper that addresses the challenges of previous approaches. A light-weight CB model-based system is developed, which incorporates explainable AI (XAI) tools to ensure precise forecasts of natural gas consumption. By combining explainability and deep learning, our model not only makes accurate predictions, but it also gives us a lot of information about what makes people use gas. This gives us a way to make decisions that is clear and easy to understand. The following briefly highlights the key contributions of this paper:

- We provide a comparative assessment of eight machine learning models for reliable forecasting of natural gas. The CB model demonstrated exceptional performance with an impressive R-squared score of 99.81%, surpassing all other methods.
- We analyze United States natural gas consumption from 2014 and 2024.
- We demonstrate how the best-performing model generates outcomes using two different explainability methods, providing insight into the decision-making procedure of the model.

The remaining content of this paper is structured as follows:

A brief overview of relevant research on the subject in question is provided in the **Literature Review** section. The **Methodology** section provides a detailed overview of the dataset and the different strategies used. Within the **Results** section, the experiment's results are examined and summarized. The **Conclusion** section ultimately provides a finale for the work.

2 Literature Review

2.1 Statistical Approaches

In four regulated industrial regions of Turkey, Cihan (2022) investigated the impact of COVID-19 lockdowns on the use of electricity and natural gas and discovered significant declines in these usage patterns. ARIMA and Holt-Winters models were developed to forecast the consumption of natural gas and electricity. The most effective models considering the data on natural gas and electricity consumption were found to be ARIMA(0,0,2)(2,1,0)$_7$ and ARIMA(0,0,2)(0,1,1)$_7$, respectively. The value for $MAPE_{Electricity}$ was 1.37%, $RMSE_{Electricity}$ was 87.2, $R^2_{Electricity}$ was 0.99, $MAPE_{Gas}$ was 5.42% and $RMSE_{Gas}$ was 50.9, R^2_{Gas} was 0.92. Using data from 2015 to 2022 acquired through the US Energy Information Administration (EIA), Bhuiyan et al. (2024) used advanced statistical methods to analyze fuel usage patterns in the production of electricity in the United States. The methodologies encompassed all four benchmark techniques, namely Mean, Naïve, Drift, and Seasonal Naïve, in addition to Seasonal and Trend Decomposition using Loess (STL), exponential smoothing (ETS), and the Autoregressive Integrated Moving Average (ARIMA) approach. The most minimal RMSE of 20,687.46 for natural gas consumption is produced by the ETS model. There are issues with the paper's dependence on historical records and forecasting methods, such as the potential for unanticipated changes in technology, economics, and policy to affect future energy trends and the requirement for more research into the capacities for regional energy production and consumption. Using historical daily natural gas consumption data from the Ghana National Gas Company spanning three years, from January 2020 to December 2022, Broni-Bediako et al. (2024) forecasted Ghana's daily natural gas consumption using both the ARIMA and SARIMA models. The results showed that both models can forecast consumption with a good degree of accuracy, with the SARIMA model slightly outperforming the ARIMA model with an RMSE of 22.25 and a Mean Absolute Percentage Error (MAPE) of 6.96%, compared to an RMSE of 23.8 and a MAPE of 7.29% for the ARIMA model. The authors did note that although the ARGIMA and SARIMA models perform well in terms of prediction, their applicability is restricted because of their short-term focus.

2.2 *Machine Learning Approaches*

Dual Convolution with Seasonal Decomposition Network (DCSDNet) is a novel technique for natural gas consumption forecasting, which was introduced by Ding et al. (2023) using actual daily city-level natural gas consumption information collected from January 2016 to June 2021. DCSDNet received 19.4977 for RMSE and 0.9063 for R^2 in terms of daily natural gas consumption forecasts. With a forecast spectrum ranging from two to seven days, DCSDNet, LSTM, CNNLSTM, and TCN execute the multi-step forecasting. With a 7-day prediction horizon, the suggested DCSDNet obtained 28.3979 for RMSE and 0.8012 for R^2. Aminu et al. (2023) used a hybrid ensemble regression machine learning approach to forecast the demand for natural gas in residential settings. Regression algorithms, which include support vector regression, decision tree regression, K-nearest neighbor, and linear regression, are combined in the hybrid ensemble approach. The Kaggle machine learning repository provided the study's dataset, which included monthly natural gas consumption data from January 1997 to August 2020. Achieved accuracy of 97.48707913, R^2 of 0.792579296, MAE of 0.721612403, and MSE of 1.164821453 are the results of the Hybrid Ensemble (HE). Gaweł and Paliński (2024) used global and local forecasting techniques to predict hierarchical long-time series of household natural gas consumption in Poland using a data set of 46,297 observations that represented natural gas consumption in Polish territorial units. With an RMSE of 4970 and a MAPE of 7.1%, MLP Global Ex produces the best results when the average performance of global models for hierarchical forecasts harmonized with the middle-out approach is analyzed.

2.3 *Approaches with eXplainable Artificial Intelligence (XAI)*

Sim et al. (2022) used data from a university building in Seoul, Republic of Korea, to present a methodology using XAI for energy consumption forecasting. With an R^2 of 0.871, MAE of 2.176, and MSE of 9.870, the prediction model demonstrated high accuracy. Based on their influence, three groups were created from the input variables using XAI analysis. When compared to other cases ($p < 0.05$ or 0.01), models that included variables from the Strong + Ambiguous or Strong groups showed better prediction performance (R^2 of 0.917, MAE of 1.859, MSE of 6.639

Table 1 Overview of literature review

Research	Year	Modeling Technique	Performance
Cihan (2022)	2022	ARIMA(0,0,2)(2,1,0)$_7$	R^2_{Gas} of 0.92
Ding et al. (2023)	2023	DCSDNet	R^2 of 0.9063 for daily natural gas consumption prediction and R^2 of 0.8012 for a 7-day prediction horizon.
Aminu et al. (2023)	2023	Hybrid Ensemble (HE)	R^2 of 0.792579296
Handayani et al. (2023)	2023	XGBoost	R^2 of 0.95
Bhuiyan et al. (2024)	2024	ETS model	Lowest RMSE of 20,687.46
Clement et al. (2024)	2024	SCAL	R^2 of 0.68649

for Strong + Ambiguous; R^2 of 0.916, MAE of 1.816, MSE of 6.663 for Strong). The Strong and Strong + Ambiguous groups did not differ significantly, indicating that concentrating on the Strong group variables (Year, E-Diff, Hour, Temp, Surface-Temp) as identified by XAI produced good prediction outcomes. With an XGBoost Regressor model that takes operational and environmental factors into account, Handayani et al. (2023) were able to predict fuel oil consumption in cargo container vessels with a high degree of predictive performance with an R^2 of 0.95 and MAE of 10.78 kg/h. By identifying the major controllable and uncontrollable factors influencing fuel consumption, the study uses SHAP analysis to provide region-specific insights for improving energy efficiency and operational strategies in the maritime industry. To improve energy consumption prediction models, Clement et al. (2024) introduced the novel SHAP Clustering-based Adaptive Learning (SCAL) technique. For the Financial Distress data set, the model's testing accuracy is 96.190. The model's testing R^2 is 0.68649 and its RMSE is 0.70856 for the Power data set. Table 1 provides the highlights of the literature review.

3 Methodology

A new methodology has been developed that encompasses data collection, preprocessing, model training and testing, performance evaluation, and model validation using explainable artificial intelligence (XAI) tools like SHAP. The goal is to determine the most accurate pipeline for predicting natural gas consumption. Each aspect of the methodology is separately illustrated in Fig. 2, which offers a top-down view of the process. The initial steps include basic preprocessing tasks such as checking for null values, scaling, and encoding. After that, the data is divided in an 80:20 ratio into training and testing sets. The models are trained using a variety of machine learning techniques, and their performance is evaluated using five metrics: R^2 score, mean absolute error (MAE), mean squared error (MSE), root mean squared error (RMSE), and mean absolute percentage error (MAPE). The results for each metric are compiled into individual tables to allow for thorough

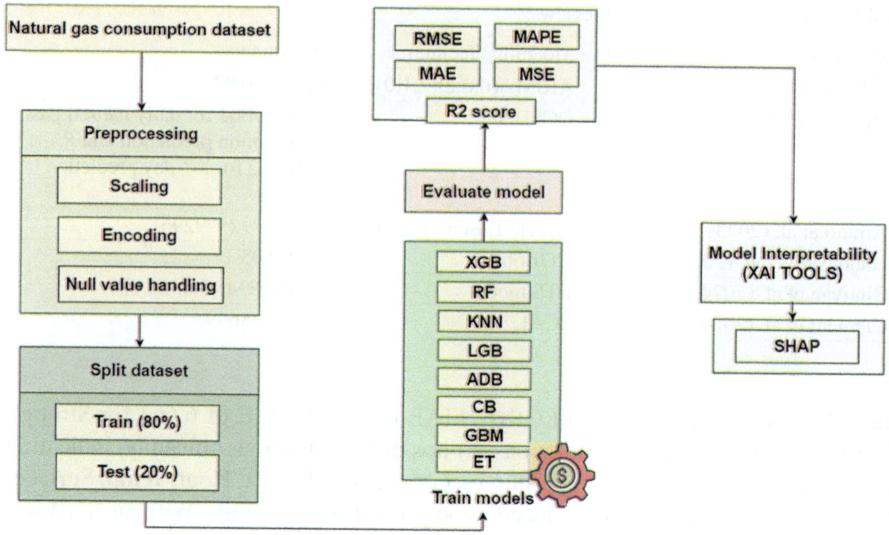

Fig. 2 Outline of the proposed methodology

comparison and analysis. The findings show that the CatBoost (CB), Extremely Randomized Trees (ERT), and Random Forest (RF) models achieved R^2 scores of 99.81%, 99.69%, and 99.60%, respectively. Due to the superior performance of the CatBoost model, an XAI tool was used on it, providing deeper insights into its underlying mechanisms and decision-making processes. A detailed overview of the models used, the methods for measuring performance, and the insights gained from the XAI tool are provided below.

3.1 Dataset Description

This dataset sourced from Kaggle (2024) includes monthly statistics on natural gas consumption for the United States from January 2014 to January 2024, segmented by state, industry (automotive fuel, commercial, industrial, residential, and electric power), and particular consumption process. The Energy Information Administration (EIA) of the United States provided the data. Table 2 provides a comprehensive overview of the structure of the dataset.

Table 2 Overview of dataset

Feature	Overview
Duoarea	State abbreviation
Area-name	State name
Product	The energy product code
Product-name	Name of the energy product
Process	The process or sector code
Process-name	Specific consumption process within the sector
Value	Consumption amount
Year	The year for the data entry
Month	The month for the data entry
Series	A unique identifier for the data series
Series description	A description of the data series
Units	Monthly consumption in millions of cubic feet (MMCF)

3.2 *Data Preprocessing*

Several critical actions are executed during the data preprocessing phase, which is essential for assuring the quality of the data and preparing it for analysis. Initially, ordinal encoding is employed to convert categorical features into numerical values. This method assigns a unique integer to each category, preserving the order, as shown in Eq. (1):

$$\text{Ordinal}\ (l_i) = o_i \tag{1}$$

where Ordinal(l_i) denotes the encoded integer for category l_i and o_i is assigned ordinal value.

The subsequent step is the standardization of numerical features, which involves adjusting the data to have a mean of 0 and a standard deviation of 1. This is mathematically represented as Eq. (2):

$$x' = \frac{x - \omega}{\varphi} \tag{2}$$

Within this equation, $x^{'}$ is the standardized feature, x is the original feature value, ω represents the mean, and φ is the standard deviation of the feature. This standardization assures that the features are consistently scaled, which is vital for the effective development of the model.

To handle missing values, any data points containing such values are removed, represented by the ruler:

$$\text{Remove } x_i \text{ if } x_i \text{ is null}$$

where x_i refers to a specific data point. Additionally, duplicate entries are eliminated to preserve data integrity, described by the following condition:

$$\text{Remove } x_i \text{ if } x_i = x_j \text{ for } i \neq j$$

These preprocessing measures are crucial for improving the dataset's quality, ensuring it is ready for subsequent analysis and modeling efforts.

3.3 *Exploratory Data Analysis*

The bar plot of Fig. 3 shows the average annual natural gas consumption from 2014 to 2024. It shows that the highest consumption rate was in 2024 and the lowest consumption rate was in 2014. From 2014 to 2018, the consumption value was between 20000 MMCF to 30000 MMCF. In 2019, 2020, 2022, and 2023 the gas consumption value was between 30000 MMCF and 35000 MMCF.

The bar plot of Fig. 4 represents the natural gas consumption from January to December. All the months are encoded serially from 1 to 12. The plot shows that the highest consumption value was in January, which is 40,000 MMCF and the lowest value was in May, which is 24,000 MMCF. In May, June, and September the consumption rate is low and their consumption value is under 25,000 MMCF. The consumption value is extremely high in January and December, their consumption values lie in 35,000 MMCF to 4000 MMCF.

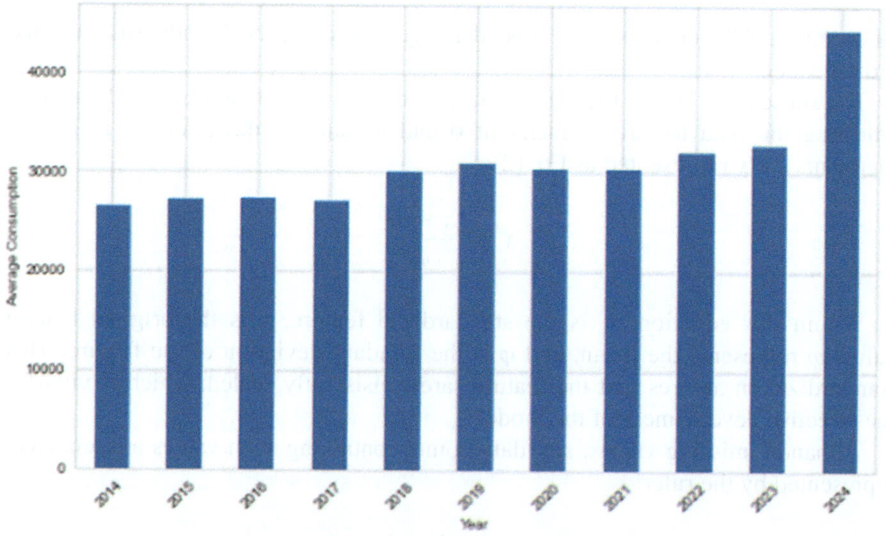

Fig. 3 Average annual natural gas consumption

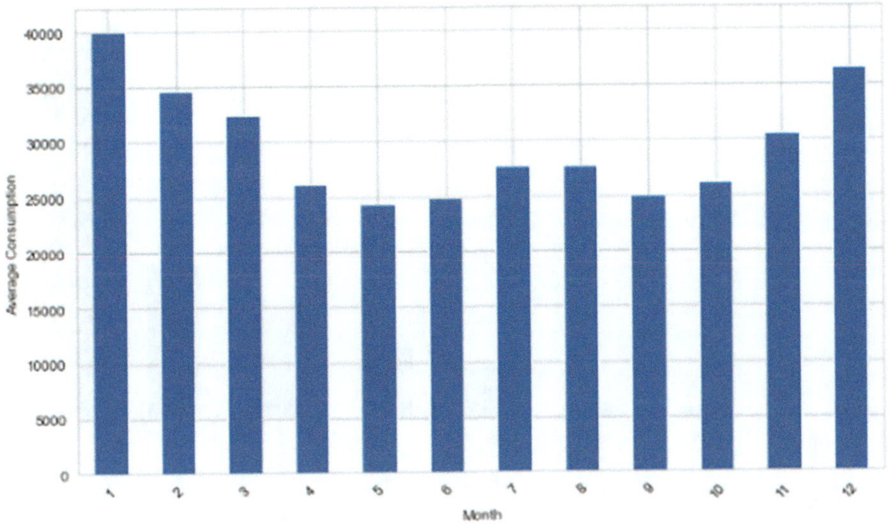

Fig. 4 Average monthly natural gas consumption

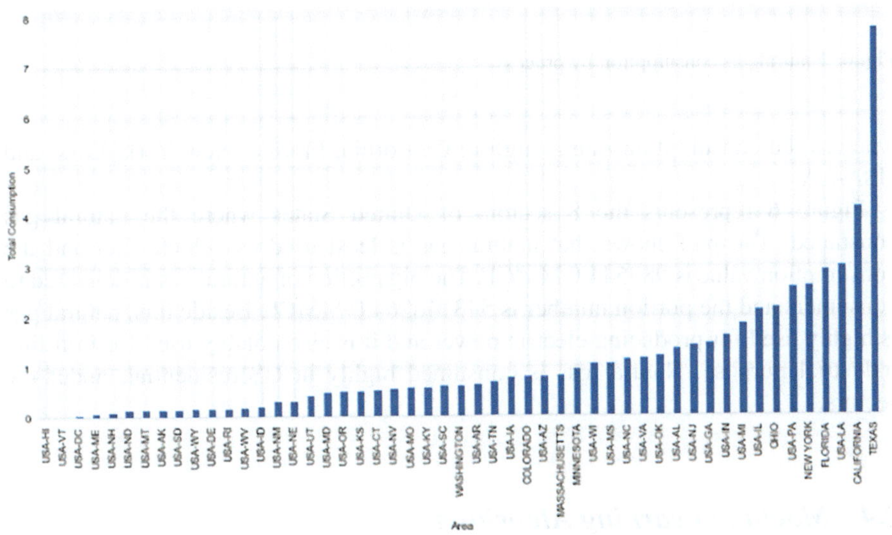

Fig. 5 Natural gas consumption in US area

Figure 5 shows the natural gas consumption in 51 areas of the United States. Among them, the natural gas consumption rate is extremely high in Texas, which is 571970195.0 MMCF. The least natural gas consumed in USA-HI and their consumption value is 59,968.0. Besides USA-HI the consumption rate is also low in USA-VT, USA-DC, USA-ME, USA-AK, USA-SD, USA-DE, and USA-RI,

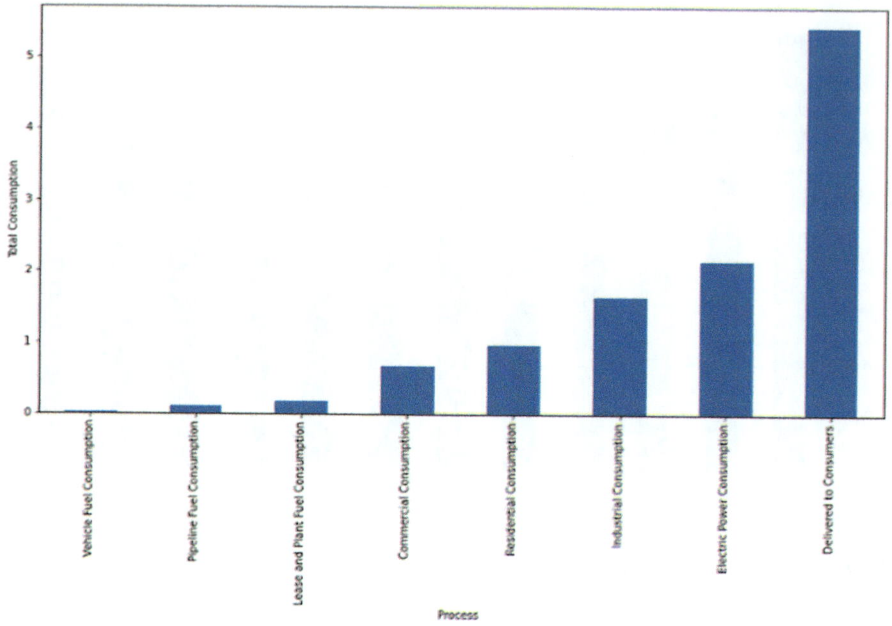

Fig. 6 Natural gas consumption by process

whereas the consumption rate is high in California, Florida, New-York, Loss, and USA-LA.

Figure 6 represents the 8 sectors of United States where the natural gas consumed. The plot shows that natural gas is least used as vehicle fuel and the consumption value is 989544.0 MMCF. The big portion of natural gas is delivered to consumers and the portion number is 543148364.0 MMCF. Beside this, natural gas is highly used for producing electric power and it is also notably used for fulfilling industrial purpose. Natural gas is consumed highly in US residential, under the sectors.

3.4 Machine Learning Algorithm

Extreme Gradient Boosting (XGB): Machine learning algorithm XGB is a member of the boosting algorithm family. It builds a sequential ensemble of weak learners (typically decision trees) such that every new learner corrects the errors of the previous ones. The culmination of all the weak learners' predictions yields the final forecast. XGB significantly outperforms the GBDT algorithm. The L1 and L2 regularization terms are introduced by XGB. Only the first derivative is utilized when the model is optimized by GBDT. The loss function undergoes a second-order Taylor expansion by XGB. To minimize computation and avoid overfitting, XGB allows column sampling. Following each iteration, XGB distributes the learning speed among the leaf nodes, lowers the weight of each tree, and improves the space available for learning that comes after (2022). The objective function of XGB can be expressed in its general form as Eq. (3):

$$Obj\,(\Theta) = \sum_{i=1}^{i=1} Loss\left(y_i, \hat{y}_i\right) + \sum_{k=1}^{k=1} \Omega\left(f_k\right) \tag{3}$$

where Θ represents the set of parameters (including both the parameters of individual weak learners and global parameters), n is the number of training examples, K is the ensemble's number of weak learners (trees), $f_k\,(x)$ denotes the prediction of the kth weak learner for input x, $\hat{y}_i$ is the predicted output of the ensemble for the ith training example, y_i is the true output for the ith training example, $Loss\left(y_i, \hat{y}_i\right)$ represents the loss function, which calculates the difference between the actual and predicted outcomes, and $\Omega(f_k)$ is a regularization term that restricts the complexity of individual weak learners to control overfitting.

Random Forest (RF): An ensemble learning technique for classification and regression problems is the RF algorithm. It generates an extensive number of decision trees throughout training, using which it extracts the class mode (classification) or the average forecast (regression). To build a regression tree, the data is separated into a series of rectangles, one after the other. A criterion (such as the residual sum of squares) is minimized for every split variable. Once split into the feature space, two regions are stored as nodes. Until a stopping condition is met—for example, the minimum number of observations in terminal nodes—these nodes are kept apart further. After that, the response variable is predicted by averaging the results for each group (2023).

For regression:

Let T be the number of trees in the forest, and (x) be the prediction of the ith tree.

The final prediction (x) for a given input x is determined by averaging over all the trees as shown in Eq. (4):

$$F(x) = \frac{1}{T} \sum_{i=1}^{T} f_i(x) \tag{4}$$

K-Nearest Neighbors (KNN): A straightforward and understandable technique for regression and classification is the KNN algorithm. It bases its predictions in the feature space on the majority class of its k nearest neighbors. This algorithm relies heavily on the distance between points; several distance metrics, including the Euclidean, Manhattan, and Minkowski distances, can be applied. The fundamental idea behind the KNN method seeks to locate the k known samples with class labels that are closest to a new sample. One may then predict or classify the new sample using the class labels associated with these k samples. When classifying a new sample in classification tasks, the KNN algorithm uses the class that emerges the most commonly among the k nearest neighbors. The KNN algorithm uses the average value of the k nearest neighbors to determine the predicted value of the new sample in regression tasks (2023). Equation (5) is used to calculate the Euclidean distance.

$$\text{Euclidean Distance} = \sqrt{\sum_{i=1}^{n} (p_i - q_i)^2} \tag{5}$$

Here, p_i and q_i is the ith dimension of points p and q, and n is the number of dimensions

Light Gradient Boosting Machine (LGB): The open-source gradient boosting framework LGB was created by Microsoft. Especially for large-scale datasets and high-dimensional features, it is made for efficient and distributed training. Written in C++, LGB offers interfaces for Python, R, and other programming languages. The LGB model performs exceptionally well when processing large amounts of data. Its primary distributed computing technique involves splitting the data into multiple parts and applying gradient operations to each part to ultimately realize the model's prediction accuracy (2024). The main goal of LGB is to minimize a particular loss function, which is commonly shown as Eq. (6):

$$\mathcal{L}(\theta) = \sum_{i=1}^{n} l(y_i, F(x_i)) + \sum\nolimits_{i=1}^{K} \Omega(f_i) \tag{6}$$

Here, $\mathcal{L}(\theta)$ is the overall objective function, l is the loss function, which quantifies the variation between the actual label y_i and the predicted value $F(x_i)$, $\Omega(f_i)$ is the regularization term that restricts complexity in the model to prevent overfitting, θ represents the parameters of the model, n is the number of samples, and K is the number of trees.

Adaptive boosting (ADB): For issues involving regression and classification, one well-known ensemble learning approach can be identified as ADB. It fuses numerous weak learners, typically decision trees, to put together a powerful classifier. ADB's basic principle is to train weak learners on the dataset iteratively, paying particular attention to the cases that were incorrectly classified in the previous iteration. The iteration concept is the foundation of the meta-ensemble ADB model. Only one weak learner is trained in a given iteration; the trained weak learner then participates in the usage of the subsequent iteration. The ultimate strong classifier is created by combining the new weak classifier and weight produced by each cycle. Additionally, a strong learner can effectively classify PV faults (2022). The equation for the final classifier is presented in Eq. (7):

$$H(x) = \text{sign}\left(\sum_{i=1}^{T} \beta_i h_i(x)\right) \tag{7}$$

Here, (x) denotes the final classifier, T indicates the total number of weak classifiers, $h(x)$ represents the weak classifier at step i, β_i denotes the weight assigned to the weak classifier $h_i(x)$, *sign* is the sign function

Categorical Boosting (CB): Yandex created the machine learning library CB, which is well-known for its efficiency when handling categorical variables. It works especially well for tasks like ranking, regression, and classification. Different data formats can be handled by this algorithm. This algorithm's ability to automatically manage categorization features means that CB can be used without the need for obvious category-number conversion preprocessing, which is one of its advantages. The algorithm also reduces overfitting, which results in more general models, which is another significant advantage (2024). An equation in this regard is illustrated in Eq. (8)

$$F(x) = F_0(x) + \sum_{t=1}^{T} \sum\nolimits_{i=1}^{N} f_t\,(x_i) \tag{8}$$

Here, $F(x)$ denotes the general prediction function that CB seeks to comprehend, $F0\,(x)$ denotes the first estimate or base prediction, T denotes entire tree count of the ensemble, N denotes the overall quantity of training samples, $\sum_{t=1}^{T}$ denotes the summation over the ensemble of trees, $\sum_{i=1}^{N}$ denotes the total of the training sample sums, and $f_t\,(x_i)$ denotes the tth tree's prediction for the ith training sample.

Gradient Boosting Machine (GBM): Strong machine learning methods such as GBM are applied to both regression and classification problems. It is a part of the ensemble learning techniques, which combine several models to increase performance as a whole. Specifically, GBM focuses on creating a sequence of weak learners, usually decision trees, and improving them iteratively by reducing the mistakes made by the earlier models. Gradient boosting combines the iterative gradient descent's optimization potential with the decision trees' flexibility. By gradually aggregating them, it seeks to improve the performance of weak learners and produce a strong and capable learner for classification and prediction tasks (2024). Equation (9) provides the equation in this regard.

$$F_m(x) = F_{m-1}(x) + \gamma_m h_m(x) \tag{9}$$

Here, $F_{m-1}(x)$ is the ensemble's prediction up to the $(m - 1)^{\text{th}}$ iteration, γ_m is the weight or contribution of the mth weak learner to the final model, and $h_m(x)$ is the weak learner added at the mth iteration.

Extremely Randomized Trees (ET): The machine learning algorithm known as ET is a member of the decision tree-based model family of ensemble learning techniques. On the basis of random subsets of the training data and features, it builds several decision trees. Similar to Random Forest, Extremely Randomized Trees uses a large number of decision trees, but it adds additional unpredictability to the process by training each tree with the entire learning sample and randomly dividing the trees top-down. It selects the division point at random rather than figuring out the best division point for each feature (for example, based on information entropy or Gini impurity). The value is chosen at random and uniformly from the empirical feature space. The division point of the node is determined by taking the division point with the highest score out of all the randomized division points (2024).

3.5 *Explainable AI*

SHapley Additive exPlanations (SHAP): SHAP, a machine learning technique for deciphering and analyzing complex model predictions. Using Shapley values from game theory, SHAP is a technique for describing the predictions of a model. This method assesses each input characteristic's influence on a machine learning model's prediction quantitatively. Shapley values, which have their roots in cooperative game theory, provide a way to divide "payoffs," or the game's prizes, equitably among several cooperating players. These "players" stand in for the characteristics or traits in the context of machine learning, and the "payoffs" are the "predicted outcomes." (2024). Equation (10) represents a linear explanation model used in SHAP:

$$h\left(x'\right) = \phi_0 + \sum_{i=1}^{T} \phi_0 x_i' \tag{10}$$

Here, h indicates the explanation model, T indicates the largest coalition size possible, x' indicates the coalition vector ($x' \in \{0, 1\}^{\mathrm{T}}$), and ϕ_i indicates the feature attribution for a feature i ($\phi_i \in \mathbb{R}$).

4 Results

4.1 *Evaluation Metrics*

Mean Squared Error (MSE): MSE is a frequently used measurement to gauge the effectiveness of a regression model. The statement conveys the average squared deviation that exists between the observed and predicted values by the model. MSE is computed using Eq. (11).

$$\text{Mean Squared Error} = \frac{1}{\mathrm{N}} \sum_{k=1}^{\mathrm{N}} \left(\Upsilon_k - \hat{\Upsilon}_k\right)^2 \tag{11}$$

Root Mean Square Error (RMSE): The precision of a prediction model is evaluated using RMSE, an often-used statistic, especially when performing regression analysis. In order to calculate it, one must extract the square root of the mean of the squared disparities between the expected and actual values. Equation (12) generates the RMSE.

$$\text{Root Mean Square Error} = \sqrt{\sum_{k=1}^{N} \frac{\left(\hat{\Upsilon}_k - \Upsilon_k\right)^2}{N}} \tag{12}$$

Mean Absolute Error (MAE): An evaluation of a regression model's performance is done using an indicator known as MAE. It determines the average absolute difference between the expected and actual values in a dataset. The formula for determining MAE is illustrated in Eq. (13) below:

$$\text{Mean Absolute Error} = \frac{\sum_{k=1}^{N} |\Upsilon_k - \mu_k|}{N} \tag{13}$$

Mean Absolute Percentage (MAPE): MAPE is an indicator that's frequently employed to evaluate the accuracy of forecasting techniques. MAPE quantifies the average absolute proportion deviation between the predicted and actual values. The computation of MAPE is demonstrated in Eq. (14).

$$\text{Mean Absolute Percentage Error} = \frac{1}{N} \sum_{k=1}^{N} \left| \frac{\Upsilon_k - \hat{\Upsilon}_k}{\Upsilon_k} \right| \times 100\% \tag{14}$$

Coefficient of Determination (R^2): R^2 score is a statistical measure which shows what proportion of the variation of the variable that is dependent can be anticipated from the independent variables. It is a variable of type binary, a value of 1 signifies a precise correspondence, whereas a value of 0 denotes the absence of a link between the independent and dependent variables. It is computed using Eq. (15).

$$\text{Coefficient of Determination} = 1 - \frac{\sum_{k=1}^{N} \left(\Upsilon_k - \hat{\Upsilon}_k\right)^2}{\sum_{k=1}^{N} \left(\Upsilon_k - \overline{\Upsilon}_k\right)^2} \tag{15}$$

In the above equations,

- Υ_k : Observed Values
- $\hat{\Upsilon}_k$: Predicted Values
- μ_k : True Value
- $\overline{\Upsilon}_k$: Mean of the Actual Values
- N: Total number of data points

Table 3 Performance comparison across multiple models

Name	MAE	MSE	RMSE	MAPE	R^2
XGB	5888.31	183314966.15	13539.39	2.38×10^{17}	99.08%
RF	2445.49	79204171.05	8899.69	1.04×10^{16}	99.6%
KNN	2484.90	1186191171.32	10891.24	5.01×10^{15}	99.4%
LGB	4622.77	151574708.29	12311.56	2.79×10^{17}	99.24%
ADB	3534.15	10714437.11	10351.05	3.79×10^{16}	99.46%
CB	1565.15	37013809.03	6083.89	4.24×10^{16}	99.81%
GBM	11264.31	716784480.06	26772.83	2.83×10^{17}	99.4%
ET	1823.70	61438319.52	7838.26	3.48×10^{14}	99.69%

4.2 Machine Learning Model Performance Analysis

As displayed in Table 3, among the models compared, CB stands out as the best-performing model, demonstrating exceptional predictive accuracy with the lowest RMSE of 6083.89 and the highest R^2 score of 99.81%. CB excels in accurately predicting outcomes, making it ideal for applications requiring precise and reliable predictions. Following CB, RF and ET perform strongly with low RMSE values of 8899.69 and 7838.26, respectively, coupled with high R^2 scores of 99.6% and 99.69%. These models provide robust predictive capabilities suitable for tasks demanding accurate modeling of complex data relationships. ADB and KNN also deliver solid performance, showing moderate prediction errors and high R^2 scores of 99.46% and 99.4%, respectively. In contrast, XGB exhibits the highest RMSE of 13539.39 and an R^2 score of 99.08%, indicating comparatively higher prediction errors among the models evaluated. While still demonstrating strong overall performance, XGB highlights areas where improvements in prediction accuracy could be advantageous.

4.3 Result of the XAI Tool: SHAP

Investigating the explainability of machine learning models is a critical endeavor in ensuring their reliability and trustworthiness. In this pursuit, the SHAP (SHapley Additive exPlanations) method emerges as a potent tool, serving to enhance the interpretability of these models by delineating the impact of individual predictor variables on model outputs. The utilization of SHAP values offers a nuanced understanding of the way each variable affects the forecasting model, thus furnishing invaluable insights into the underlying mechanisms governing predictions. Figure 7 encapsulates the culmination of this investigative process, presenting summary plots that meticulously elucidate the relationship between model inputs and SHAP values. Within these plots, each characteristic is symbolized with a vertical bar, positioned along the x-axis in accordance with its corresponding SHAP value. A pivotal

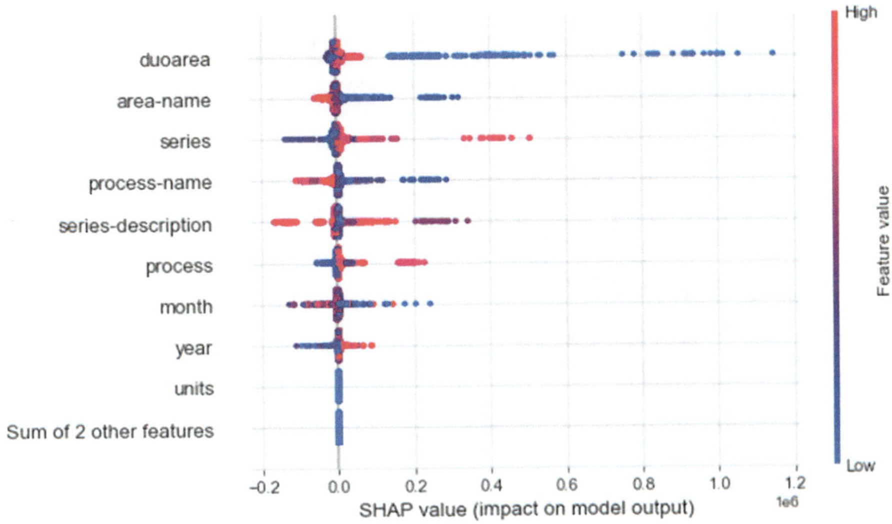

Fig. 7 The SHAP values on the model for every feature

measure of a feature's impact on the model output is the polarity of these values, where positive indicates an augmentative effect and negative indicates a diminutive influence. Moreover, the magnitude of SHAP values provides a quantitative measure of the strength of this influence, furnishing researchers with a comprehensive gauge of variable importance. Crucially, the incorporation of a color gradient into these visualizations further enriches their interpretability, with blue hues connoting lower feature values or adverse effects, and red hues signifying higher values or beneficial contributions. This color scheme not only accentuates the relative significance of each feature within the dataset but also facilitates a nuanced understanding of their respective roles in shaping model predictions. Thus, by imbuing the SHAP plots with both quantitative and qualitative insights, researchers are empowered to discern the intricate interplay between feature values and predictive outcomes. Figure 7 not only displays a graphic depiction of the SHAP value distribution across different features but also affords a hierarchical ranking of these features based on their mean absolute SHAP values. Through this dual perspective, researchers are equipped to discern both the relative importance of individual features and the magnitude of their impact on forecasting accuracy. Notably, the findings underscore the preeminent influence of the "duoarea" variable on prediction outcomes, followed closely by "area-name" and "series" as the second and third most influential features, respectively.

Figure 8 displays a more basic feature importance plot. The model was more affected by the variables that were at the top than by the ones at the bottom. Attributes, namely duoarea, area-name, series, and process-name, showed substantial influence on the model outcome, as indicated by the depiction of SHAP values in Fig. 8. The series description, procedure, month, and year were the four attributes

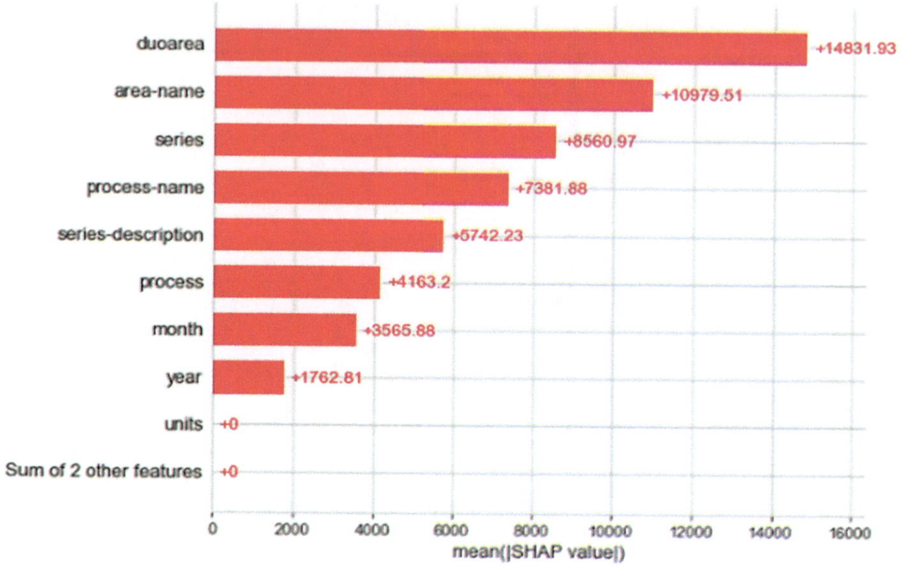

Fig. 8 The importance of each feature to the prediction result

with moderate importance. The least significant features, which had little to no impact on the forecasting model, were likewise highlighted by the SHAP values in Fig. 8, as opposed to the high-importance features. Three features, namely units, product, and product-name, had no bearing on the forecasting model, as can be observed from the graphic in Fig. 8.

5 Conclusion

In summary, our in-depth study has shown how powerful advanced machine learning techniques can be applied to resolve important forecasting problems in the energy industry. Following a thorough evaluation of a wide range of models using data on natural gas usage, the study has shown that the CB algorithm is the best method, with an outstanding R^2 score of 99.81%. Additionally, the study advances beyond merely summarizing the numerical outcomes by offering insight into the CB model's decision-making procedure and utilizing two different explainability techniques to better understand the complex connections and patterns the model discovered. The findings of the research have a significant impact on policymaking, optimizing operations, and energy planning as they can help the industry make better decisions about how to allocate resources, manage risk, and choose how much natural gas to use.

References

Aminu, A., Zambuk, F. U., Gital, A. Y., Lawal, M. A., Pyelshak, Y., & Yakubu, I. Z. (2023). Application of hybrid ensemble machine learning approach for prediction of residential natural gas demand and consumption. *International Journal of Innovative Science and Research Technology (IJISRT), 8*(9), 158–165. https://doi.org/10.5281/zenodo.8348415

Bhuiyan, M. M., Hossain, A. N., Sakib, S. I., & Alawee, and Talayeh Razzaghi. (2024). Fueling the future: A comprehensive analysis and forecast of fuel consumption trends in US electricity generation. *Sustainability, 16*(6), 2388.

Broni-Bediako, E., Buabeng, A., & Allotey, P. (2024). Predicting Ghana's daily natural gas consumption using time series models. *Petroleum Science and Engineering, 8*(1), 27–37. https://doi.org/10.11648/j.pse.20240801.14

Cattani, G. (2023). Combining data envelopment analysis and Random Forest for selecting optimal locations of solar PV plants. *Energy and AI, 11*, 100222. https://doi.org/10.1016/j.egyai.2023.100222

Chen, D., & Liu, B. 2024. Temperature prediction for stored grain: A multi-model fusion strategy based on machine learning. *arXiv preprint arXiv:2404.07175*

Cihan, P. (2022). Impact of the COVID-19 lockdowns on electricity and natural gas consumption in the different industrial zones and forecasting consumption amounts: Turkey case study. *International Journal of Electrical Power & Energy Systems, 134*, 107369.

Clement, T., Nguyen, H. T. T., Kemmerzell, N., Abdelaal, M., & Stjelja, D. 2024. Beyond explaining: XAI-based adaptive learning with SHAP clustering for energy consumption prediction. *arXiv preprint arXiv:2402.04982*.https://arxiv.org/abs/2402.04982

Ding, J., Zhao, Y., & Jin, J. (2023). Forecasting natural gas consumption with multiple seasonal patterns. *Applied Energy, 337*, 120911.

Gadagi, A., Sivaprakash, B., Adake, C., Deshannavar, U., Hegde, P. G., Santhosh, P., Rajamohan, N., & Osman, A. I. (2024). Epoxy composite reinforced with jute/basalt hybrid–Characterisation and performance evaluation using machine learning techniques. *Composites Part C: Open Access, 14*, 100453.

Gao, Q. (2024). Multi-temporal scale wind power forecasting based on Lasso-CNN-LSTM-LightGBM. *EAI Endorsed Transactions on Energy Web, 11*.

Gaweł, B., & Paliński, A. (2024). Global and local approaches for forecasting of long-term natural gas consumption in Poland based on hierarchical short time series. *Energies, 17*(2), 347. https://doi.org/10.3390/en17020347

2024. Global Gas Demand Set for Stronger Growth in 2024 despite Heightened Geopolitical Uncertainty. January 26. https://www.iea.org/news/global-gas-demand-set-for-stronger-growth-in-2024-despite-heightened-geopolitical-uncertainty

Handayani, M. P., Kim, H., Lee, S., & Lee, J. (2023). Navigating energy efficiency: A multifaceted interpretability of fuel oil consumption prediction in cargo container vessel considering the operational and environmental factors. *Journal of Marine Science and Engineering, 11*(11), 2165. https://doi.org/10.3390/jmse11112165

Hu, Y., & Man, Y. (2023). Energy consumption and carbon emissions forecasting for industrial processes: Status, challenges and perspectives. *Renewable and Sustainable Energy Reviews, 182*, 113405.

King, A. (2024). *Natural Gas Usage (2014-Now)*. March 31. https://www.kaggle.com/datasets/alistairking/natural-gas-usage

Lee, J., Eom, J., Park, J., Jo, J., & Kim, S. (2024). The development of a machine learning-based carbon emission prediction method for a multi-fuel-propelled smart ship by using onboard measurement data. *Sustainability, 16*(6), 2381. https://doi.org/10.3390/su16062381

Li, X., Ma, L., Chen, P., Xu, H., Xing, Q., Yan, J., Lu, S., Fan, H., Yang, L., & Cheng, Y. (2022). Probabilistic solar irradiance forecasting based on XGBoost. *Energy Reports, 8*, 1087–1095. https://doi.org/10.1016/j.egyr.2022.05.013

Liu, Z., Luo, Z., Gong, P., & Guo, M. 2013. The research of character recognition algorithm for the automatic verification of digital instrument. In: *Proceedings of 2013 2nd International Conference on Measurement, Information and Control* (IEEE) 01: 177–181. https://doi.org/10.1109/MIC.2013.6757941.

Lodhi, E., Wang, F.-Y., Xiong, G., Dilawar, A., Tamir, T. S., & Ali, H. (2022). An AdaBoost ensemble model for fault detection and classification in photovoltaic arrays. *IEEE Journal of Radio Frequency Identification, 6*, 794–800.

Qu, X., & Liu, Z. (2024). Forecasting the total building energy based on its architectural features using a combination of CatBoost and meta-heuristic algorithms. *Energy & Environment*.

Shaikh, M., Qureshi, Z., Sheikh, H. H., & Sami, M. (2022). The adoption of mobile app for automating gas meter reading in industry 4.0. *Engineering Proceedings, 20*(1), 32.

Sim, T., Choi, S., Kim, Y., Youn, S. H., Jang, D.-J., Lee, S., & Chun, C.-J. (2022). eXplainable AI (XAI)-based input variable selection methodology for forecasting energy consumption. *Electronics, 11*(18), 2947. https://doi.org/10.3390/electronics11182947

Song, Y., Zhou, Z., Xiang, P., & Fang, S. 2020. Instrument recognition in transformer substation based on image recognition algorithm. In: *2020 5th Asia-Pacific Conference on Intelligent Robot Systems (ACIRS)* (IEEE) 7–11.

2024. *U.S. natural gas consumption 2023.* June 25. http://www.statista.com/statistics/184329/energy-consumption-from-natural-gas-in-the-us-from-1995/.

Yang, H., Ran, M., & Feng, H. (2023). Improved data-driven building daily energy consumption prediction models based on balance point temperature. *Buildings, 13*(6), 1423. https://doi.org/10.3390/buildings13061423

Farhana Sultana Eshita graduated with a Bachelor's degree in Computer Science and Engineering from the University of Asia Pacific in July 2024. Throughout her academic journey, she distinguished herself as a dedicated competitive programmer, participating in numerous programming contests. Eshita's research interests are broad and impactful, encompassing cutting-edge fields such as Computer Vision, Natural Language Processing (NLP), Financial Technology (FinTech), and Sustainable Energy. She has contributed to esteemed journals such as *Data in Brief* and has presented her work at notable conferences, including NIDS, STI, and ICEEICT.

Tasnim Jahin Mowla is a highly motivated graduate student in the Department of Computer Science and Engineering at the University of Asia Pacific, Dhaka, Bangladesh. Her consistent academic excellence is highlighted by her standing as the top student in her cohort and securing a coveted spot on the Vice Chancellor's Honor List for five consecutive terms. Her research is at the forefront of Machine Learning and Deep Learning. With a profound commitment to exploring the transformative potential of artificial intelligence, she is dedicated to pushing the boundaries of knowledge and aims to make substantial, meaningful contributions through her scholarly work. Her dedication and accomplishments reflect her unwavering pursuit of excellence and her passion for advancing the field of computer science.

Abu Bakar Siddique Mahi graduated with a degree in Computer Science and Engineering from the University of Asia Pacific in July 2024. He is currently employed as a Teaching Assistant at the same institution. During his academic career, he served as a Research Assistant on a project with the Bangladesh Accreditation Council and previously worked remotely as a Research Assistant at Swansea University, UK. Mahi has also held the position of Machine Learning Instructor at the University of Asia Pacific. His research interests encompass Computer Vision, Natural Language Processing (NLP), Financial Technology (FinTech), and Sustainable Energy. He has contributed to esteemed journals such as *Data in Brief* and has presented at notable conferences, including NIDS, STI, and ICEEICT.

An Extensive Statistical Analysis of Time Series Modeling and Forecasting of Crude Oil Prices

Mahmudul Hasan, Md. Iftekhar Hossain Tushar, Most Mozakkera Jahan, Touhida Sultana Ety, and Md. Palash Uddin

1 Introduction

Crude oil, often called "black gold", is an innate, unprocessed petroleum product derived from deposits of hydrocarbons and other organic matter. This versatile resource fuels most vehicles, heats houses, and provides electricity for the planet. Besides the energy sector, crude oil represents an irreplaceable raw material in all related industries: plastics, pharmaceuticals, and chemicals (Hasan et al., 2023). Crude oil is of immense economic importance; it fuels the world economy. It is the primary fuel in the world, and its supply and price determine essential effects on economic activity. This is because crude oil is a vital component in so many industries, affecting profit and output. Also, the oil industry is among the largest employers around the world: from upstream activities through midstream to marketing. Finally, crude oil is one of the most actively traded commodities in

M. Hasan (✉) · M. P. Uddin
Department of Computer Science and Engineering, Hajee Mohammad Danesh Science and Technology University, Geelong, Dinajpur, Bangladesh

School of Information Technology, Deakin University, Melbourne, VIC, Australia
e-mail: palash_cse@hstu.ac.bd

M. I. H. Tushar
Department of Computer Science and Engineering, Hajee Mohammad Danesh Science and Technology University, Geelong, Dinajpur, Bangladesh

M. M. Jahan
Department of Economics, Begum Rokeya University, Rangpur, Bangladesh

T. S. Ety
Department of Management, University of Dhaka, Dhaka, Bangladesh

M. Z. Abedin, W. Yong (eds.), *Machine Learning Technologies on Energy Economics and Finance*, International Series in Operations Research & Management Science 367, https://doi.org/10.1007/978-3-031-94862-6_4

the world, with its price movements affecting trade balances and national economic health (Sajid et al., 2023).

As an energy source, crude oil provides the most inelastic component of energy supply. It contributes over a third to the world's energy consumption and, thus, serves as an essential component toward realizing global energy security (Moon et al., 2019). The massive infrastructure of the oil sector in terms of thousands of miles of pipelines, storage facilities, and refineries constitutes a multitrillion dollar sector that is in a position to influence global economic dynamics (Zhang et al., 2022). In this view, predicting crude oil prices is of paramount importance because of the pervasive impact that the commodity exercises upon the global economy (Hasan et al., 2024). Fluctuations in the price of oil exert huge impacts on the transportation, manufacturing, agriculture, and consumer goods sectors. The capability to provide exact price forecasts for the managers of businesses, investors, and even policymakers opens up room for well-informed decisions and strategies that also manage risks. Forecasting oil prices is also crucial for energy-dependent countries' budget planning, managing foreign exchange reserves, and forming economic policies.

The necessity of involving statistical models in crude oil price forecasting is because many variables are interlinked and together impact the oil market: world supply and demand patterns, geopolitical events, economic growth forms, technological progress, and ecological policy. They provide a systematic way of analyzing past data to possibly detect some pattern through which a forecast can be made from this multifaceted influence. This gives a quantitative framework to process vast amounts of information and derive insights upon which action can be taken.

Statistical analysis, to a great extent, influences the forecasting of crude oil prices. This will enable the researcher and analyst to find trends and patterns based on historical price data, measure relationships among several factors that influence oil prices, estimate the relative importance of different variables in price determination, generate probabilistic forecasts that account for uncertainty, and appraise the accuracy and reliability of various forecasting methods. This includes numerous aspects: Forecasters can utilize different statistical techniques, time series analysis, and regression models, among others, with machine learning algorithms to enable the development of more prosperous and more predictive forecast models of crude oil prices. The development of these techniques also allows the involvement of multiple variables and the understanding of the nonlinear relationships that need to be there in oil markets.

Further, the development of statistical analysis lays out the framework for comparing the strengths and the weaknesses of different forecasting models. In the accurate comparison of the predicted values and the outcome produced by them, the researcher, with time, improves the model and overall ability of predictions. Only through such iterative procedures, researchers can develop and validate models to find a much better prediction of crude oil prices. The need for statistical analysis is also related to risk management strategies for investment decisions and policy formulation, in addition to price level predictions. For instance, statistics-based prediction can help oil-producing countries optimize their production levels, as well as assist them in price risk management. On the other hand, these areas are where energy-intensive industries may use to reduce cost overruns and plan their business investments. The role of crude oil as the engine of the global economy dictates that there is evident importance to predicting its price in the most accurate way possible. Statistical models and analysis provide a powerful toolkit to work out the multiplicities of the oil market and insights of vital significance to economic planning, risk management, and decision-making in different sectors. As the world continuously depends on crude oil, there is an ongoing shift toward cleaner energy sources, and predicting oil prices remains a crucial competency for economists, policymakers, and industry leaders. The technical contributions of this chapter are:

- We design a methodology for time series modeling and forecasting of crude oil prices using statistical models.
- We handle the missing values and generate stationary data using Augmented Dickey-Fuller (ADF) test to make the data more suitable for analysis.
- We perform residual analysis to evaluate the adequacy of the model by examining the residuals to ensure the random, normally distributed, and exhibit no autocorrelation, thereby validating the model's assumptions and accuracy.
- We perform a comparative statistical analysis to find the most suitable statistical model for predicting crude oil price.

The structure of the remaining sections of this chapter is outlined as follows. The related works are outlined in Sect. 2. Section 3 is dedicated to presenting our proposed methodology and the experimental setup. We detail the approach we have taken to address the research problem, including the methods, techniques, and tools employed in our study. Within Sect. 4, we present the outcomes of our experiments. The chapter concludes in Sect. 5 with a summary of our findings and their significance. Additionally, we outline avenues for future research and development in this domain, emphasizing the potential directions for further exploration and enhancement.

2 Literature Review

Tianxiang Wang proposed a model (Wang, 2024) for predicting the price of WTI crude oil for the next year. In this work the author used the daily data of WTI crude oil price in the range of the first day of January 2019 to the end of September 2023. The motive was to find the best projection of the next year 2024. Initially she used the auto ARIMA, an autonomous stepwise search, for identifying the best parameters p, d, and q to get high performance on the measurement scale. Secondly she used ARIMA on the training data using specified parameters and projected the outcome for the year 2024 where ARIMA model avoided the small fluctuations of time series. The model forecasted almost a constant value of 100 USD/Bbl. In the measurement of correctness, applied procedure represents approximately 0.04 for

MSE, 0.18 for MAE, and 0.2 for RMSE. The measurement scale justifies ARIMA as a model of high-level accuracy. In another work, Alfaki and Masih (2015) introduced the Box-Jenkins method for designing the model and forecasting the monthly sales for Naphtha. The dataset was collected from Azzawiya Oil Refining Company, Libya as the monthly sales of Naphtha. They used ARIMA models for the iterative process of Box-Jenkins to forecast both stationary and nonstationary time series. In this model, ADF test was used to test the unit root and stationarity in the data, and differencing order was chosen for the integrated component of ARIMA to make the time series stationary. The parameter value for AR and MA components was selected from the graphical presentation of Autocorrelation Function (ACF) and Partial Autocorrelation Function (PACF). The researchers fitted the time series into the ARIMA model with parameter tuples of (1, 1, 1), (3, 1, 3), and (6, 1, 6). They found ARIMA (1, 1, 1) was the best having 0.3506, 1.8629, and 1.9510 for MSE, AIC, and SC in the measurement scales. At the end, the researchers forecasted monthly sales of Naphtha for 6 years from January 2015 to December 2020.

Suleiman et al. (2023) also followed the iterative process of four steps such as identification, estimation or model fitting, diagnostic checking, and model refinement of Box-Jenkins method to analyze the time series. It is a country-based work that used the monthly time series dataset of crude oil price in Nigeria from 2006 to 2020. Firstly, the time series dataset was observed by the ACF and PACF and was identified as nonstationary and having unit root with the examination of Kwiatkowski-Phillips-Schmidt-Shin (KPSS), ADF, and Phillips-Perron (PP) test. The nonstationary behavior in time series was removed after performing the first difference. Secondly, in the model fitting step they fitted ARIMA model with different parameter tuples to find the optimized estimation. Researchers had found two optimal ARIMA with tuples of (2, 1, 1) and (3, 1, 1) based on HQC, AIC, and BIC information criteria. Finally, they identified the ARIMA (3, 1, 1) model performed better and suitable among these two of different parameter tuples in forecasting the monthly price of crude oil considering MSE, RMSE, MAE, MPE, and MAPE predictive measures. A comparison-based work (Tularam & Saeed, 2016) of statistical models on time series dataset was proposed by Tularam et al. in 2016. In this researchers discussed about three univariate models of statistics which were ARIMA, ES, and Holt-Winters. They collected the time series dataset of regular crude oil prices from West Texas Intermediate. The time series was fitted to each univariate model merging with the best hyper parameters to get high performance from them. The outcome showed that the Holt-Winters model performed better with a confidence interval of 95% than the ES model and ARIMA model with parameter tuple of (2, 1, 2) results best among three by considering six measurements in model selection such as MSE, RMSE, MAE, MAPE, and Theil's U statistic (Theil's statistics were implemented and defined as U1 and U2). In a different comparative study (Ning et al., 2022), Ning et al. (2022) presented two statistical models and a Recurrent Neural Network (RNN) model namely ARIMA, Prophet, and LSTM. Those models were fitted with a time series to extract the remarkable behaviors and fluctuations of historical data and forecast values of a future time sequence. The oil production data of 65 wells, a reservoir located in

Denver-Julesburg (DJ) Basin, was used as the time series data for this analysis. The 65 wells' data was divided into four pads, and 70% of data was used to train the models and rest of the data to evaluate the performance of those models for each pad. Though Prophet captures the fluctuations of winter season more preciously, DJ Basin's time series showed that ARIMA and LSTM performed better due to not all pads were not facing seasonal impacts, and considering all the measurement scales they observed that ARIMA (0, 1, 1) was more appropriate in shorter time predictions, i.e., next 1-year period.

Rangsan Nochai and Titida Nochai worked (Nochai & Nochai, 2006) to find the best parameter tuple for ARIMA model in different time series. They used palm oil prices of Thailand in three formats as farm prices, wholesale prices, and pure oil prices. The goal of researchers is to obtain the parameters for these three different time series while applying the ARIMA model. They used MAPE measurement technique to judge the ARIMA with every parameter tuple, and they had found the effective parameter tuples of that model as ARIMA (2, 1, 0) for farm prices, ARIMA (1, 0, 1) or ARMA (1, 1) for wholesale prices, and ARIMA (3, 0, 0) or AR (3) for pure oil prices of the palm oil. Caspah Lidiema used (Lidiema, 2017) two statistical models in modeling and predicting the inflation rate in Kenya. He implemented the Box-Jenkins method with SARIMA and triple ES of Holt-Winter. The time series of Consumer Price Index was collected in a range of November 2011 to October 2016 which was published by Kenya National Bureau of Statistics (KNBS). The SARIMA model was trained with parameter tuple (1, 1, 0) for nonseasonal impacts and (1, 0, 0) for the seasonal impact which can be defined as SARIMA (1, 1, 0) (1, 0, 0). On the other hand, the triple ES of Holt-Winter was trained with smoothing parameters such as $\alpha = 0.9999$ and $\beta = 0.0001$, and γ confirms that the model did not contain any seasonal component. The researcher analyzed the performance of two models with measurement scales including MAE, MAPE, and MASE. The SARIMA resulted 0.0036, 0.073, and 0.059 for MAE, MAPE, and MASE, respectively, and the triple ES of Holt-Winter resulted 0.595, 0.400, and 0.643 for MAE, MAPE, and MASE, respectively. Comparing the measurement scale's results, the SARIMA model was chosen as the best model to forecast the inflation rate correctly. In a different investigation (Mardiana et al., 2020) of comparing statistical models in between additive model of Holt-Winter and ARIMA was performed by Mardiana et al. in 2020. They used the gasoline time series dataset and its three components' time series which are gasoline 88, gasoline 90, and gasoline 92 spanning a range of 2017 to 2019 period. The goal of researchers is to forecast the total demand of gasoline from 2020 to 2022. In this study, it was shown that the additive model of Holt-Winter outperformed ARIMA model and the joined application of Holt-Winters model with a neural network resulted lower error in predicting the demand of gasoline 92. Though the components of gasoline were in different trends, the forecasted result showed the increasing behavior in total gasoline demand.

3 Methodology

3.1 Approach Overview

To analyze the time series crude oil price, we proposed a statistical extensive analysis system. After collecting the raw data from online, we preprocess the data by handling the missing values and consistent data and differencing the data. We apply AR, MA, ARIMA, SARIMA, ES, and VAR models for the analysis and forecasting of the crude oil price. We show the model coefficients, ACF and PACF plots, simple moving average (SMA) and EMA values, and residual analysis and finally measure the key statistics of the models. The overview of the methodology is in Fig. 1.

3.2 Description of Dataset

We collect the daily crude oil data from online marketwatch.com. The dataset contains data from May 20, 1987 to June 05, 2024. We consider the open market price of the opening days for the analysis. The trend of crude oil price is in Fig. 2.

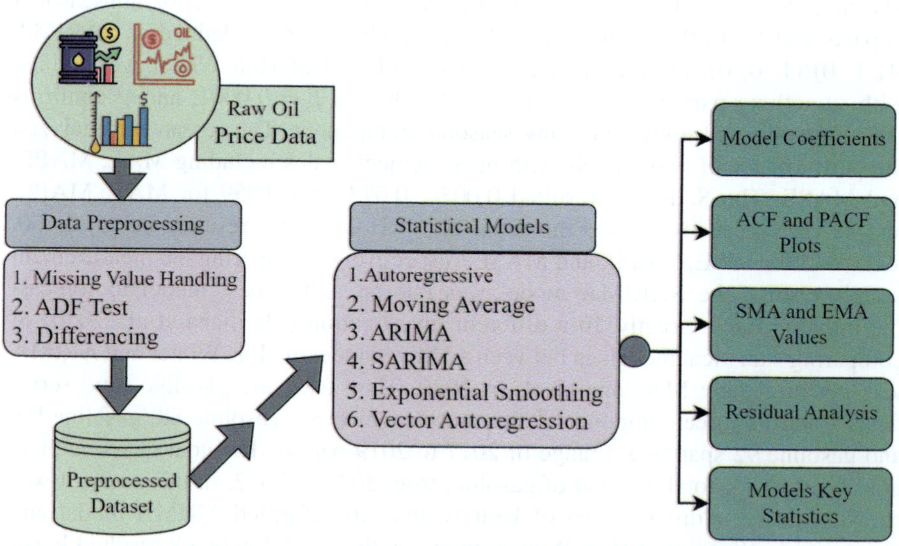

Fig. 1 Overview of the proposed statistical analysis-based forecasting system

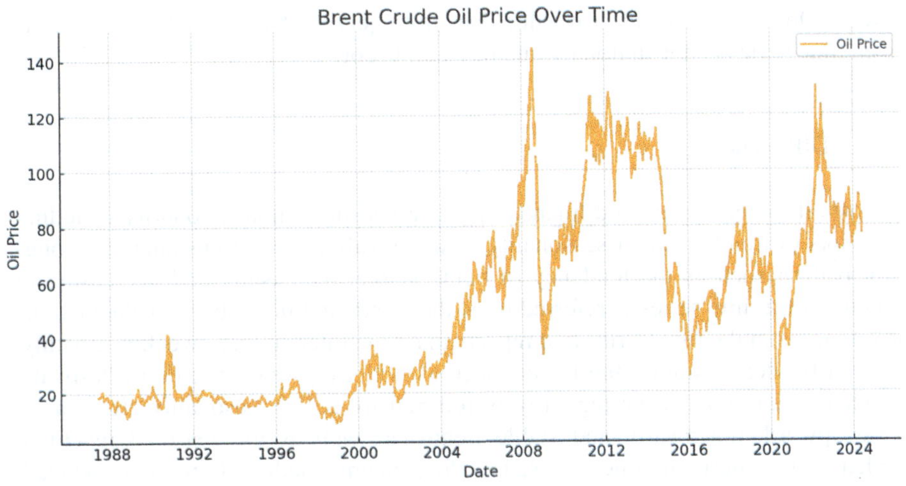

Fig. 2 Change of the crude oil price from 1988 to 2024

3.3 Data Preprocessing Techniques

3.3.1 Missing Value Handling

We handle the missing values using mean imputation process. We consider the mean of previous 3 days and after 3 days to put a suitable value in the missing prices. It provides the consistency and prevents information loss from the dataset.

3.3.2 Stationarity Check Using ADF Test

Checking the stationarity of a time series is an important factor in some time series forecasting statistical models such as ARIMA, SARIMA, and ARCH. Checking stationarity means checking the statistical properties of a time series whether they are varying or not with time. In this purpose, the ADF test, a type of unit root test in statistics, is a most commonly used statistical testing model which is an extended representation of the Dickey-Fuller test (Demetrescu, 2010). To find the stationarity of the time series ADF initially defines two hypotheses. The Null Hypothesis (H0) states that the time series is not stationary or the time series has a unit root, and the Alternate Hypothesis (HA) states that the time series is stationary or the time series has no unit root. The ADF test finds a critical value named as p-value which determines whether the test rejects the null hypothesis or not. If it finds the critical p-value less than the significant level (considering 5%), then the ADF test rejects the Null Hypothesis, which means the given time series is stationary and there is no unit root; more elaborately it can be said that the statistical properties such as mean, variance, covariance, and standard deviation of that time series are not the function of time. On the other hand, if p-value does not reject the Null Hypothesis, then the given time series is not stationary and has a unit root.

3.3.3 Differencing

Almost all the economic and financial time series shows nonstationarity including behaviors of trend, cycle or seasonality, random walking, etc. to its data. The long-term pattern such as the tendency of being stable, increasing, and decreasing in direction of a time series is referred to as the trend, and the pattern of fluctuations or variations in the time series that repeats in a time interval (e.g., weekly, monthly, and yearly) depending on the factors such as holidays, weather, cultural festivals, or other events that occur regularly is referred to as the seasonality. To remove these upward, downward, and stable trends, as well as the seasonal repetitive tendency in pattern of time series, the differencing methodology is introduced. It makes the time series to stationary series from nonstationary. The technique of differencing calculates the differences between consecutive instances in the time series data. Three most impactful differencing methods named as first-order differencing, second-order differencing, and seasonal differencing are exhibited here. The first-order difference is effective in removing the linearly changing trend over time, and it is also called random walk model which can be written as

$$y'_t = y_t - y_{t-1}$$

The first-order differencing results $t - 1$ values because there is no differencing value for the first instance of time series data so it is eliminated from the dataset. Sometimes the first-order differences do not provide the time series as stationary so the second-order involves on the first-order differences and calculates the differences of consecutive first-order instances shows calculation as

$$y''_t = y'_t - y'_{t-1} = (y_t - y_{t-1}) - (y_{t-1} - y_{t-2})$$

The second-order differencing results $t - 2$ values because there is no differencing value for the initial two instances of time series data, so these are eliminated from the dataset. First-order and second-order differences remove the trend in the time series data, but they cannot handle the seasonality of time series where patterns found in repetitive structure in a time interval. Considering this problem, the seasonal differences are introduced. Seasonal difference calculates the differences not between the consecutive instances, but it calculates the differences between two instances which are seasonally related or have similar fluctuations over a time interval. The equation can be written as

$$y'_t = y_t - y_{t-m}$$

where m is the time interval of similar fluctuations which is also called "the lag-m differences." The subtraction in between instances which have a lag of m periods. The seasonal differencing results $t - m$ values because there is no differencing value for the initial m number of instances of time series data, so these are eliminated from the dataset.

3.4 *Description of the Statistical Models*

Statistical models are the mathematical frameworks that capture the components (e.g., trend, seasonality, random fluctuations, and irregularities) of time series data points, understand the data's behavior, identify the relationship among them, and predict the future values. There are several models in the shadow of statistics which works on time series data to forecast data point. Common models of them are AR, MA, ARIMA, SARIMA, VAR, Generalized AutoRegressive Conditional Heteroskedasticity (GARCH), and ES models. Each model works effectively in various circumstances by maintaining many characteristics such as randomness or stochastic nature in the time series, temporal dependency of one-time data point to its previous data points, variations in time series components, parametric or nonparametric description, selection of model and then evaluating them, and finally the inference and prediction. The right model is chosen analyzing all these characteristics. Sometimes model selection is performed with the guidance of different diagnostic measurements including ACF, PACF, Akaike Information Criterion (AIC), Bayesian Information Criterion (BIC), residual analysis, and so on.

3.4.1 Autoregressive (AR) Model

AR is the fundamental model in time series analysis and forecasting. It has grandiose applications in the fields of economics, finance, climate science, and more. Its simplicity in understanding and implementation, interpretability to make relationships between past and future values, and solid foundation of building blocks in time series analysis provides efficacious result in forecasting but it has limitations on nonstationarity, complex trends, and external factors (Nassar et al., 2004). The AR model is a type of regression model in the time series analysis which refers to that the interested value can be predicted from the linear combination of previous values together with an error term. For example, to forecast tomorrow's values of any time series, it might consider today's, yesterday's, and so many past values of that time series. So, the AR model with order p defined as the current value considers p numbers of past values in AR. The equation can be written as

$$Y_t = P_0 + P_1 Y_{t-1} + P_2 Y_{t-2} + \cdots + P_p Y_{t-p} + E_t$$

where $P_1, P_2, \ldots, P_p$ are the changing parameters, E_t is noise, and Y_t is the forecasting value depending on $Y_{t-1}, Y_{t-2}, \ldots, Y_{t-p}$.

The value of p can be defined with the help of PACF and ACF.

3.4.2 Moving Average (MA) Model

MA is also a fundamental model in time series analysis and forecasting as like AR model. It has widespread applications on predicting stock prices, analyzing market trend, managing risk, forecasting demand of product, predicting sales based on seasonal variations, and modeling the weather patterns. Its simple and interpretable structure helps to identify trends and random errors on past observations and forecast the intended value proficiently (Akrami et al., 2014). The MA model is a type of regression model in the time series analysis which refers to that the interested value can be fluctuated over past forecasting errors. The MA model with order q defines as the current value considers q number of past errors. The equation can be written as

$$Y_t = E_0 + E_t + Q_1 E_{t-1} + Q_2 E_{t-2} + \cdots + Q_q E_{t-q}$$

where $Q_1, Q_2, \ldots, Q_q$ are the changing parameters, E_t is considered as the weighted MA of past forecast errors, and $E_0, E_t, E_{t-1}, \ldots, E_{t-q}$ are the errors of forecasted values.

The value of q can be defined with the help of ACF and PACF.

3.4.3 Autoregressive Integrated Moving Average (ARIMA) Model

ARIMA model (Newbold, 1983) is a popular, widely used, and versatile statistical forecasting model on time series data which has three different components referred to as AR, Integrated (I), MA. It works on identifying patterns and trends of historical data and forecasts values of the time series based on past values by handling seasonality, trends, and fluctuations in data. Combining three components it merges the facilities of three independent models into one to predict the interested value effectively (Valipour et al., 2012). Combining integrated or differencing with autoregression and moving average the ARIMA (p, d, q) model forms the following equation:

$$Y'_t = P_0 + P_1 Y'_{t-1} + \cdots + P_p Y'_{t-p} + Q_1 E_{t-1} + \cdots + Q_q E_{t-q} + E_t$$

where p is the order of AR part, q is the order of MA part, and d is the differencing order of integrated part.

Though ARIMA is a powerful tool in time series forecasting, but it has drawbacks on nonlinear trends or patterns, external factors, focusing on long-term periods, selecting parameters, and handling outliers.

3.4.4 Seasonal Autoregressive Integrated Moving Average (SARIMA) Model

To overcome the limitation of ARIMA model in forecasting the time series which fluctuates over a fixed interval or seasonal manner, the SARIMA model comes in Alharbi and Csala (2022). ARIMA model only captures the trend and patterns in the time series in its learning period. On the other hand, SARIMA not only includes the working process of forecasting time series but also captures the seasonal fluctuations in the time series and gives more accurate result than ARIMA in forecasting the interested value. SARIMA joins two sets of parameter in its seasonality modeling. One set represents the nonseasonal effects with parameters of AR, I, and MA which are denoted in lowercase letters, and other set represents the seasonal effects with parameters of AR, I, and MA which are denoted in uppercase letters. The functionality of SARIMA can be formed as

$$SARIMA(p, d, q)(P, D, Q)m$$

where p, d, q parameters represent the nonseasonal effects for AR, I, MA, P, D, Q parameters represent the seasonal effects for AR, I, MA, and m parameter represents the number of observations in one season. In a weekly time series there would be 54 observations.

In SARIMA the parameter selection and handling the overfitting risk is a challenging task, but its performance on time series having both trend (e.g., upward trend, downward trend, and stable trend) and seasonality makes this model very valuable.

3.4.5 Exponential Smoothing (ES) Model

ES is a forecasting model for univariate time series data where the interested value is calculated as the weighted linear summation of lags or past observations. Using the exponential window function, it assigns weights to the past observations which are exponentially decreasing (Gardner Jr., 2006). The idea behind this strategy is giving more importance to the recent observations and decreasing the importance exponentially smaller to the older observations. The equations for calculating the ES are:

1. Simple Exponential Smoothing (SES):

$$F_{t+1} = S_t = \alpha X_t + (1 - \alpha)S_{t-1}$$

   where t represents the time period, X_t is the current observation, S_t is the smoothed value representing the weighted average for X_t, α is the smoothing factor in a range of (0, 1), and F_{t+1} is the forecasting value for the next period $t + 1$.

2. Holt's Linear Trend Smoothing:

$$F_{t+1} = L_t + T_t$$

$$L_t = \alpha X_t + (1 - \alpha)(L_{t-1} + T_{t-1})$$

$$T_t = \beta(L_t + L_{t-1}) + (1 - \beta)T_{t-1}$$

where t represents the time period, X_t is the current observation, α is the smoothing factor in a range of $(0, 1)$, β is the trend smoothing factor in a range of $(0, 1)$, L_t is the smoothed value for level of X_t in time period t, T_t is the smoothed value for trend in time period t, and F_{t+1} is the forecasting value for the next period $t + 1$.

Holt's Linear Trend Smoothing combines the smoothed value for level and smoothed value for trend, represented as Lt and Tt, to forecast the interested value. The level of equation is almost the same to the SES, but it includes the previous trend.

The ES model is comparatively simple than ARIMA and other models to understand and implement. Its computational efficiency in calculating only the weighted sum requires minimal processing power which is very suitable for real-time forecasting with quick turnaround. But, the model focuses more on recent data than past observations as a result long-term prediction may result less reliable when the time series contains highly volatility, sudden changes, and intricate patterns.

3.4.6 Vector Autoregression (VAR) Model

VAR is an extended version of simpler AR model in single time series. VAR is used for multivariate time series where it captures the interdependent characteristics of multiple time series and explains each time series variable with its own past observations or lags and past observations of other variables (Lütkepohl, 2013). If there are N number of time series variables, then there will be N equations, one for one variable. The function of VAR is written as

$$VAR(p)$$

where $VAR(p)$ model has n equations for all the time series variables, and p is the number of past observations in each equation.

VAR performs very efficiently in analyzing the relationships between several patterns and forecasting the value which is dependent on the behavior of multiple time series.

4 Result Analysis

4.1 Stationarity Check

We use the ADF test to check if the time series is stationary. A stationary time series has a constant mean and variance over time. Table 1 shows the ADF test statistic and p-value of the dataset before differencing and after differencing. We consider a null hypothesis as follows:

H_0 : *The Series Is Nonstationary*
Before differencing the p-value is 0.2147 that is greater than 0.05. We fail to reject the null hypothesis. This means the series is not stationary and requires differencing to make it stationary. After differencing, we get p-value as 1.09×10^{-26} that is less than 0.05. We reject the null hypothesis and conclude that the differenced series is stationary.

4.2 Results of the ARIMA Model

After getting the stationary series, we fit it into an ARIMA model. To do this, we need to determine the appropriate order of the ARIMA model (p, d, q). We use the ACF and PACF plots in Fig. 3 to determine the values of p and q.

Table 1 The ADF test values and p-values of the dataset

State of dataset	ADF test statistic	P-value
Before differencing	−2.1771	0.2147
After differencing	−14.3356	1.09×10^{-26}

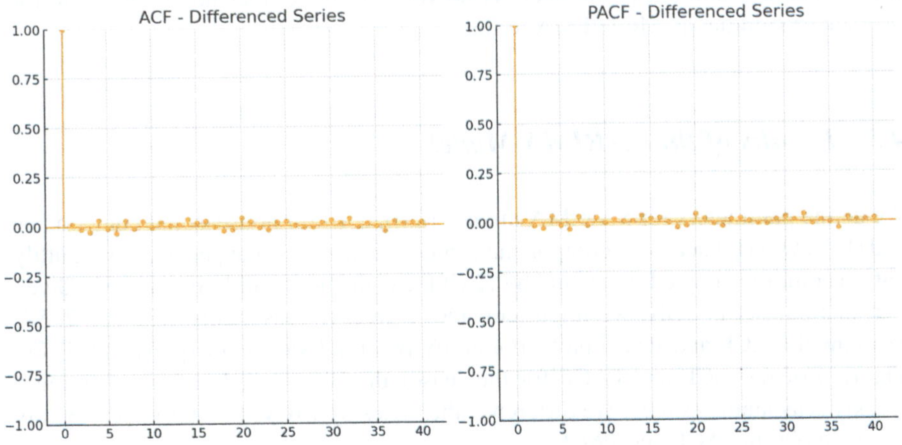

Fig. 3 ACF and PACF plots to determine the values of p and q in ARIMA

Table 2 Model coefficients in ARIMA model

Model coefficient	Coefficient	Standard error	z-Value	P>|z|
ar.L1 (AR term at lag 1)	0.6681	0.870	0.768	0.443
ma.L1 (moving average term at lag 1)	−0.6706	0.867	−0.773	0.439
sigma2 (variance of the residuals)	1.4565	0.007	197.825	0.000

The ACF and PACF plots provide insight into the potential values for the ARIMA model parameters. The ACF plot shows the correlation between the time series and its lagged values. The significant lags can help to determine the q parameter. The PACF plot shows the partial correlation between the time series and its lagged values after removing the effects of intermediate lags. The significant lags can help to determine the p parameter.

From the plots, we can observe that the ACF plot shows significant spikes at lag 1, indicating $q = 1$. The PACF plot shows significant spikes at lag 1, indicating $p = 1$. Given that we applied first differencing ($d = 1$), we can fit an ARIMA (1, 1, 1) model to the data.

Table 2 summarizes the coefficients of an ARIMA (1, 1, 1) model, providing details on the AR, MA, and variance components. In terms of statistical significance both the AR term (ar.L1) and the MA term (ma.L1) have high p-values (> 0.05), indicating that they are not statistically significant. This suggests that neither the past values nor the past errors significantly impact the current values in the model. The variance of the residuals (sigma2) is highly significant with a very low p-value (< 0.05), indicating that the variability in the residuals is a critical component of the model. The lack of statistical significance in the AR and MA terms suggests that the model may not effectively capture the relationships in the data. This could result in a model that does not adequately predict future values based on past values and errors. The significant sigma2 value indicates that the model's residuals have a consistent and measurable amount of variance. However, this alone does not compensate for the lack of significant AR and MA terms.

4.3 *Results of the SARIMA Model*

We define parameters for SARIMA models as (p, d, q) for nonseasonal components and (P, D, Q, s) for seasonal components, where s is the seasonal period. For monthly data, a common choice for s would be 12 (if the data had monthly frequency). Given that this dataset is daily, we might consider a weekly seasonality with s = 7. We examine the ACF and PACF plots to identify potential values for (p, q) and (P, Q). Figure 4 of the ACF and PACF for the seasonally differenced series, assuming a weekly seasonality (s = 7). We difference the series by 7 days to remove seasonality and then plot the ACF and PACF.

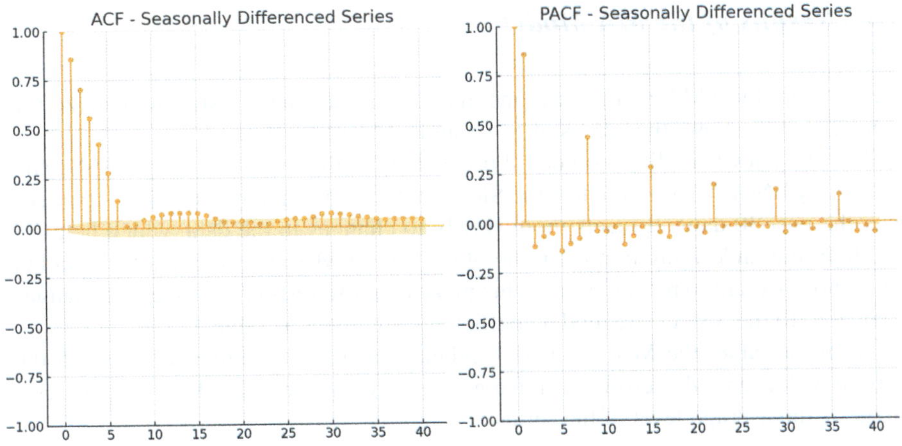

Fig. 4 ACF and PACF plots to determine the values of (p, q) and (P, Q) in SARIMA

Table 3 Model coefficients of the SARIMA model

State	Components	Coefficient	Standard Error	z-Value	P>|z|
Nonseasonal	ar.L1	−0.1452	0.348	−0.417	0.677
	ma.L1	0.1597	0.347	0.460	0.645
Seasonal	ar.S.L7	0.0290	0.006	4.860	0.000
	ma.S.L7	−1.0000	0.044	−22.808	0.000
Common	sigma2	1.4554	0.064	22.71	0.000

Based on these plots, we make the following observations. For nonseasonal components (p, d, q), the ACF plot of the original series shows significant spikes at lag 1, indicating $q = 1$. The PACF plot of the original series shows significant spikes at lag 1, indicating $p = 1$. For seasonal components (P, D, Q, s), the ACF plot of the seasonally differenced series shows significant spikes at lag 7, indicating a potential seasonal AR or MA component. Given the seasonal differencing applied (D = 1) and assuming a weekly seasonality ($s = 7$), we start with $p = 1$ and $q = 1$. The final SARIMA model contain parameters: $(p, d, q) = (1, 1, 1)$ and $(P, D, Q, s) = (1, 1, 1, 7)$. This model includes one nonseasonal autoregressive term (AR (1)), one nonseasonal difference (d = 1), one nonseasonal moving average term (MA (1)), seasonal components with one seasonal autoregressive term (SAR (1)), one seasonal difference (D = 1), one seasonal moving average term (SMA (1)), and a seasonal period of 7. The model coefficients are given in Table 3.

From Table 3, we get the nonseasonal components (ar.L1 and ma.L1) which are not statistically significant, but the seasonal components (ar.S.L7 and ma.S.L7) are statistically significant, indicating the importance of considering seasonal effects in the model.

4.4 Results of the MA Model

We compute the SMA and the exponential moving average (EMA) to analyze the trend. Figure 5 shows the values of SMA and EMA of the dataset.

The plot shows the Brent crude oil price along with the 30-day simple SMA and the 30-day EMA. We observe that the SMA smooths the data by averaging the prices over the past 30 days. It reacts more slowly to changes in the data, which helps to highlight the underlying trend. On the other hand, EMA also smooths the data but gives more weight to recent prices. This makes it more responsive to recent changes compared to the SMA. The model coefficients are given in Table 4.

From the table, the MA (1) model indicates a strong relationship between the current value and the error term from the previous period. The constant term suggests the average level of the series.

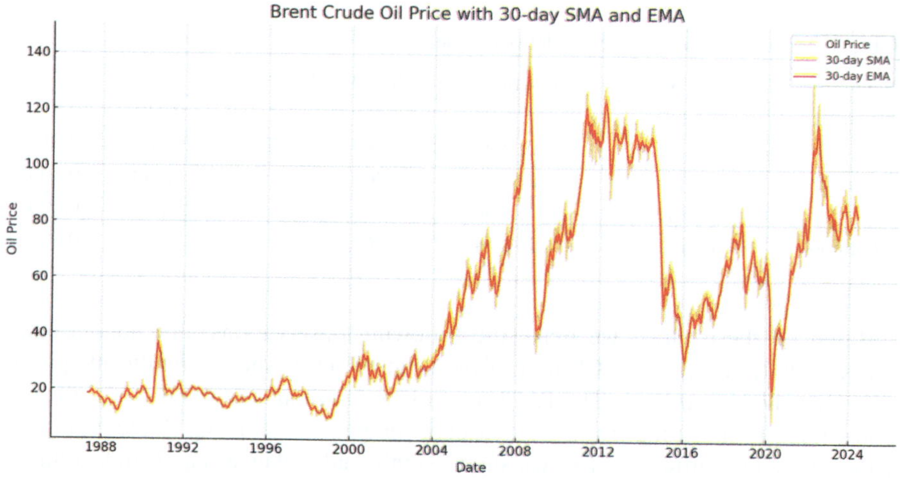

Fig. 5 The 30-day SMA and EMA values of crude oil price data

Table 4 Model coefficients of the MA model

Model coefficient	Coefficient	Standard error	z-Value	P>|z|
Const. (constant term)	49.6502	0.424	117.045	0.000
ma.L1	0.9749	0.002	580.984	0.000
sigma2	283.8572	6.324	44.889	0.000

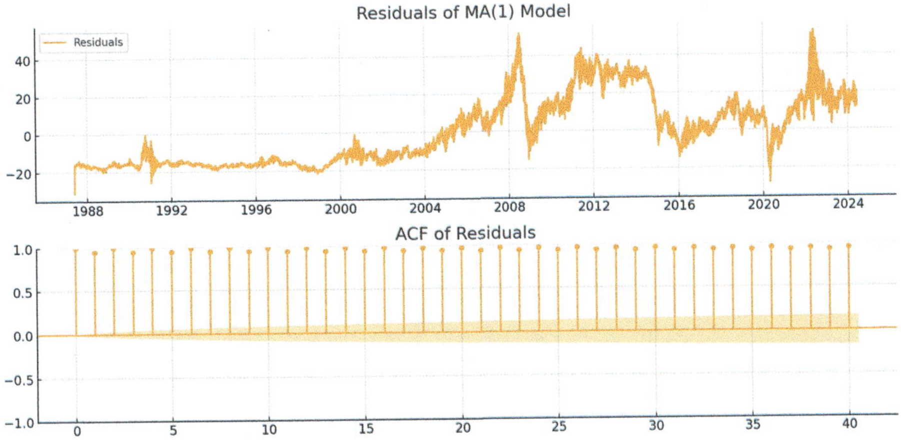

Fig. 6 Residuals and ACF of residuals of MA model

4.5 Residual Analysis of the Models

4.5.1 MA

The result of the residual analysis is in Fig. 6. The residuals fluctuate around zero, but there are noticeable patterns, indicating that the MA (1) model may not have captured all the underlying structure of the data. On the other hand, the ACF plot shows significant spikes, suggesting that there is still some autocorrelation left in the residuals. This implies that the MA (1) model has not fully accounted for the time dependence in the data.

4.5.2 AR

The result of the residual analysis is in Fig. 7. The residuals fluctuate around zero, indicating that the AR (1) model has captured the main structure of the data. However, there are noticeable patterns, suggesting that the model may not have captured all the underlying dependencies. The ACF plot shows significant spikes, suggesting that there is still some autocorrelation left in the residuals. This implies that the AR (1) model has not fully accounted for all the time dependence in the data.

4.5.3 ES

We plot the residuals and their ACF to inspect any patterns or anomalies both for adaptive seasonal and for multiplicative seasonal model. From Fig. 8, we find that the residuals fluctuate around zero, indicating that the additive seasonality model

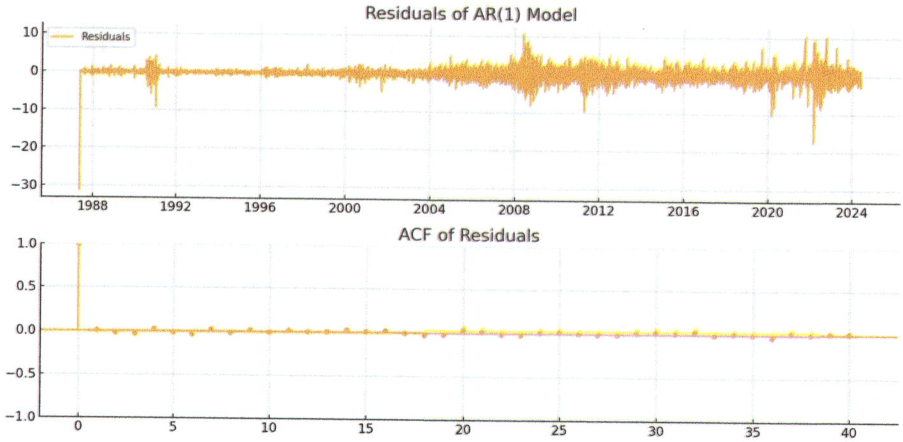

Fig. 7 Residuals and ACF of residuals of AR model

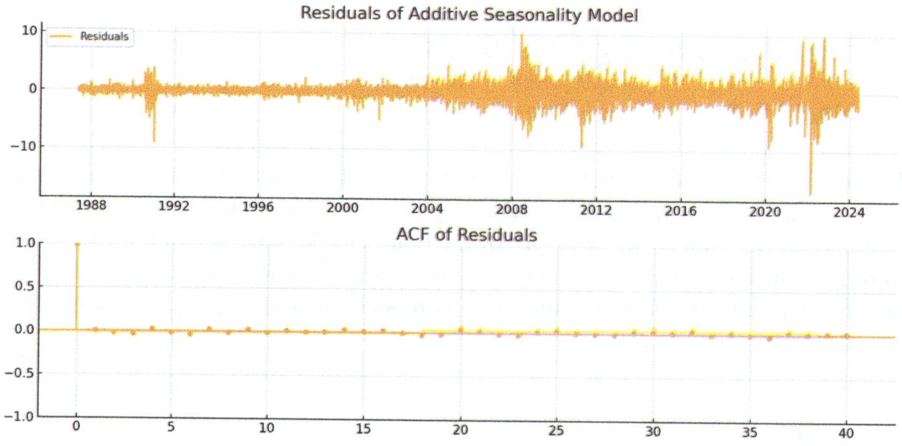

Fig. 8 Residuals and ACF of residuals of adaptive seasonal ES

captures the main structure of the data. However, there are noticeable patterns, suggesting that the model may not have captured all the underlying dependencies. On the other hand, the ACF plot shows significant spikes, suggesting that there is still some autocorrelation left in the residuals. This implies that the additive seasonality model has not fully accounted for all the time dependence in the data.

From Fig. 9, the residuals fluctuate around zero, indicating that the multiplicative seasonality model captures the main structure of the data. Similar to the additive model, there are noticeable patterns, suggesting that the model may not have captured all the underlying dependencies. On the other hand, the ACF plot shows significant spikes, suggesting that there is still some autocorrelation left in the residuals. This implies that the multiplicative seasonality model has not fully accounted for all the time dependence in the data.

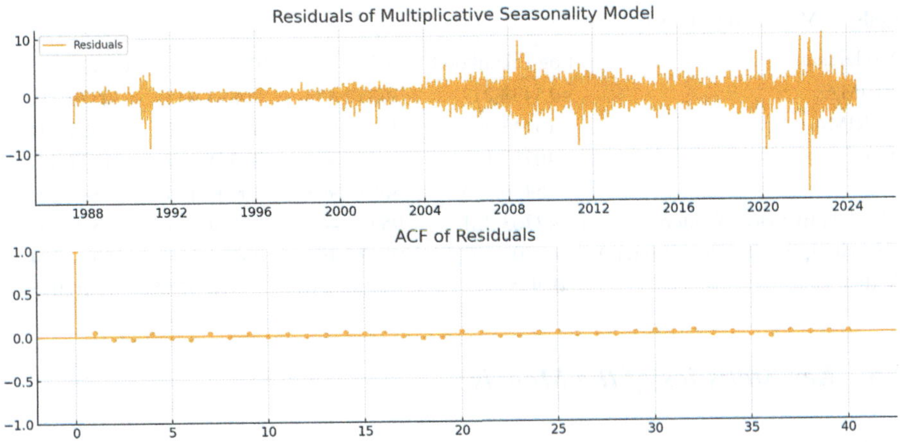

Fig. 9 Residuals and ACF of residuals of multiplicative seasonal ES

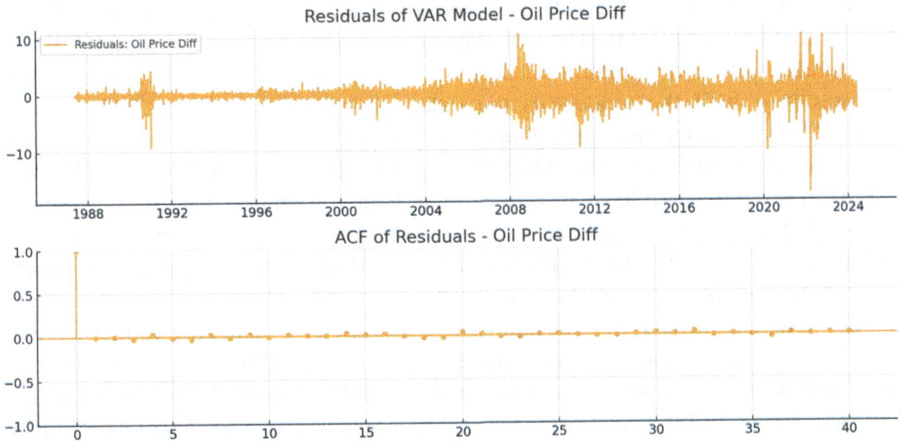

Fig. 10 Residuals and ACF of residuals of VAR

4.5.4 VAR

Figure 10 indicates that the residuals of the “Oil Price Diff” series fluctuate around zero, indicating that the VAR model has captured the main structure of the data. However, there are still noticeable patterns, suggesting that the model may not have captured all underlying dependencies. On the other hand, the ACF plot shows some significant spikes, suggesting that there is still some autocorrelation left in the residuals. This implies that the VAR model has not fully accounted for all the time dependence in the data.

Table 5 Models key statistics

Model name	Log Likelihood	AIC	BIC	HQIC
ARIMA	−15433.212	30872.423	30893.933	30879.719
SARIMA	−15443.843	30897.685	30933.531	30909.844
MA	−40757.192	81520.383	81541.894	81527.679
AR	−15436.454	30878.908	30900.418	30886.204
ES (Additive Holt-Winters)	−47703.221	95428.442	95674.654	95519.942
ES (Multiplicative Holt-Winters)	−47703.202	95428.404	95674.616	95519.904
VAR	308955	−70.0267	−70.0222	−70.0252

4.6 Key Statistics of the Models

Table 5 presents a comparative analysis of key statistics for various time series models applied to the Brent crude oil price data. The models under consideration are ARIMA, SARIMA, MA, AR, ES using both additive and multiplicative Holt-Winters methods, and VAR. The key statistics include Log Likelihood, AIC, BIC, and Hannan-Quinn Information Criterion (HQIC). In the Log Likelihood value indicates how well the model fits the data. Higher values suggest a better fit. VAR (308955) model has the highest Log Likelihood, indicating an exceptional fit compared to other models. ARIMA (−15433.212), SARIMA (−15443.843), and AR (−15436.454) have mostly similar Log Likelihood values, suggesting comparable model fits among these models. MA (−40757.192) and ES models (Additive: −47703.221 and Multiplicative: −47703.202) have significantly lower Log Likelihood values, indicating poorer fits.

AIC penalizes models for the number of parameters, balancing model fit and complexity. Lower values indicate a better model. VAR (−70.0267) has the lowest AIC, reinforcing its strong performance and model fit. Among the ARIMA family, ARIMA (30872.423) has a slightly lower AIC compared to SARIMA (30897.685) and AR (30878.908). MA (81520.383) and ES models (Additive: 95428.442 and Multiplicative: 95428.404) have much higher AIC values, indicating poorer performance. BIC is like AIC but imposes a heavier penalty for the number of parameters. Lower values are preferred. VAR (−70.0222) again shows the best performance with the lowest BIC. ARIMA (30893.933) has a lower BIC compared to SARIMA (30933.531) and AR (30900.418), suggesting a better fit among these models. MA (81541.894) and ES models (Additive: 95674.654 and Multiplicative: 95674.616) have higher BIC values, indicating a lesser fit.

HQIC also penalizes for model complexity, though less severely than BIC. Lower values are preferred. VAR (−70.0252) shows the best performance with the lowest HQIC. ARIMA (30879.719) has a lower HQIC compared to SARIMA (30909.844) and AR (30886.204), indicating a better fit. MA (81527.679) and ES models (Additive: 95519.942 and Multiplicative: 95519.904) have higher HQIC values, indicating poorer performance.

The VAR model consistently outperforms all other models in terms of Log Likelihood, AIC, BIC, and HQIC, indicating it provides the best fit for the Brent crude oil price data. ARIMA Family Models: Among the ARIMA family, the ARIMA model shows a slightly better performance compared to SARIMA and AR based on AIC, BIC, and HQIC values. Both the additive and multiplicative Holt-Winters models perform similarly, but they do not fare well compared to the other models, indicated by their higher AIC, BIC, and HQIC values. The MA model shows the poorest performance among the models considered, with the highest AIC, BIC, and HQIC values. In summary, the VAR model emerges as the most suitable for capturing the underlying structure and dependencies in the Brent crude oil price data, followed by the ARIMA model. Both the ES and MA models show significant limitations in terms of model fit and complexity.

4.7 *Diagnostic Tests of the Models*

Table 6 presents a comparative analysis of the diagnostic tests performed on various time series models applied to the Brent crude oil price data. The models under consideration are ARIMA, SARIMA, MA, AR, ES using both additive and multiplicative Holt-Winters methods, and VAR. The diagnostic tests include the Ljung-Box test for autocorrelation, the Jarque-Bera test for normality of residuals, and the Breusch-Pagan test for heteroskedasticity.

In Ljung-Box test for autocorrelation, ARIMA (0.16) and AR (0.23) models have p-values greater than 0.05, indicating no significant autocorrelation in the residuals, suggesting that these models have adequately captured the temporal dependencies in the data. SARIMA (0.92) shows the highest p-value, strongly indicating no significant autocorrelation and suggesting it is the best model in terms of capturing autocorrelation. MA (0.00), ES (Additive Holt-Winters) (0.00), ES (Multiplicative Holt-Winters) (0.00), and VAR (0.00) all show significant autocorrelation in the residuals, suggesting these models may have omitted some temporal dependencies.

Table 6 Models diagnostic tests

Model name	Ljung-Box (prob.)	Jarque-Bera (prob.)	Heteroskedasticity (Prob)
ARIMA	1.98(0.16)	101197.49(0.00)	10.91(0.00)
SARIMA	0.01(0.92)	103701.56(0.00)	10.90(0.00)
MA	8676.88(0.00)	943.33(0.00)	1.30(0.00)
AR	1.45(0.23)	100672.88(0.00)	10.87(0.00)
ES (Additive Holt-Winters)	1429.45(0.00)	187332.81(0.00)	1338.23(0.00)
ES (Multiplicative Holt-Winters)	1429.67(0.00)	188135.32(0.00)	1350.67(0.00)
VAR	216.45(0.00)	22740.57(0.00)	1338.23(0.00)

For Jarque-Bera test for normality, all models, including ARIMA (0.00), SARIMA (0.00), MA (0.00), AR (0.00), ES (Additive Holt-Winters) (0.00), ES (Multiplicative Holt-Winters) (0.00), and VAR (0.00), show p-values of 0.00, indicating significant deviation from normality in the residuals. This suggests that none of the models' residuals are normally distributed, which can affect the reliability of statistical inferences made from these models.

In Heteroskedasticity (Breusch-Pagan test) for Nonconstant Variance, all models, including ARIMA (0.00), SARIMA (0.00), MA (0.00), AR (0.00), ES (Additive Holt-Winters) (0.00), ES (Multiplicative Holt-Winters) (0.00), and VAR (0.00), have p-values of 0.00, indicating significant heteroskedasticity in the residuals. This means that the residuals of all these models exhibit nonconstant variance over time, suggesting that the models may not fully capture the variability in the data.

SARIMA emerges as the most effective model in capturing autocorrelation with the highest Ljung-Box p-value (0.92), indicating no significant autocorrelation in the residuals. ARIMA and AR models also show no significant autocorrelation, but to a lesser extent. None of the models pass the Jarque-Bera test for normality, indicating that all models have residuals that significantly deviate from normal distribution. All models show significant heteroskedasticity, suggesting that the models do not fully account for varying variance in the data.

Overall, while the SARIMA model performs best in terms of capturing autocorrelation, all models exhibit significant issues with normality and heteroskedasticity in their residuals. These findings suggest that while SARIMA might be preferable for capturing autocorrelation, additional model refinement or alternative modeling approaches are necessary to address the non-normality and heteroskedasticity observed in the residuals.

4.8 Discussion

The analysis of Brent crude oil prices using various time series models reveals significant insights into their relative performance and suitability for forecasting. The models examined include ARIMA, SARIMA, MA, AR, ES using both additive and multiplicative Holt-Winters methods, and VAR.

The VAR model consistently outperformed all others, evidenced by its highest Log Likelihood and lowest AIC, BIC, and HQIC values. This indicates that the VAR model provides the best fit for the data, effectively capturing the underlying structure and dependencies. The strong performance of the VAR model suggests its robustness in handling complex temporal dependencies and interactions within the data.

Among the ARIMA family, the ARIMA (1, 1, 1) model showed slightly better performance compared to SARIMA (1, 1, 1) (1, 1, 1, 7) and AR(1), based on AIC, BIC, and HQIC values. This indicates that while seasonality is an important factor, the simpler ARIMA model can still provide a reasonably good fit, potentially due to the relatively stable seasonal patterns in the data. ES models, both additive and multiplicative Holt-Winters, displayed significantly higher AIC, BIC, and HQIC values, indicating poorer fits. Their residuals exhibited substantial autocorrelation, non-normality, and heteroskedasticity, suggesting that these models are less capable of capturing the data's complexity. The MA model performed the worst, with the highest AIC, BIC, and HQIC values, indicating significant limitations in its ability to model the data effectively.

Finally, we find that the VAR model is the most suitable for forecasting Brent crude oil prices, followed by the ARIMA model. Both ES and MA models demonstrate considerable limitations, highlighting the need for more sophisticated modeling approaches to improve forecasting accuracy. This comprehensive evaluation underscores the importance of selecting appropriate models based on key statistical metrics to achieve reliable forecasts.

5 Conclusion and Future Work

The aim of this research is to analyze and forecast crude oil prices using daily time series data. We have collected daily oil prices from online sources and performed some basic preprocessing to make the data more suitable for analysis using statistical models. Since the data was not stationary, we employed the ADF test to determine its state. By using differencing, we achieved stationarity in the dataset. We applied various statistical models, and although the performance of these models was generally good, the VR model showed superiority over the others. It achieved the highest forecasting capability, which was confirmed by the residual analysis of the model. This research also forecasts the next 30 days, aiding in various decision-making processes to achieve a sustainable world with fair economic activities. Additionally, it helps in maintaining the supply and demand of crucial energy resources like crude oil and assessing its impact on society. Future work will focus on designing a real-time forecasting system with long-term forecasting capabilities to inform both short-term and long-term policy decisions.

Acknowledgments We extend our deepest gratitude to the Center for Multidisciplinary Research and Development (CeMRD) for their resources, guidance, and support. The expertise and encouragement from CeMRD members have been invaluable. This research was made possible by the collaborative environment and cutting-edge facilities at CeMRD.

References

Akrami, S. A., El-Shafie, A., Naseri, M. & Santos, C. A. (2014). Rainfall data analyzing using moving average (MA) model and wavelet multi-resolution intelligent model for noise evaluation to improve the forecasting accuracy. *Neural Computing and Applications, 25*, 1853–1861.

Alfaki, M. M. A. & Masih, S. B. (2015). Modeling and forecasting by using time series ARIMA models. *International Journal of Engineering Research & Technology (IJERT), 4*, 2278–0181.

Alharbi, F. R. & Csala, D. (2022). A seasonal autoregressive integrated moving average with exogenous factors (SARIMAX) forecasting model-based time series approach. *Inventions, 7*, 94.

Demetrescu, M. (2010). On the dickey–fuller test with white standard errors. *Statistical Papers, 51*, 11–25.

Gardner Jr, E. S. (2006). Exponential smoothing: The state of the art—part ii. *International Journal of Forecasting, 22*, 637–666.

Hasan, M., Das, U., Datta, R. K. & Abedin, M. Z. (2023). *Model development for predicting the crude oil price: Comparative evaluation of ensemble and machine learning methods* (pp. 167–179). Springer.

Hasan, M. et al. (2024). A blending ensemble learning model for crude oil price forecasting. *Annals of Operations Research*, 1–31.

Lütkepohl, H. (2013). *Vector autoregressive models* (pp. 139–164). Edward Elgar Publishing.

Lidiema, C. (2017). Modelling and forecasting inflation rate in Kenya using SARIMA and holt-winters triple exponential smoothing. *American Journal of Theoretical and Applied Statistics, 6*, 161–169.

Mardiana, S., Saragih, F. & Huseini, M. (2020). Forecasting gasoline demand in Indonesia using time series. *International Journal of Energy Economics and Policy, 10*, 132–145.

Moon, M. H. et al. (2019). *A cryptographic algorithm based on ASCII and number system conversions along with a cyclic mathematical function* (pp. 1–5). IEEE.

Nassar, S., Schwarz, K.-P., El-Sheimy, N. & Noureldin, A. (2004). Modeling inertial sensor errors using autoregressive (AR) models. *Navigation, 51*, 259–268.

Newbold, P. (1983). ARIMA model building and the time series analysis approach to forecasting. *Journal of Forecasting, 2*, 23–35.

Ning, Y., Kazemi, H. & Tahmasebi, P. (2022). A comparative machine learning study for time series oil production forecasting: ARIMA, LSTM, and prophet. *Computers & Geosciences, 164*, 105126.

Nochai, R. & Nochai, T. (2006). *ARIMA model for forecasting oil palm price* (pp. 13–15). Academia Penang.

Sajid, S. W., Hasan, M., Rabbi, M. F. & Abedin, M. Z. (2023). *An ensemble LGBM (light gradient boosting machine) approach for crude oil price prediction* (pp. 153–165). Springer.

Suleiman, M. et al. (2023). Modelling Nigeria crude oil prices using ARIMA time series models. *NIPES-Journal of Science and Technology Research, 5*, 230–241.

Tularam, G. A. & Saeed, T. (2016). Oil-price forecasting based on various univariate time-series models. *American Journal of Operations Research, 6*, 226–235.

Valipour, M., Banihabib, M. E. & Behbahani, S. M. R. (2012). Parameters estimate of autoregressive moving average and autoregressive integrated moving average models and compare their ability for inflow forecasting. *Journal of Mathematics and Statistics, 8*, 330–338.

Wang, T. (2024). WTI crude oil price forecasting for the next year using ARIMA model. *Highlights in Science, Engineering and Technology, 88*, 804–809.

Zhang, K., Hong, M., Zhang, K. & Hong, M. (2022). Forecasting crude oil price using LSTM neural networks. *Data Science in Finance and Economics, 2*, 163–180.

Mahmudul Hasan is currently pursuing a PhD in Information Technology (IT) at Deakin University, Melbourne, Australia. He earned his BSc (Eng.) and MSc (Eng.) degrees in Computer Science and Engineering (CSE) from Hajee Mohammad Danesh Science and Technology University, Dinajpur, Bangladesh, in 2021 and 2023, respectively. He previously served as a lecturer in the Department of CSE at the University of Creative Technology, Chittagong (UCTC), Bangladesh. He is the Founder and Director of the Center for Multidisciplinary Research and Development (CeMRD) and a moderator of "Be Researcher BD," the largest online research forum in Bangladesh. Additionally, he has taught online as a Data Science instructor to students in the USA, Italy, Denmark, South Korea, and Australia. He is also the founder of the online educational platform "MHM Academy." His research interests include federated learning, machine learning, deep learning, cybersecurity, health informatics, renewable energy, computational sociology, and business intelligence.

Md. Iftekhar Hossain Tushar is currently pursuing his BSc (Eng.) degree in CSE from Department of Computer Science and Engineering in Hajee Mohammad Danesh Science and Technology University, Dinajpur, Bangladesh. He completed his Higher Secondary degree from Notre Dame College, Dhaka in 2018. Currently, he is working as a Research Assistance at Center for Multidisciplinary Research and Development (CeMRD). During his bachelor's degree, including his academics, he participated in various programs and achieved a best paper award in his academic life which was organized by EEE club of HSTU in his university. His research interests include machine learning, deep learning, and time series analysis.

Most Mozakkera Jahan is currently pursuing her BSS (Honours) degree in Economics in Department of Economics from Begum Rokeya University, Rangpur, Bangladesh. She is also working as a Research Assistant (RA) in Center for Multidisciplinary Research and Development (CeMRD). She is a moderator on the online educational platform "MHM Academy." Her research interests include data analytics, econometrics, machine learning, and data mining.

Touhida Sultana Ety is currently learning Japanese language at Hotsuma International School, Gifu, Japan. She completed her Bachelor of Business Studies in Management from the Department of Management at Eden Mohila College affiliated with University of Dhaka, Dhaka, Bangladesh, in 2020. Her research interests include Advanced Forecasting, Entrepreneurship and Innovation, Marketing Management, Financial Management, Business Intelligence, and Business Analytics.

Md Palash Uddin is currently working as a Postdoctoral Research Fellow at the School of Information Technology, Deakin University, Australia. He received a PhD degree in Information Technology from Deakin University, Australia, in 2023. He also received a BSc degree in Computer Science and Engineering from Hajee Mohammad Danesh Science and Technology University (HSTU), Bangladesh and an MSc degree in Computer Science and Engineering from Rajshahi University of Engineering & Technology, Bangladesh. He is also an academic faculty member at HSTU, Bangladesh. His research interests include machine learning, federated learning, blockchain, and remote sensing image analysis.

Comparative Analysis of Selected Emerging Economies Energy Transition Scenario: A Transition Pathway for the Continental Neighbors

Dip Bindu Bhattachergee, Zarin Tasnim, Md Nafeez Hasan, Rakib Hasan, and Mahmudul Hasan

1 Introduction

The world's health and human well-being are at stake due to climate change, and there is a confined period of time left to ensure a livable and sustainable future for all creatures (IPCC, 2023). In 2015, Paris agreement (Process and meetings: UNFCCC, 2015) is propagated with an aim to alleviate the impact of climate change upon the world. Under the Paris agreement, COP28 was particularly notable as it concluded the first "Global Stock take" of the worldwide effort to combat climate change (Process and meetings: Conferences: UN Climate Change Conference - United Arab Emirates, 2023). Traditional energy or fossil fuel-based energy is inevitably linked with climate change due to immense greenhouse gas emissions (Elias, 2018). Consequently, renewable energy became the center of attention all over the world replacing fossil fuel in order to gain energy efficiency.

Renewable resource enriched nations are usually emerging and middle-income economies located in the sunbelt (Mühlbauer et al., 2023), and many of them are swiftly employing their enormous resources (Manish Ram, 2022). However, some are facing economic, social, and technological barriers to efficiently utilize their resources (Moorthy et al., 2019). One major constraint that an emerging economy generally encounters while exploiting RE resources is the lack of availability of finance (Anthony, 2021). Additional barriers include inadequate knowledge of the advantages of renewable energy (Moorthy et al., 2019), inexperienced

D. B. Bhattachergee · Z. Tasnim · M. N. Hasan · R. Hasan
Department of Accounting and Information Systems, Hajee Mohammad Danesh Science and Technology University, Dinajpur, Bangladesh

M. Hasan (✉)
Department of Computer Science and Engineering, Hajee Mohammad Danesh Science and Technology University, Dinajpur, Bangladesh

M. Z. Abedin, W. Yong (eds.), *Machine Learning Technologies on Energy Economics and Finance*, International Series in Operations Research & Management Science 367, https://doi.org/10.1007/978-3-031-94862-6_5

technical experts and a shortage of training facilities (Ansari et al., 2013), a lack of infrastructure availability, inadequate research and development, ineffective operational and maintenance expertise initiatives (Zhao et al., 2016), and so on. Not much research has been conducted on the consideration of emerging or developing economies in the past decades (Hansen et al., 2019). Those who focused on the emerging economy's energy transition mostly concentrated their studies either on specific countries, for instance, Indonesia (Reyseliani & Purwanto, 2021), China (Xu, 2020), Egypt (Mühlbauer et al., 2023), Brazil (Dranka & Ferreira, 2018), etc., or into regions, for instance South Asia (Breyer et al., 2023), Latin America (de Souza Noel Simas et al., 2017), the MENA (Bogdanov et al., 2020), and sub-Saharan Africa (Barasa et al., 2018). However, an aggregated study on the emerging sunbelt economies regarding RE transition with consideration of their cultural, demographic, and geographic dimensions is scarce. A combined study pertaining to RE transition in developing economies would comprise country- or region-specific energy production, consumption, storage capacity, share of renewables in the energy portfolio, policies regarding net zero emissions by 2050, interim targets, etc. in order that countries with minimal development in this sector but with colossal resources can follow their superior continental neighboring states, who are leading the world in terms of decarbonizing the energy sector. Hence, this study selects some countries from each sunbelt region on the basis of continental supremacy regarding RE, i.e., those that comprise the most renewable share in their portfolio and are looking forward to a sustainable transition to a fully renewable energy sector and aim to provide their contiguous states as a pathway to their energy transition.

Selected countries are India from South Asia, Vietnam from Southeast Asia, Morocco from the Middle East and North Africa, Brazil from South America, South Africa from sub-Saharan Africa, and Mexico from North America.

- This research work is intended to find out a pathway toward energy sustainability for the emerging sunbelt countries across the world.
- After studying vast amount of literature and observing database the results are drawn.
- Most of the countries of specified continents may pursue the findings to widely adopt renewable energy omitting fossil or oil-based energy.

2 Literature Review

The documents from the IPCC, UNFCCC primarily construct the introduction section of the study. The data conducted into the scenario of the countries are acquired from the documents of IRENA, IEA, ITA, World Bank, BBC portal, NDC partnership, World Economics, European Commission, Green Hydrogen Organization, Our World in Data, and more. Statistical information is extracted from Statista, Global Carbon Atlas provided the facts regarding carbon emissions, and the data regarding RE capacity and generation are assembled from CLIMATESCOPE and

IRENA's Renewable energy statistics. Moreover, few other supporting information are aggregated from different dedicated sites such as Climate Action Tracker, Global Energy Monitor, Greencare, Global Methane Pledge, and World Meteorological Organization.

Websites and published articles from country-specific and continental authority of renewable energy and environment, such as MNRE, SENER, INSAMER, DMRE, MME, etc., exhibit the current position of the country in the light of climate change. Research articles that are utilized to assert real-time data contain the available current information of 2022, and all the referred articles in this study are from rated journals and published by expert author in the field.

3 Methodology

3.1 Overview

3.2 Comparative Analysis

A comparative analysis of production, generation, consumption, storage capacity, regulatory framework, and policy terms regarding renewable energy among emerging countries of sunbelt regions, namely India, Brazil, Mexico, Vietnam, Morocco, and South Africa, have been studied to justify their strategy applicability into their contiguous states.

The contextual variables for the analysis are total RE production (including biofuels & heating system), total generation of electricity from RE, total electricity consumption generated by RE, total storage capacity, regulations established & policy undertaken by the selected countries.

The overall comparative ratio of these variables is competent to indicate a country's position in deploying RE resources. Strategies or theories that have worked out for them regarding the exploitation of the resources can well be derived from the analysis. We have provided an overview in Fig. 1.

3.2.1 Literature Selection & Analysis

The literature used for the comparative analysis is collected mostly from secondary sources (published article, secondary database) and some are from primary sources (government authorized websites, primary database).

Firstly, keyword search (renewables, energy, emerging, sunbelt) has been applied to obtain a handful of potential research articles. Secondary screening comprises the exclusion of articles which does not meet the following criteria:

- Articles with respect to the RE transition scenario of the above-named countries.
- Articles containing up-to-date real-time information.

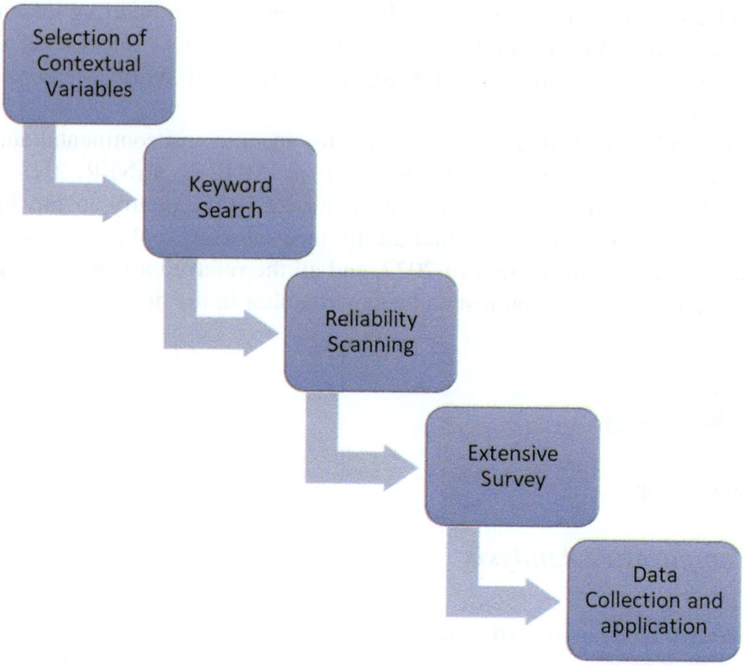

Fig. 1 Overall process of methodology

- Articles published by renowned authors and in peer-reviewed journals.
- Findings of the article must be broadly useful.

Finally, endured articles are revised according to the pre-determined selection criteria. Subsequently, data from official and valid websites, including IEA, IRENA, UNFCCC, etc., are complemented with the previously mentioned information.

Considering all these information, a range of strategy and policy implication path is eventually outlined for the adjoining states of studied countries.

3.3 Appropriateness

The method employed above to arrive at the answer is an exoteric and established way of carrying out qualitative research, and it has been proven to be particularly effective for cross-national comparative analysis. Given the wealth of information the database offers on renewable energy generation, consumption, storage, public awareness campaigns, local and national policies, certifications, statements from international organizations, and other related topics, it is reasonable to disclose several solutions to the most pressing issues and recommendations for those neighboring states' policymakers, subject to a thorough and exploratory investigation.

3.3.1 Scenario of India

With 1.417 billion people, India is the second-most populous country in the world and one of the biggest users of fossil fuels, which contribute to global warming. With the rapid growth of population and industrialization, energy demand is soaring in the country. India's overall energy consumption is expected to double by 2030, with power demand rising to three times current levels (Energyworld, 2022). The government of India formed MNRE (Ministry of New and Renewable Energy) to detect and implement new sources of energy generation and has already marked several progresses through uplifting RE resources. The energy crisis, the drastically rising level of environmental pollution, and the growing population are the main causes of the emphasis on switching to renewable energy sources. India has already achieved the target of 40% power generation from non-fossil or renewable sources, which was committed to achieve by 2030 (PIB Delhi, 2023), and through these achievements, India became one of the largest manufacturers of renewable energy. According to Climatescope's latest ranking, India is the most attractive spot for renewable energy investment.

As of 2022, India's share of renewable energy in total energy generation is 22.4%, according to the IEA, which amounts to 343,138.7 GWh (IRENA, 2022a). The total installed renewable energy capacity of the country, excluding large hydro, was 150.27 GW by 2022 (MNRE, 2022a), which accounts for 33.52% of total energy storage capacity (IRENA, 2022b).

Hydropower leads India's total RE generation mix in 2022, with 165,715 GWh of generation including small and large hydro. Total electricity generation from solar PV accounts for 102,010 GWh, while wind and biomass generate 71,814 GWh and 17905.4 GWh of electricity, respectively (Bloomberg NEF, 2023a).

India ranks 5th in terms of solar PV deployment and 4th in terms of wind power installed capacity in the world (MNRE, 2022b). It has experienced a drastic increase in the growth of installed solar energy storage capacity of 24.07 times the 2014 rate, standing at 63.30 GW in 2022 (Sarraju Narasinga Rao, 2023). Solar PV capacity storage as of December 2022 is 63,048 MW, whereas CSP stands at 343e MW in India. Wind energy is no exception; it has augmented to 41,930 MW in terms of its installed capacity until 2022. Also, storage capacity installed for biomass and hydropower reaches 10,669 MW and 47,220 MW, respectively, in 2022 (Whiteman et al., 2023). According to the MNRE annual report 2022–2023, installed storage capacity for waste to energy arrived at 522.42 MW eq.

The incessantly rising population of India undoubtedly resulted in unremitting consumption of energy throughout the country. In 2022, India invests a total of $11015.37 million in renewable energy, which is 12.42% higher compared to the previous year (Bloomberg NEF, 2023a). At the 2019 UN Climate Summit, it committed to achieving 450 GW of renewable energy (RE) by 2030 (PIB Delhi, 2021).

The profitability view of RE indicates that in comparison with domestic coal-fired power plants, which charge between 3.5 and 5 INR/kWh (43.7 and 62.5 €/MWh), solar PV-based electricity generation ranges between 1.99 and

2.36 INR/kWh (24.8–29.5 €/MWh) (Gulagi et al., 2022). Average price of electricity for the country stands at 94.78 USD/MWh in 2022 (Bloomberg NEF, 2023a).

The country ranks 3rd in terms of global carbon emissions, followed by China and the United States, respectively, and accounts for 2830 MtCO2 of carbon emissions (Global Carbon Atlas, 2022).

However, according to Gulagi et al., GHG emissions from the electricity sector are declining quickly, nearly to zero, prior to 2050.

India's RE Policies & Projects

- The framework for the Renewable Energy Certificate (REC), which can be called the currency of the renewable energy market, was established in 2010 by the Central Electricity Regulatory Commission. All types of renewable energy generators now have the chance to take advantage of the benefits without having to worry about the terms of the power purchase agreement for the trade of renewable power, thanks to the renewable energy certificate system (Elavarasan et al., 2017).
- The Jawaharlal Nehru National Solar Mission (JNNSM), a significant energy mission, was launched in 2010 under the National Action Plan on Climate Change (NAPCC) with a view to increasing the generation of electricity through solar energy within 2022. It has the current target of generating 22,000 MW of power combining on-grid and off-grid plants (Elavarasan et al., 2017).
- In a few places, solar and wind power plants were installed on agricultural lands, which is lucrative for both crops and power plants. Plants can benefit from indirect sunlight, which can be created by installing solar panels above crops and other vegetation. By reducing the humidity and moisture level below the panels, it also lessens the heating effect of the solar panels (Patel et al., 2018). Innovative ideas like that should be facilitated.
- Tariff policy in India is revised multiple times to simplify the purchase of RE. It is coordinated in a way that is beneficial for both the distributor and the consumer of renewables. A minimum amount for purchasing energy is fixed, taking into consideration the perspective of distribution companies, and creates ample capacity and robust infrastructure to ensure improved services to consumers (Elavarasan et al., 2017).
- Several ground-breaking ongoing projects, including the development of a high-efficiency (21%/ 19%) PERC type of c-Si/mc-Si solar cell, green hydrogen mobility projects, met-ocean measurements at the Gulf of Khambhat and Gulf of Mannar, and biomass gasification through plasma pyrolysis technology, are acknowledging them as a future world leader in this field (MNRE. *Home: Ongoing Projects*. [Online]. Available: https://mnre.gov.in/; NTPC Renewables. *Verticals: Green-hydrogen* [Online]. Available: https://ntpcrel.co.in/; NIWE. *Department: Offshore Wind Development: Met-ocean Measurements*. [Online]. Available: https://niwe.res.in/; CMERI, 2022).

- In 2016, India and France jointly formed The International Solar Alliance with the aim of expanding solar energy globally (Nguyen et al., 2021).

3.3.2 Scenario of Vietnam

Vietnam is another highly climate-vulnerable emerging economy of South East Asia (Nguyen et al., 2021), with a population of 98.19 million. Vietnam is moderately positioned in terms of equality distribution, with a Gini coefficient of 36.1 in 2022 (World Bank Group, 2022a). Vietnam is one of those countries which witnessed some unprecedented surge in energy sector within a short period. In 2015, total installed capacity for solar energy was only 4 MW in the country, and till 2019 it is uplifted to 7.4 GW followed by a massive investment in this sector (Tachev, 2024), as the government of Vietnam found that sustainable energy development and energy security are inevitable components of its strategic plan to achieving sustainability in near future (Nguyen et al., 2021).

Harnessing renewable energy sources including hydro power, wind power, solar power, and biomass power is especially advantageous for Vietnam. Although hydropower plants are the main RE producer of the country (Polo et al., 2015), solar and wind generation accounts for 69% of the total RE generation in ASEAN region (Rosalia et al., 2024).

In 2022, total installed renewable energy capacity in Vietnam is 44,691 MW, followed by an increment of 3.79% from the previous year, and 64% increase from the year 2014. Installed capacity for renewable hydropower accounts for 22,535 MW in 2022, while 5065 MW indicates the capacity for wind energy (4071 MW onshore wind, 994 MW offshore wind). Solar PV and bioenergy stand at 16,698 MW and 393 MW, respectively (Prime et al., 2024). Concentrated Solar Power (CSP) is not available in Vietnam. Solar PV experienced the most expansion in installed capacity in 2022—it jumped from 24.14 in 2021 to 24.54% in 2022 (Bloomberg NEF, 2023b).

Electricity generation from hydropower accounts for 99,370 GWh in 2022 and mounts at the top among renewable technologies, within which generation from large hydro stands at 73,844 GWh and small hydro at 25,526 GWh. Wind generates a total of 8852 GWh of electricity in this year, whereas solar PV and biomass generates 25,526 GWh and 379 GWh, respectively (Bloomberg NEF, 2023b). Large hydro constitutes 27.93% of total energy generation, which is the second largest among the energy mix after coal (Bloomberg NEF, 2023b).

According to CLIMATESCOPE, Vietnam's 2022 clean energy investment was approximately \$559.53 million, a 93.19% drop from 2021 (\$8221.73 million). The year 2020 recorded the most investment in sustainable energy, at \$10815.62 million. In Vietnam, the average cost of energy witnesses a drop from 98.48 USD/MWh in 2021 to 96.4 USD/MWh in 2022. Vietnam's average power cost has varied from 82.98 USD/MWh in 2017 to 98.48 USD/MWh in 2021 (Bloomberg NEF, 2023b).

Vietnam emits 344 $MtCO_2$ in 2022 and rank 17th in the list of worldwide carbon emissions (Global Carbon Atlas, 2022). Vietnam is one of the largest emitters of carbon among the emerging economies. According to the Global Carbon Atlas, it experienced a slight fall of 2.5% in 2022 in terms of carbon emissions (Global Carbon Atlas, 2022).

Vietnam's RE Policies and Projects & Plans

Vietnam is one of those countries, who implemented policies to encourage shifting toward renewable energy. In order to develop the sector, the Vietnamese government counts on the national power development plans, which project demand growth and outline the entire shift in the power industry to satisfy demand for the next 10 years (ITA, 2024). We provide the details in Table 1.

- The government of Vietnam issued and implemented a sustainable energy development strategy through 2030 and a vision to 2045 (Climate Action Tracker. Countries: Vietnam. [Online]. Available: https://climateactiontracker.org/).
- Viet Nam joined the Just Energy Transition Partnership (JETP) in December 2022 with the aim of achieving net-zero emissions by 2050. Viet Nam will be granted USD 15.5 billion until 2026–2028 (Climate Action Tracker. Countries: Vietnam. [Online]. Available: https://climateactiontracker.org/).
- In May 2023, Vietnam adopted the much-expected Power Development Plan 8 (PDP8), which sent contrasting messages to the country's power industry, following the signing of the JETP.

Table 1 Comparison of the scenario of Vietnam

Type of renewable energy	Type of technology	Selling price (excluding vat)
Small Hydropower (Under 30 MW)	Power production	According to the announcement of Ministry of Industry and Trade
Wind power (projects came into operation before November 2021)	Project on land	8.5 UScents/kWh
	Offshore project	9.8 UScents/kWh
Biomass	Cogeneration of heat and electricity	7.03 UScents/kWh
	Not heat-electricity cogeneration	8.47 UScents/kWh
Electricity from waste	Burn	10.05 UScents/kWh
	Bury	7.28 UScents/kWh
Solar power	Floating solar power	7.69 UScents/kWh
	Ground solar power	7.09 UScents/kWh
	Rooftop solar power	8.38 UScents/kWh

Source: National Steering Committee for Electricity Development (Vietnam)

- Vietnam pledges at COP26 to halt erecting new coal-fired power plants and to gradually phase out coal-fired power generating (Climate Action Tracker. Countries: Vietnam. [Online]. Available: https://climateactiontracker.org/).
- In November 2022, Vietnam revised its Paris Agreement goal. In comparison with the previous NDC, the target was 39 MtCO2e (excluding LULUCF), and sectors coverage and accountability have both increased (Climate Action Tracker. Countries: Vietnam. [Online]. Available: https://climateactiontracker.org/).
- One crucial component of policies for cleaner energy is intergovernmental cooperation. Vietnam is pursuing collaboration with Switzerland to assist them in the clean energy transition (Prime et al., 2024).

All feed in tariff above will be available for the first 20 years of operation of the technologies, except small hydropower, which is cost tariff exempted.

- Other incentives comprise tax incentives (income tax, import tax), land use, low interest finance, etc.
- The Politburo of Vietnam adopted Resolution 55, which calls for changing the Electricity Regulation to permit private sector participation in electricity infrastructure, which will attract abundant investment (Central Committe of the Vietnam, 2020).

3.3.3 Scenario of Mexico

Mexico is one of the most uneven countries in the world, with a Gini coefficient of 43.5 (World Bank Group, 2022b). It has a population of approximately 129 million people, with half of them living below the poverty line (De La Peña et al., 2021). This country's economy is $2.87 trillion, which is the 11th largest in the world (World Economics, 2023). It is also one of the most climate-vulnerable emerging economies in the world. The electricity sector in Mexico had previously experienced monopoly control by the Federal Electricity Commission, followed by a reform in 2013, which granted private parties the opportunity to partake in the electricity market. This energy reform ensured that the government would continue to place emphasis on clean or renewable energy (Diezmartínez, 2020). The Energy Transition Law, which was passed in 2015, mandates that 35% of the electricity generated by 2024 must originate from renewable sources (De La Peña et al., 2021). The 2013 energy reform served as the basis for the development of various laws, strategies, programs, and initiatives (Castrejon-Campos, 2022), which are mentioned in Mexico's policy section.

Mexico terms its sustainable energy resource as clean energy. The generation of total clean energy is 106,302.45 GWh (including non-renewable nuclear) for the country in 2022, which constitutes 31.2% of the total electricity generation (ITA, 2023a). Generation from large hydropower accounts for 30,390.9 GWh in the year, while generation from wind and solar remains closer at 20,528.8 GWh and 20,338.3 GWh respectively. Small hydro generates 5168 GWh electricity in 2022, and 4412.7 GWh and 2141.3 GWh of electricity is generated by geothermal and biomass technologies, respectively, although natural gas leads the total energy generation mix, generating 56.58% of total generation (192,508 GWh) (Bloomberg NEF, 2023c). Total renewable energy supply by source stands at 209.73 TWh in 2022, or 9.72% of the total supply mix of the year (IEA, 2023a). Mexico's total installed renewable energy capacity witnesses an increment of 4.36% in 2022 compared to the previous year, which stood at 31.95 GW. Hydropower leads the installed capacity portfolio, followed by solar and wind in the 2nd and 3rd positions, respectively, and it is also to be noted that Mexico comprehends fifth largest geothermal power capacity in the world after the USA, Philippines, Indonesia, and New Zealand (Castrejon-Campos, 2022).

Solar energy installed capacity is 9364 MW, of which solar PV is 9347e MW and concentrated solar power is 17e MW. The total installed capacity for hydropower is 13,304 MW in the year. Wind (onshore) installed capacity remains at 7318 MW, and bioenergy and geothermal installed capacity are 966 MW and 999o MW, respectively, as of the year 2022 (Whiteman et al., 2023).

Mexico's 2022 clean energy investment was approximately $717.81 million, down 8.56% from 2021 ($784.98 million). This is the lowest amount of investment since 2017. In Mexico, the average price of electricity is 119.52 USD/MWh in 2022, which was 128.5 USD/MWh in 2021. The price generally ranges between 111 USD and 132 USD per MWh (Bloomberg NEF, 2023c). Mexico ranks 12th in the world in terms of greenhouse gas emissions since the country still mostly relies on fossil fuel-based energy (Global Carbon Atlas, 2022). However, to comply with the Paris Agreement 1.5 target, shifting the energy sector from fossil fuel-based to renewable energy is vital since globally increasing GHG emissions are dominated by FF-based energy use. Mexico accounts for 512 MtCO2 GHG emissions in 2022, comprising gas and oil in the 1st and 2nd positions, respectively (Global Carbon Atlas, 2022).

Mexico's RE Policies and Projects

- The energy reform of 2013 simultaneously focuses on reducing GHG emissions by boosting the proportion of clean energy sources to achieve climate sustainability and opening the formerly closed oil, gas, and power sectors in the country (IEA, 2017).
- Some policies and programs predominantly devised from the reform are: laws include: Energy Transition Law 2015, Geothermal Energy Law 2014, etc., strategies include: Transition Strategy to Promote the Use of Cleaner Technologies and Fuels 2016 & others, programs include: National Program for Sustainable Use of Energy (2014–2018), Special Program for the Energy Transition (2017–2018), and so on, and initiatives include: Energy Transition Fund, and Energy Sustainability Fund (Castrejon-Campos, 2022).
- In order to strengthening the innovation capability, Mexico formed a fund called "Energy Sustainability Fund" in 2008. It generally seeks to facilitate research & development, sustainable energy development, technological advancement, etc. (OECD/IEA, 2017).

- The Sonora Plan, which is unveiled at the COP27 climate summit, consists of several green infrastructure projects in the state's northern region with the goal of increasing the nation's capacity for manufacturing and renewable energy (Godoy, 2022).
- President of Mexico announced an ongoing project concerning generation of green hydrogen through water electrolysis (Parkes, 2024).
- In order to electrify whole territory, including remote parts, new generation plants are erected, new dams are restored, transmission channels are dilated, and also solar panels are installed.

3.3.4 Scenario of Brazil

Brazil is the country, who is effectively transiting its energy sector to renewable based system among the sunbelt countries. It is a country with 215.3 million of people and a significant amount of disparity in income with a Gini coefficient of 52 in 2022 (World Bank Group, 2022c). Brazil has been moving forward with its shift to a low-carbon and more sustainable energy system. The nation has adopted renewable energy sources including hydropower, solar electricity, and wind power in an effort to minimize its reliance on fossil fuels. Through the rapid adoption of off-grid solar technology, Brazil is making progress in solar and wind energy.

Brazil boasts the seventh-largest power generation capacity in the world and the sixth-largest consumer electricity market worldwide (ITA, 2023b). It ranks second in terms of hydropower as well as bioenergy generation in the world. It also held 7% of global renewable energy generation. In Brazil, hydroelectric electricity is the second most often utilized primary energy fuel after oil.

Renewable energy comprises 84% of Brazil's total capacity mix and 87% of total generation mix (Bloomberg NEF, 2023d). Its total installed renewable energy capacity is 176,709 MW as of 2022, among which hydropower accounts for 109,802 MW, solar stands at 25,520 MW. Wind and Bioenergy held 24,165 MW and 17,224 MW of installed capacity, respectively, in 2022 (Whiteman et al., 2023). According to ITA, over 44 GW of installed wind generating capacity is anticipated in Brazil by 2028, making up 13.2% of the country's total electricity mix.

Hydropower also dominates the generation mix of Brazil with 409,890 GWh in 2022, while wind generates 81,631 GWh of electricity in the year. Solar and bioenergy accounts for 64,235 GWh and 52,046 GWh of electricity generation, respectively, as of December 2022 (Bloomberg NEF, 2023d).

Brazil's average price per MWh of electricity rose from 151.1 USD in 2021 to 159.95 USD in 2022. It is fluctuating from 138.62 USD/MWh to 183.34 USD/MWh within the period of 2017–2022. Investment in clean energy for Brazil amounts $11502.35 million in 2022, an increase of 52.92% compared to the previous year ($7521.92 million) (Bloomberg NEF, 2023d). By 2029, projected investment in the Brazilian energy sector would be around $100 billion, encompassing transmission, distributed generation, and utility-scale generation projects (ITA, 2023b).

Brazil's energy sector is among the least carbon-intensive in the world, although it emits 484 MtCO2 of carbon and ranks 13th globally in terms of emissions in 2022 (Global Carbon Atlas, 2022). 320 MtCO2 of it accounts for oil emissions. Deforestation in Brazil expanded in 2020 and 2021 and expected to increase as well in the coming years due to cattle rearing and illegal mining. That might cause the recent unexpected surge in carbon emissions of the country (Climate Action Tracker, 2023a).

Brazil's RE Policies, Programs, and Plans

Brazil has been paving the way in renewable energy, enacting laws, and programs to support sustainable energy sources. Brazil is determined to a sustainable energy future that benefits its people and the planet globally, as seen by its regulations and investment.

- Law 9478 of 1997 established Brazil's national energy policy, with a focus on the exploitation of renewable energy sources as a fundamental element (Ministerio De Minas Energia, 1997).
- The New Legislative Framework for Solar Energy encourages photovoltaic development by exempting individuals who commence producing solar energy from paying taxes until 2045 (Government of Brazil, 2022a).
- The Program of Incentives for Alternative Electricity Sources supported in the development of local manufacturing abilities for wind turbines and their components (IEA, 2015).
- Implemented in 2020, the "**RenovaBio**" policy establishes emission targets for transportation and promotes the production of biofuel by means of decarbonization credits (MME, 2021).
- Brazil is pursuing new frontiers in energy innovation, one of which is hydrogen. Recent initiatives in the National Hydrogen program are focused on research and development (R&D) (Government of Brazil, 2022b).
- An initiative named "The Fuel of the Future," which prioritizes R & D to advance low-emission technologies (Fick, 2023).
- To establish guidelines for businesses interested in establishing offshore wind farms in Brazil, the Brazilian government established a working group (Enerdata, 2022).
- Brazil has tremendous capability of producing off shore wind energy due to its vast coastline of 7400 K.M, steady wind, and relatively shallow ocean (Government of Brazil, 2022c).
- Ministry of Mines and Energy formed The Ten-Year Energy Expansion Plan, named PDE 2019–2029, followed by due consultation with stakeholders (Ministerio De Minas Energia. *Ten-Year Energy Expansion Plan 2029*. [Online]. Available: https://www.epe.gov.br/).

According to the former president of Brazil Fernando Henrique Cardoso, consolidation of a low-carbon matrix is a requirement of the twenty-first century, and Brazil has all the necessary resources to progress in this direction, given its abundance of renewable energy. Nonetheless, the nation must constantly innovate and foresee what is going on in the rest of the world (Fundacao, 2019).

3.3.5 Scenario of Morocco

Morocco is a developing country with high growth rate of population and soaring demand of energy consumption. It has a population of 37,457,971 crores in 2022, and a Gini coefficient of 50.6 in 2019 (most recently measured) (World Economics, London, 2019). 11.9% of active people of the country are unemployed, which is approximately 6.8% higher than the regional average (Focus Economics, 2022).

Morocco started to develop renewable energy in the 2010s, followed by a slowdown in the later period, because of price volatility, environmental issues, and scarcity of resources (Bloomberg NEF, 2023e). Given the national policy and the locally acquired expertise of the national and international operators, presently Morocco is actively participating in the renewable energy sector (El Ghazi et al., 2021). According to World Bank 2018, Morocco has the commanding position in its region in terms of incorporating renewables in its energy mix. It aims to boost its dependence on renewable energy sources from about 20% of its current power generation to 52% by 2030 (Roscoe, 2017), and further 80% in 2050 (Bloomberg NEF, 2023e).

Morocco has a total renewable capacity of 3725 MW for renewables in 2022, which is 2.34% higher compared to the previous year (Whiteman et al., 2023). Solar PV capacity witnessed biggest leapfrog in that year from 7.14% in 2021 to 10.24% in 2022.

Capacity installed for hydropower, wind energy, and bioenergy in Morocco stands at 1306 MW, 1558 MW, 7 MW, respectively, in 2022. Solar PV accounts for 854 MW of capacity installed, in which 314 MW implies solar PV and 540 MW indicates CSP capacity in 2022 (Whiteman et al., 2023). Morocco's dependency on coal is diminishing gradually which stands at 35.81% of total capacity installed in 2022.

Morocco's total electricity generation from renewables is led by wind energy with 6504 GWh of electricity generation, followed by large hydro and solar PV with 1432 GWh and 1300 GWh of electricity generation, respectively, in 2022. Its largest annual increase in generation in 2022 also comes with wind energy at 14.86% up from 13.92% in 2021. Generation from solar thermal reaches at 1252.8 GWh at the end of 2022. In the meantime, generation from small hydro mounts at 540 GWh. (Bloomberg NEF, 2023e). According to BBC, Morocco has tremendous natural resources to produce solar, wind, and hydropower, comprehending that the country is heading toward sustainable energy sector and dilating as an energy investment hub of Northern Africa (Alami, 2021).

Morocco's clean energy investment amounts $553.12 million in 2022, up 12.72% from $490.73 million in 2021. In the recent periods, Morocco makes its largest renewable energy investment in 2018, which was \$2930.38 million (Bloomberg NEF, 2023e).

The average price of residential, industrial, and commercial use electricity in Morocco decreased from 108.48 USD/MWh in 2021 to 97.54 USD/MWh in 2022 (Bloomberg NEF, 2023e).

Morocco is one of the most climate-vulnerable countries in the region, where the average temperature is thriving despite not being a large emitter of carbon. In 2022, Morocco emits 68 MtCO2 of carbon and ranks 49th as a country in terms of carbon emission, constituting only 0.183% of global carbon emission. Per capita emission of the country in 2022 stands at 1.8 ton (M. R. a. P. R. Hannah Ritchie, 2022).

Morocco's Policies and Projects

According to INSAMER, Morocco has undergone an enormous transition to renewable energy and energy efficiency, placing it at the forefront of this industry in various aspects for the continent of Africa. However, there are still difficulties with putting policies into practice (Mhamed, 2022).

- During COP26, Morocco signed the methane agreement. Methane is mostly found in the waste and agriculture sectors and accounted for 17% of global GHG emissions. (Europian Commission, 2023). Overall methane emissions are anticipated to be impacted by mitigation strategies in the agriculture sector.
- Morocco committed to halt issuing permits and building new coal-fired plants when it approved clauses 1, 3, and 4 of the coal exits at the UN Climate Change Conference (COP26), 2021 (Climate Action Tracker, 2023b).
- Morocco committed to expediting the spread and uptake of electric zero-emission vehicles (Nabil Samir, 2022).
- The Moroccan government pledged to raise the proportion of renewable energy in the country's electricity mix to 80% by 2050 in its long-term strategy, which was released on December 21, 2021 (Climate Action Tracker, 2023b).
- The 2021 NDC incorporates the 2030 National Solar Plan, which sets a new goal of achieving a 4 GW total capacity by 2030. Also, Morocco currently plans to reach a total wind power capacity of 2.2 GW by 2030 as part of the 2030 National Wind Plan. Further, the government stated in its revised NDC that it intended to add 1.1 GW of hydropower capacity by 2030 (Climate Action Tracker, 2023b).
- Morocco's green hydrogen sector and its derivatives are predicted to be able to satisfy the country's demand between 13.9 TWh and 30.1 TWh in 2030, and between 153.9 TWh and 307.1 TWh in 2050, according to the country's roadmap, and it aspires to become a global leader of green hydrogen (Moroccan Ministry of Energy, Mines and Environment and IRESEN. Countries: Morocco. [Online]. Available: https://gh2.org).
- The Noor-Ouarzazate complex, located in Morocco, is the largest concentrated solar power plant in the world. It is made up of a massive network of curved

mirrors that cover 3000 hectares (11.6 square miles), concentrating sunlight onto tubes of fluid that are then heated to generate electricity (Josephs, 2023).

- Moroccan Youth Center for Sustainable Energy founded by Rachid Ennassiri, a Moroccan environmentalist, works on several climate change projects, i.e. project of making sustainable mosques using solar panel (Alami, 2021).

3.3.6 Scenario of South Africa

South Africa is yet another unequal country with a Gini coefficient of 63 and with a population of 59.89 million people (Dyvik, 2024). 86% of country's total wealth is held by the richest 10% people of the country (Anekwe et al., 2024). 40% of the youth are unemployed in the country. Since 2021, the country's electrification rate is 89.30%, followed by a national electrification campaign. The coal industry has been the backbone of South Africa's energy supply for the past few decades, offering relatively well-paid jobs for workers with lower skill levels as well as a local fossil fuel that serves as the country's main source of electricity (Hanto et al., 2022). Coal industry utterly dominates the share of installed capacity and electricity generation with 72.19% of total installed capacity and 84.54% of total electricity generation in 2022 (Bloomberg NEF, 2023f).

The Renewable Energy Independent Power Producer Procurement Program (REIPPPP), which has so far secured approximately 10 GW capacity in six bidding windows, is crucial in obtaining RES through its bidding process. REIPPPP is introduced by the South African government in 2011 for the sake of attracting private investment in the renewable energy sector (Anton Eberhard, 2016). Solar photovoltaic and onshore wind are to be considered as qualifying technology at the 7th bid submission phase of the REIPP procurement program 2024 (DMRE, 2023a). Now the government is targeting widespread phase out of coal and deployment of RE to reduce GHG emissions and achieve carbon neutrality within 2050. It is worth to be mentioned that South Africa is the 14th largest GHG emitter of the world (Robert McSweeney, 2018).

South Africa's total installed renewable energy capacity accounts for 10,505 MW in 2022, which is 7% higher compared to the previous year, also the largest capacity among sub-Saharan Africa region (Cowling, 2024). However, the country seeks to uptake its renewable energy capacity to 19 GW within 2030 (NDC Partnership. *Making renewable energy affordable: The South African Renewables Initiative*. [Online]. Available: https://ndcpartnership.org/). It comprises 17.69% of total energy installed capacity. Among all the renewable energy sector capacity, hydropower remained steady over the years and accounts for 752 MW in 2022. Wind energy (Onshore) stands at 3163u MW in 2022, after witnessing a slight decrease in the previous year. A total of 6326 MW of installed capacity for solar energy is available at the country in 2022, in which 5826 MW stands for solar PV and the rest indicates CSP. Installed capacity regarding bioenergy 265 MW, solid biofuel and renewable waste is 242e and biogas 23e MW (Whiteman et al., 2023). South Africa is not a country with geothermal resource available. The highest increase in terms of capacity installed is 9.46% for solar PV in 2022, up from 8.05% in the previous year (Bloomberg NEF, 2023f).

Wind energy is the mainstay of renewable's share in electricity generation with 9640.9 GWh, followed by solar PV and large hydro, which amounts 4962.7 GWh and 3022.7 GWh, respectively, in 2022. Solar thermal generation for the year is 1589.5 GWh, whereas small hydro and biomass waste stand at 280.4 GWh and 201.4 GWh, respectively (Bloomberg NEF, 2023f).

According to Statista, SA's renewable energy held 18.26% share in the total final energy consumption (TFEC) in 2022 (Degenhard, 2024). Country's average Price of energy decreased from 101.25 USD/MWh in 2021 to 99.94 USD/MWh in 2022. Total investment in renewable energy in the year is $4787.14 million, a threefold increment from $1576.62 million in 2021 (Bloomberg NEF, 2023f). As previously mentioned, South Africa is the 14th largest GHG emitter, alongside it ranks 15th in terms of emitting CO2. In 2022, it emits 404 MtCO2 carbon, 338 MtCO2 came from coal sector therein. 38 MtCO2 of carbon is released from the oil sector this year, which stands for the second largest carbon emitting sector in South Africa (Global Carbon Atlas, 2022).

SA's Policies and Programs

- One of the policy papers that established the ground work for the development of renewable energy technologies, including solar, hydro, biomass, and wind, was the White Paper on Renewable Energy, 2003 (DMRE, 2003).
- Integrated Resource Plan (IRP) 2019 is the national electricity strategy by the country's government from 2018 to 2030, which indicates how the specific demand will be supplied (Hanto et al., 2022).
- Renewable Energy Independent Power Producer Procurement Program (REIPPPP) is a unique initiative by the country's government to entice the private sector investment to the several renewable energy sectors like solar PV, CSP, wind, etc.
- Climate change mitigation is the explicit focus of legislative measures and policy instruments including the Climate Change Bill, the Carbon Tax, and offsetting schemes (Hanto et al., 2022).
- DMRE is carrying out an initiative to connect the youth and women of their country to the energy sector to make them vigilant about the gradual coal phase-out and decarbonization of the sector (DMRE. *Energy Resources: Programmes and Projects: Programmes and Projects Management Office: Women Empowerment* [Online]. Available: https://www.dmre.gov.za; DMRE. *Energy Resources: Programmes and Projects: Programmes and Projects Management Office: Youth Empowerment* [Online]. Available: https://www.dmre.gov.za).
- Integrated Energy Center (IEC), an initiative undertaken by the government, seeks to improve rural enterprise development, reduce poverty, and increase access to energy (DMRE. *Energy Resources: Programmes and Projects: Pro-*

grammes and Projects Management Office: Integrated Energy Centre [Online]. Available: https://www.dmre.gov.za).
- Request for proposal (RFP) from municipalities in preparation for the 2024/25–2027/28 is a project that is underway by the Department of Minerals and Energy, South Africa (DMRE, 2023b).
- The DMRE's new licensing regulations for plants <100 MW are another significant recent policy change that is anticipated to promote the uptake of RES. These regulations are expected to release approximately 5 GW of additional industrial and mining capacity in the coming years (Hanto et al., 2022).
- South Africa launched an initiative called South African Renewable initiative (SARi) to push the electricity generation from renewable technologies (NDC Partnership. *Making renewable energy affordable: The South African Renewables Initiative*. [Online]. Available: https://ndcpartnership.org/).

Regulation and legislative changes are gradually fostering a policy environment that is more RES-friendly (Hanto et al., 2022).

4 Comparative Analysis

4.1 Discussion

Among the emerging economies of the sunbelt considered in the study, Brazil peaked in respect of renewable energy capacity and generation, as well as the respective mix percentage, setting a benchmark for many other developing and developed nations. India is also pursuing through the way and on the verge of stretching the Paris agreement target as well. India has the lowest per kWh price of electricity and the 2nd highest annual investment in RE followed by Brazil among the countries by which it devised as one of the RE hubs of the world. Morocco, one of the least carbon-intensive economies in the world, stands as the lowest carbon emitter in the study and is projected to attain the net-zero emission target soon. Vietnam has one of the fastest growing renewable energy sectors in the world. The surge in Vietnam's renewables is the transcendent among the studied economies. South Africa and Mexico are also competitively utilizing their RE resources and minimizing their reliance on fossil fuels and carbon-intensive technologies. A comparative analysis is in Table 2.

Table 2 Comparative analysis of six countries' renewable energy indicators according to the data of 2022

Countries	Installed capacity (MW)	Generation (GWh)	Investment (million) USD ($)	Price/MWh USD ($)	Emissions (MtCO2)	Capacity (%) of total mix	Generation (%) of total mix
India	177418	357444	11015.37	94.78	2830	36.43	19.42
Brazil	181446	625025	11502.35	159.95	484	84	87.84
Mexico	34952	82977	717.81	119.52	512	35	24.39
South Africa	10573	19700	4787.14	99.94	404	17.69	8.42
Vietnam	46840	129811	559.53	96.4	344	57.91	49.1
Morocco	4592	11029	553.12	97.54	68	40	25.19

Note: Data from (CLIMATESCOPE) are applied in the comparison

5 Implications

5.1 India (South Asia)

India is superiorly positioned among the South Asian region concerning renewable resource deployment. Various factors guided the country to attain the position including economy, geography and certainly policies, projects and plans following their implications. India's neighboring states, who are geographically and most often economically similar can replicate its policies and strategies and implement it in their respective country to emerge as a renewable intensive country and decarbonize the energy sector sooner.

Sri Lanka, Pakistan, Nepal, and Bangladesh are India's neighboring subordinates with regard to the utilization of RE technologies. In India, the diffusion of REC has been propagated, especially since 2017, following a lackluster demand for it (Sawhney, 2022). REC has been contributing to the development of RE and raising awareness throughout the country. Other countries of the region, including Nepal, Pakistan also have introduced REC but not propagated that much yet. Remarkable diffusion of REC can result in improved awareness and eventually in accomplishment of sustainability goals.

Moreover, as a result of being Agri-based country, India is widely spreading its strategy to install solar panels on crop fields. Adjoining states of India can certainly imitate this exercise as most of them rely on agriculture.

India simplified its feed in tariff policy followed by multiple time revision to facilitate renewables purchase. Some other countries of South Asia, including Nepal and Sri Lanka also captivatingly implemented FIT policy (Elavarasan et al., 2017). Other SA countries except Nepal and Sri Lanka might naturalize their renewable tariff policy to sustain their energy transition and fulfill their drastic energy demand through renewables, i.e. in Bangladesh for electricity produced by renewable energy sources, an incentive tariff that is 10% greater than the utility's maximum purchase price from private producers may be taken into consideration (Dastagir, 2018). While India undertakes strategically different approach, India's MoP has made it clear that, in order to encourage the use of renewable energy, the green tariff cannot, under any circumstances, be greater than the total of the average power purchase costs of renewable energy, plus a surcharge equal to 20% of the average cost of supply (Ministry of Power, 2023).

Most of the South Asian countries except Nepal and Bhutan are surrounded by waterbody like Bay of Bengal, Indian Ocean, and Arabian Sea. Hence, they possess a great advantage of producing offshore wind energy. India is already working on offshore wind project at the gulf of Khambhat and the gulf of Mannar.

Further, international collaboration (regional and outside of the region) pertaining to renewable technologies with resourceful countries would be beneficial for South Asian countries, such as India.

5.2 *Brazil (South America)*

South America is one of the cleanest regions in the world regarding electricity sectors. Apart from Brazil, Chile, Argentina, Colombia also is rapidly proceeding toward a carbon-free energy sector (IEA, 2023b). Other countries are also initiating plans and policies for the sake of impact of climate change.

Brazil is one of the renewable superpowers of the world, hence, would pursue by all over the world. According to Brazil's Solar Energy legislation, tax would be exempted for those, producing solar energy newly until 2045. Such tax exemption of 15–20 years on RE, according to the state's economic convenience, greatly instigates private investor and producer to commence producing renewables.

Another Brazil's masterstroke is RenovaBio policy. Its objective is to lower the carbon intensity of Brazil's transportation system by increasing the use of biofuels and developing a market for carbon credits to offset greenhouse gas emissions from the burning of fossil fuels, and to be included into the country's NDC. In this policy, producers of biofuel proactively verify their output, earning them points for energy-environmental sufficiency, which results in the decarbonization credit, that can be commercialized (MME, 2021).

Thus, not only other country of the continent but also countries all over the world, infested with bioenergy resources may replicate the strategy to decarbonize their transportation sector.

In Brazil, a bill called Fuel of the Future, which promotes the manufacture of sustainable fuels like biodiesel, biomethane, and sustainable aviation fuel (SAF), was approved by Parliament in 2024. Such policies are crucial for carbon-intensive countries, whose energy sector are mostly relying on fossil fuel. Countries including Argentina, Peru, and Ecuador heavily rely on diesel and natural gas, therefore, these countries may reduce their dependability on fossil fuels and increase the number of renewables in their energy mix by imitating the law, since the law establishes initiatives for the decarbonization of natural gas and the manufacturing of sustainable aviation fuel, as well as a 20% increase in the blend content of biodiesel in diesel. Moreover, annual, or multiyear plan concerning RE exploitation in accordance with respective NDC of the countries will also provide pathway toward sustainable energy.

5.3 *Vietnam (South East Asia)*

South East Asia is one of the world's renewable-intensive regions, consisting of countries that are more or less equally charged with renewable resources. Apart from Vietnam, Indonesia, Philippines, Thailand are also similarly concentrated on producing renewable energy. Because of Vietnam's rapidly growing RE sector, it is taken into consideration for the study. Now focusing on the Vietnam-related issues that can be pursuit by its neighbors.

A Just Energy Transition Partnership (JETP) between Vietnam and the International Partners Group (IPG) was announced with the goal of securing funding to assist Vietnam in implementing a fair and sustainable energy transition in 2022 (Eurpean Union, 2023). Indonesia also entered into this partnership on the same year (UNDP). This partnership program helps a country with resource mobilization, policy frameworks and implementation, financial incentives, and so on. Therefore, all the other countries of the region who seek a swift transition of energy sector would pursue this partnership strategy. Moreover, international collaboration with the countries with similar contemplation of renewables enhances the capacity, generation, and policy of renewables.

Vietnam also halted the erection of new coal-fired power plants to reduce the emission of carbon and achieve sustainability, Philippines and Indonesia also did the same. Philippines announced a moratorium on the permit of new coal plant, and Indonesia postponed its scheduled power facilities up to 15 GW (Global Eergy Monitor, 2020). Other SEA countries may also exert the same to expand the usage of renewables and alleviate reliance on fossil fuels.

Remediation of greenhouse gas is not possible overnight. It requires long-term provident planning and roadmap along with proper execution. Considering this, Vietnam conferred a much-anticipated roadmap to implement National Electricity Development Plan also known as Power Development Plan 8 (PDP8) in May 2023.

It includes an aggregate depiction of a country's energy sector, therefore, comparatively backward countries of SEA, such as Cambodia, Myanmar, and Malaysia may adopt this type of roadmap and can imitate Vietnam's to a great extent, since they are geographically, economically, and culturally similar. With some changes according to their energy targets, financial allocations, preferences, and positions, these countries may effectively adopt this type of programs.

Other lucrative initiatives by Vietnam government, including, alluring tariff policy, import tax exemption, incentivize investor and most importantly allowing private sector participation in energy industry to attract large investment. Simulation of these actions may certainly support other countries to further approach toward sustainability.

5.4 *South Africa (Sub-Saharan Africa)*

South Africa is the country with maximum renewable energy in sub-Saharan Africa region, although its economy is greatly dependent on coal. All the countries of the region contain numerous RE resources, while some are lagging behind in terms of utilization of these resources. South Africa is rapidly implementing policies and procedures to utilize its RE resources and reducing reliance on coal and other carbon-intensive technologies. Hence, it is needless to say that replicating South Africa's resource utilization policy and strategy must facilitate its neighboring countries' energy sustainability actions.

Now focus on how other sub-Saharan countries can emulate SA's environmental sustainability strategies and policies. South Africa's latest Integrated Resource Plan (IRP) 2019, under its National Development Program (NDP) is a comprehensive plan that outlines the country's electricity infrastructure development goals, including integrated renewable energy generation and capacity target, electricity tariff, investment trends, R&D, regional integration, technology usage, plan performance, and so on (DMRE, 2019). Economically competitive neighbors of South Africa, such as Nigeria, Botswana, Namibia, Zimbabwe may easily replicate such comprehensive plan with some changes according to their recent positions in utilization and preference.

South Africa's most vital effort to diversify its energy mix and reduce reliance on fossil fuel is the Renewable Energy Independent Power Producer Procurement Program (REIPPPP), which is designed to procure RE generation capacity from private companies followed by tendering and bid submission.

This idiosyncratic strategy of the country is highly suggested to pursue by other countries of the region. A country operating under this strategy can invite IPP's to submit bids to develop RE project in accordance with their requirement, evaluate the bids under the light of different criteria, enter into negotiations with the bidders, control and maintain the overall project. Thus, devices like REIPPPP may stimulate a country's overall economic growth, job creation, and environmental sustainability.

Everyone is aware of the fact that youth are the major changemaker of a nation, and no nation can progress without the advancement of their women. Keeping this in mind, Department of Mineral and Renewable Energy is connecting the nation's youth and women to their energy sector to make them concern about climate and need of decarbonization of the sector. This is also such an initiative that other country may contemplate to imply.

Another considerable program is the South African Renewables initiative (SARi), established by SA government to help accelerate and aggressively scale up renewable energy in South Africa in a way that will benefit the country's economy, society, and environment (SARi, 2011). Pursuing this initiative may help a country to channelize or mobilize its public finance to its green energy activities.

5.5 *Mexico (North America)*

Mexico is often considered as a country of North America along with the USA and Canada. Other two countries are developed and not located in sunbelt completely, since some parts of the USA included in the sunbelt region. Therefore, if not the country's regional neighbors, then its geographically adjoining countries can pursue its energy sustainability strategy to a great extent, such as Panama and Costa Rica.

Mexico's root to the energy transition and climate sustainability is its energy reform program in 2013, which prioritizes reduction of GHG emissions and boosting the proportion of clean energy sources into their energy mix in order to achieve climate sustainability (IEA, 2017). Prior to 2013, Mexico's energy sector completely relied on fossil fuels, such as coal, natural gas, and diesel. The impact of climate change made the country concern regarding the climate sustainability and forced it to reform the energy policy. Hence, all the countries that are adversely impacted due to climate change, but still have not taken any progressive measure to deploy renewable energy, may find such reform policy beneficial for its future. Some specific programs, policies, and initiatives devised from this energy reform, such as Energy Transition Law 2015, Geothermal Energy Law 2014, Strategy to Use Cleaner Technologies and Fuels, and National Program for Sustainable Use of Energy, are noteworthy to depict its potential significance to the countries.

One of the Mexico's presidential initiatives named, Sonora Sustainable Energy Plan, which aims to create a sustainable ecosystem, encouraging the expansion of vital sectors includes semiconductor, automation, and electromobility (NDC Partnership, 2023). The four primary foundations of this ecosystem are the development of human skills, clean energy generation, key minerals, and strategic infrastructure.

Sonora's renewable energy plan aims to receive an investment of US$1.64 billion for its photovoltaic plant to generate 1000 MW of electricity and to facilitate 1.6 million consumers. Consequently, it goes without saying that such a city-based plan or turning a city into an energy hub on the basis of the infrastructure, transportation, onshore and offshore access of the city is conducive to a country's energy transition.

The effectiveness of hydrogen in electricity and transportation sector is immense, which includes fueling vehicles and aircrafts, power plant fuel, fuel cell power generation, etc. Therefore, use of green hydrogen is benignant solution in terms of climate impact. The president of Mexico announced a project regarding generation of green hydrogen through water electrolysis. Countries with abundant water resources may imitate this technique as well to reduce reliance on conventional hydrogen and enhance clean energy.

5.6 Morocco (MENA)

Morocco is one of the most carbon-intensive countries in Middle East and North Africa region. There are some other countries that are competitive to Morocco in terms of RE generation and capacity. But, the primary reason behind considering Morocco is its overall profile including carbon emissions flexibility. Along with Morocco's regional neighbors, other countries with higher carbon emission ratio may pursue its strategy and replicate its carbon lowering initiatives.

The impact of climate change by drastic increase in worldwide greenhouse gas emissions forced Morocco to decarbonize all the sectors possible, as it is one of the world most adversely affected countries by GHG.

In COP26, an agreement regarding Methane is signed by Morocco, which aims to reduce the emission methane mostly from waste and agricultural sector. Morocco also ceased the issuance of permits for building new coal-based power plant to reduce the use of coal in energy. These are the primary actions to pursue by a nation who anticipates decarbonization, although many MENA nations also have joined the methane pledge except Iran, Syria, and Algeria.

Morocco is also determined to disseminate the purchasing and operation of zero-emission vehicles, which further demonstrates their resolve to create a carbon free nation, and yet again fosters other country to contemplate the same by indicating appropriate pathways.

Moreover, country's renewable energy strategies and targets are tempting. Morocco seeks to enhance its RE proportion to 80% in its electricity mix within 2050. According to Morocco's 2021 NDC, it aims to achieve 4 GW of solar capacity and 2.2 GW of wind capacity within 2030. In the meantime, the country is targeting to generate 14–30 GW of green hydrogen.

Individual initiatives from the people of Morocco further exhibit public responsibilities toward sustainability. Rachid Ennassiri, a Moroccan environmentalist established an organization named Youth Center for Sustainable Energy, currently working on building sustainable mosques by installing solar panel. Other countries can exemplify these initiatives to enhance public integrity to climate and sustainability.

6 Conclusion

In this era of constantly deteriorating climate, energy sustainability is the key to frame a protest to the impact of it. Hence, this study aims to suggest several pathways, experienced from regionally supreme countries for reducing the reliance on fossil fuels and gain energy sustainability of EMMIE's of those regions. These formulas are proved to be effective and vary according to the geography, economy, impact, and preferences. This study verdicts that most of the policies, plans, and programs implemented by studied countries are pursuable by their neighboring states. It pinpoints a number of experiences that are shared across countries, especially those dealing with how such renewable energy initiatives are structured, which may be adopted with minimum modifications to different national contexts. Shared geographical characteristics include a rich resource base in terms of solar potential, an important predisposing advantage. The immediate neighbors will have to factor into their peculiar economic capacity and governance system a manner of adopting the strategies. Success will come with regional cooperation, sharing resources, and a commitment to investing in renewable energy infrastructure. This study is the first to cover entire sunbelt regions to support them achieving renewable based energy sector on the basis of their geographical location. Economic conditions of the neighboring countries are a limitation of the study, although economic condition was never considered as a factor under climate change setting and its impacts. As it still emerges as a vital factor in a viable way, further conducting of research is suggested to investigate countries' RE adoption opportunities in accordance with their economic conditions and preferences.

References

Alami, A. (2021, November 19). *How Morocco went big on solar energy*. [Online]. Available: https://www.bbc.com

Anekwe, I. M. S., Akpasi, S. O., Mkhize, M. M., Zhou, H., Moyo, R. T., & Gaza, L. (2024). Renewable energy investments in South Africa: Potentials and challenges for a sustainable transition - a review. *Science Progress, 107*(2), 368504241237347.

Ansari, M. F., Kharb, R. K., Luthra, S., Shimmi, S. L., & Chatterji, S. (2013). Analysis of barriers to implement solar power installations in india using interpretive structural modeling technique. *Renewable and Sustainable Energy Reviews, 27*, 163–174.

Anthony, A.-D. (2021). *Global 100% energy transition by 2050: A fiction in developing economies?* Cell Press.

Anton Eberhard, R. N. (2016). The South African renewable energy independent power producer procurement programme: A review and lessons learned. *Journal of Energy in Southern Africa.*

Barasa, M., Bogdanov, D., Oyewo, A. S., & Breyer, C. (2018). *A cost optimal resolution for Sub-Saharan Africa powered by 100% renewables in 2030*. Elsevier.

Bloomberg NEF. (2023a). Markets: in. [Online]. Available: https://www.global-climatescope.org

Bloomberg NEF. (2023b). Markets: vn. [Online]. Available: https://www.global-climatescope.org.

Bloomberg NEF. (2023c). Markets: mx. [Online]. Available: https://www.global-climatescope.org

Bloomberg NEF. (2023d). Markets: br. [Online]. Available: https://www.global-climatescope.org/

Bloomberg NEF. (2023e). Markets/ma. [Online]. Available: https://www.global-climatescope.org/

Bloomberg NEF. (2023f). Markets/za. [Online]. Available: https://www.global-climatescope.org

Bogdanov, D., Breyer, C., & Aghahosseini, A. (2020). *Towards sustainable development in the MENA region: Analysing the feasibility of a 100% renewable electricity system in 2030*. Elsevier.

Breyer, C., Solomon, O. A., Gulagi, A., & Keiner, D. (2023). *Renewable energy enabling pathways towards prosperity in Africa and South Asia*. Elsevier.

Castrejon-Campos, O. (2022). Evolution of clean energy technologies in Mexico: A multi-perspective analysis. *Energy for Sustainable Development, 67*, 29–53.

Central Committe of the Vietnam. (2020). *Resolution of the Politburo On Orientations of the Viet Nam's National Energy Development Strategy to 2030 and outlook to 2045*. Central Committe of the Vietnam.

Climate Action Tracker. (2023a, December 5). *Countries: Brazil*. [Online]. Available: https://climateactiontracker.org/.

Climate Action Tracker. (2023b, April 17). *Countries: Morocco*. [Online]. Available: https://climateactiontracker.org

Climate Action Tracker. Countries: Vietnam. [Online]. Available: https://climateactiontracker.org/

CMERI. (2022, November). *Projects: Biomass-gasification-through-plasma-pyrolysis-technology-chemicals-production*. [Online]. Available: https://cmeri.res.in/

Cowling, N. (2024, April 4). *Total renewable energy capacity in South Africa from 2013 to 2023*. [Online]. Available: https://www.statista.com

Dastagir, R. U. M. (2018). *Renewble Energy in Bangladesh*. Ministry of Power, Energy & Mineral Resources.

De La Peña, L., Guo, R., Cao, X., Ni, X., & Zhang, W. (2021). Accelerating the energy transition to achieve carbon neutrality. *Elsevier, Resources, Conservation & Recycling, 177*, 105957.

de Souza Noel Simas Barbosa, L., Bogdanov, D., Vainikka, P., & Breyer, C. (2017). Hydro, wind and solar power as a base for a 100% renewable energy supply for South and Central America. *PLoS One, 12*(3), e0173820.

Degenhard, J. (2024, January 30). *Share in the total final energy consumption of renewable energy in South Africa from 2014 to 2029*. [Online]. Available: https://www.statista.com/

Diezmartínez, C. (2020). *Clean energy transition in Mexico: Policy recommendations for the deployment of energy storage technologies*. Elsevier.

DMRE. (2003). *White paper on Renewable Energy*. Department of Mineral Resources and Energy.

DMRE. (2019). *Integrated Resource Plan (IRP)*. Department of Mineral Resources and Energy.
DMRE. (2023a). *Overview of the request for qualification and proposals for new generation capacity under Seventh Bid submission phase of the REIPPPP*. Department of Mineral Resources and Energy.
DMRE. (2023b, August 15). *Energy Resources: Programmes and Projects: Energy Efficiency and Environment: Projects* [Online]. Available: https://www.dmre.gov.za/
DMRE. *Energy Resources: Programmes and Projects: Programmes and Projects Management Office: Women Empowerment* [Online]. Available: https://www.dmre.gov.za
DMRE. *Energy Resources: Programmes and Projects: Programmes and Projects Management Office: Youth Empowerment* [Online]. Available: https://www.dmre.gov.za
DMRE. *Energy Resources: Programmes and Projects: Programmes and Projects Management Office: Integrated Energy Centre* [Online]. Available: https://www.dmre.gov.za
Dranka, G. G., & Ferreira, P. (2018). *Planning for a renewable future in the Brazilian power system*. Elsevier.
Dyvik, E. H. (2024, July 4). *20 countries with the biggest inequality in income distribution worldwide in 2023, based on the Gini index*. [Online]. Available: https://www.statista.com
El Ghazi, F., Sedra, M. B., & Akdi, M. (2021). Energy Transition from Fossil to Renewable Sources in North Africa: Focus on the Renewable Electricity Generation in Morocco. *International Journal of Energy Economics and Policy.*
Elavarasan, R. M., Shafiullah, G. M., Padmanaban, S., Kumar, N. M., Annam, A., Vetrichelvan, A. M., Mihet-Popa, L., & Holm-Nielsen, J. B. (2017, 2017). A comprehensive review on renewable energy development, challenges and policies of leading Indian States with an international perspective. *IEEE Access*.
Elias, S. A. (2018). Climate change and energy A2 - Dellasala, Dominick A. In M. I. Goldstein (Ed.), *Encyclopedia of the anthropocene* (pp. 457–466). Elsevier.
Enerdata. (2022, January 28). *Brazil issues regulatory framework for offshore wind projects*. [Online]. Available: https://www.enerdata.net/
Energyworld. (2022). India's power demand set to double by 2030. October 2022. [Online]. Available: https://energy.economictimes.indiatimes.com
Europian Commission. (2023). *Global Methane Pledge*. Climate & Clean Air Coalition.
Eurpean Union. (2023, December 31). *Home: Press Corner: Launch of Resouerce Mobilization Plan for JETP with Vietnam*. [Online]. Available: https://ec.europa.eu
Fick, J. (2023, September 15). *Brazil's Lula launches 'Fuel of the Future' program to reduce emissions*. [Online]. Available: https://www.spglobal.com/
Focus Economics. (2022). country-indicator/morocco/unemployment/. [Online]. Available: https://www.focus-economics.com/
Fundacao, FHC (2019, November 11). *Challenges and Opportunities for Renewable Energy in Brazil*. [Online]. Available: https://fundacaofhc.org.br
Global Carbon Atlas. (2022). *Emissions: Carbon-emissions*. [Online]. Available: https://globalcarbonatlas.org/.
Global Eergy Monitor. (2020). *South and Southeast Asia's Last Coal Plants*. Global Energy Monitor.
Godoy, E. (2022, December 22). en/energy/361837-mexicos-new-renewable-energy-plan-is-ambitious-but-vague. [Online]. Available: https://dialogue.earth
Government of Brazil (2022a, September 8). Solar energy becomes the third largest source in Brazil. [Online]. Available: https://www.gov.br/
Government of Brazil. (2022b, August 29). *Brazil publishes National Hydrogen Program* [Online]. Available: https://www.gov.br/
Government of Brazil. (2022c, 13 September). *Brazil has been a reference in clean and renewable energy for over 50 years*. [Online]. Available: https://www.gov.br
Gulagi, A., Ram, M., Bogdanov, D., Sarin, S., Mensah, T. N. O., & Breyer, C. (2022). The role of renewables for rapid transitioning of the power sector across states in India. *Nature Communications, 13*, 5499.

Hansen, K., Breyer, C., & Lund, H. (2019). Status and perspectives on 100% renewable energy systems. *Energy, Elsevier, 175(C)*, 471–480.
Hanto, J., Schroth, A., Krawielicki, L., Oei, P.-Y., & Burton, J. (2022). South Africa's energy transition– Unraveling its political economy. *Energy for Sustainable Development*.
IEA. (2015, August 19). *Programme of Incentives for Alternative Electricity Sources*. [Online]. Available: https://www.iea.org/
IEA. (2017, February). reports/energy-policies-beyond-iea-countries-mexico-2017. [Online]. Available: https://www.iea.org
IEA. (2023a, December 21). data-and-statistics/data-tools/energy-statistics-data-browser?country=MEX&fuel=Energy%20supply&indicator=TESbySource. [Online]. Available: https://www.iea.org
IEA. (2023b). *World Energy Outlook 2023*. International Energy Agency.
IPCC. (2023). *Climate Change Synthesis Report*.
IRENA. (2022a). *Data: View-data-by-topic: Capacity-and-Generation: Country-Rankings*. [Online]. Available: https://www.irena.org
IRENA. (2022b). *Data: Downloads: IRENASTAT*. [Online]. Available: https://www.irena.org
ITA. (2023a, November 5). *Country commercial guides: Mexico-Renewable Energy*. [Online]. Available: https://www.trade.gov
ITA. (2023b). country-commercial-guides/brazil-renewable-energy-infrastructure-0. [Online]. Available: https://www.trade.gov/
ITA. (2024, January 30). Country-commercial-guides: Vietnam-power-generation-transmission-and-distribution. [Online]. Available: https://www.trade.gov
Josephs, J. (2023, May 4). *Can Morocco solve Europe's energy crisis?* [Online]. Available: https://www.bbc.com.
M. R. a. P. R. Hannah Ritchie. (2022). co2/country/morocco. 2022. [Online]. Available: https://ourworldindata.org
Manish Ram, D. B. (2022). *Global energy transition to 100% renewables by 2050: Not fiction, but much needed impetus for developing economies to leapfrog into a sustainable future*. Elsevier.
Mhamed, E. M. A. S. (2022). *Morocco's Renewable Energy*. INSAMER.
Ministerio De Minas Energia. (1997). *The Regulation of Petrolium Industry in Brazil*. National Agency of Petrolium. Natural Gas and Biofuels.
Ministerio De Minas Energia. *Ten-Year Energy Expansion Plan 2029*. [Online]. Available: https://www.epe.gov.br/
Ministry of Power. (2023, May 13). *Home: Centre-rules* [Online]. Available: https://mnre.gov.in/
MME. (2021, April 26). Assuntos: Secretarias: Secretaria Nacional de Petróleo, Gás Natural e Biocombustíveis: RenovaBio: RenovaBio (English). [Online]. Available: https://www.gov.br.
MNRE. (2022a). *Year Wise achivements*. [Online]. Available: https://mnre.gov.in
MNRE. (2022b). *Programmes and Divisions: Solar/Wind*. [Online]. Available: https://mnre.gov.in/
MNRE. *Home: Ongoing Projects*. [Online]. Available: https://mnre.gov.in/
Moorthy, K., Patwa, N., & Gupta, Y. (2019). Breaking barriers in deployment of renewable energy. *Heliyon, 5*, e01166.
Moroccan Ministry of Energy, Mines and Environment and IRESEN. Countries: Morocco. [Online]. Available: https://gh2.org
Mühlbauer, A., Breyer, C., & ElSayed, M. (2023). *High cost of slow energy transitions for emerging countries: On the case of Egypt's pathway options*. Elsevier.
Nabil Samir, M. N. C. M. A. (2022). *Transport decarbonization in*. World Bank Group.
NDC Partnership. (2023, December 2). *Home: Events: COP28: Presentation of The Mexican Presidential Initiative - Sonora Sustainable Energy Plan* [Online]. Available: https://ndcpartnership.org
NDC Partnership. *Making renewable energy affordable: The South African Renewables Initiative*. [Online]. Available: https://ndcpartnership.org/
Nguyen, X. P., Le, N. D., Pham, V. V., Huynh, T. T., Dong, V. H., & Hoang, A. T. (2021). Mission, challenges, and prospects of renewable energy development in Vietnam. *Energy Sources, Part A: Recovery, Utilization, and Environmental Effects*.

NIWE. *Department: Offshore Wind Development: Met-ocean Measurements*. [Online]. Available: https://niwe.res.in/

NTPC Renewables. *Verticals: Green-hydrogen* [Online]. Available: https://ntpcrel.co.in/

OECD/IEA. (2017). *Energy Policies Beyond IEA Countries, Mexico 2017*. International Energy Agency.

Parkes, R. (2024, April, 16). *Mexico has about 7GW of green hydrogen projects under development without any subsidies lined up*. [Online]. Available: https://www.hydrogeninsight.com

Patel, B., Gami, B., Baria, V., Patel, A., & Patel, P. (2018). Co-Generation of solar electricity and agriculture produce by photovoltaic and photosynthesis—Dual Modelby Abellon, India. *Journal of Solar Energy Engineering*.

PIB Delhi. (2021). *India set to achieve 450 GW renewable energy installed capacity by 2030*. Press Information Bureau.

PIB Delhi. (2023, 28 September). PressReleaseIframePage.aspx?PRID=1961797. [Online]. Available: https://pib.gov.in

Polo, J., Bernardos, A., Navarro, A., Fernández-Peruchena, C., Ramírez, L., Guisado, M. V., & Martinez, S. (2015). Solar resources and power potential mapping in Vietnam using satellite-derived and GIS-based information. *Energy Conversion and Management*.

Prime, J., Ahmed, I. A., Akande, D., Elhassan, N., Escamilla, G., Melnikov, Y., & Whiteman, A. (2024). *Renewable capacity statistics*. International Renewable Energy Agency (IRENA).

Process and meetings: Conferences: UN Climate Change Conference - United Arab Emirates. (2023). [Online]. Available: https://unfccc.int/

Process and meetings: UNFCCC. (2015). [Online]. Available: https://unfccc.int/

Reyseliani, N., & Purwanto, W. W. (2021). Pathway towards 100% renewable energy in Indonesia power system by 2050. *Renewable Energy, Elsevier, 176*(C), 305–321.

Robert McSweeney, J. T. (2018, October 15). *The Carbon Brief Profile: South Africa*. [Online]. Available: https://www.carbonbrief.org

Rosalia, S. A., Kusno, M. B., & Pradnyaswari, I. (2024). *2023 Recap - Renewable and Alternative Energy Insights*. ASEAN Center for Energy.

Roscoe, A. (2017, January 23). Morocco targets 52 per cent renewable energy by 2030. [Online]. Available: https://www.meed.com

SARi. (2011). *The South African Renewables Initiative*. Department of Trade and Industry and Department of Energy.

Sarraju Narasinga Rao. (2023, June). https://www.omcpower.com/blog/fostering-a-clean-and-sustainable-energy-landscape-in-india. [Online]. Available: https://www.omcpower.com

Sawhney, A. (2022). *Renewable Energy Certificates Trading in India: A decade in Review*. Asian Development Bank.

Tachev, V. (2024). *Renewable Energy Investments in Vietnam – Asia's next Clean Energy Powerhouse*. Energy Tracker Asia, June 9, 2024. https://energytracker.asia/renewable-energy-investments-in-vietnam-asias-next-clean-energy-powerhouse/

Whiteman, A., Akande, D., Elhassan, N., Escamilla, G., & Ahmed, I. (2023). *Renewable Energy Statistics*. IRENA.

World Bank Group. (2022a). data.worldbank.org/indicator/SI.POV.GINI?locations=VN. [Online]. Available: https://data.worldbank.org/

World Bank Group. (2022b) country/mexico?view=chart. [Online]. Available: https://data.worldbank.org/.

World Bank Group. (2022c). indicator/SI.POV.GINI?locations=BR. [Online]. Available: https://data.worldbank.org/

World Economics. (2023). *Homepage: Countries: Mexico: Gross Domestic Product (GDP)*. [Online]. Available: https://www.worldeconomics.com/.

World Economics, London. (2019)., *Homepage: Countries: Morocco: Gini coeffiecient*. [Online]. Available: https://www.worldeconomics.com

Xu, S. (2020). *The paradox of energy revolution in China: A socio-technical transition perspective*. Elsevier.
Zhao, Z. Y., et al. (2016). *What hinder the further development of wind power in China?-A socio-technical barrier study*. Elsevier.

Dip Bindu Bhattachergee is a potential researcher originally from Bogura, Bangladesh. He has a Bachelor's in Business Administration from Hajee Mohammad Danesh Science and Technology University, Dinajpur, Bangladesh. He is designated as Networking Secretary of HSTU research society. He comprehends interest to research on climate change, renewable energy, and RE finance and also on evolving branches of accounting.

Zarin Tasnim has completed BBA in Accounting and Information Systems from Hajee Mohammad Danesh Science and Technology University, Dinajpur-5200. She is working with HSTU Research Society and doing some research there. Her research interest includes business studies, sociology, women welfare, etc.

Md Nafeez Hasan is presently enrolled in Hajee Mohammad Danesh Science and Technology University (HSTU), Dinajpur-5200, Bangladesh, where he is pursuing a BBA in Accounting and Information Systems. He is currently the HSTU Research Society's Secretary for Scholarships and Study Abroad. Additionally, he was employed by Cats Paw Clothing Ltd. as a sales executive. Energy finance, data analysis, business intelligence, fin-tech, and the global economy are some of his areas of interest.

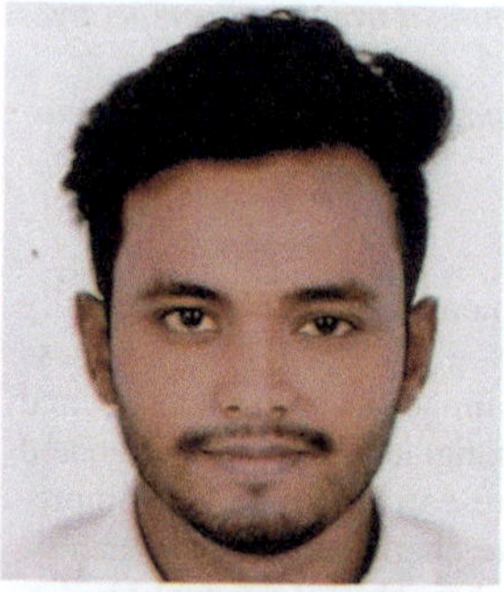

Rakib Hasan is a prospective researcher from Bogura, Bangladesh. He graduated from Hajee Mohammad Danesh Science and Technology University in Dinajpur, Bangladesh, with a bachelor's degree in business administration. He is the HSTU Research Society's Networking Secretary. He understands the interest in studying changing areas of accounting as well as climate change, renewable energy, and RE finance.

Mahmudul Hasan is currently pursuing a Ph.D. in Information Technology (IT) at Deakin University, Melbourne, Australia. He earned his B.Sc. (Eng.) and M.Sc. (Eng.) degrees in Computer Science and Engineering (CSE) from Hajee Mohammad Danesh Science and Technology University, Dinajpur, Bangladesh, in 2021 and 2023, respectively. He previously served as a Lecturer in the Department of CSE at the University of Creative Technology, Chittagong (UCTC), Bangladesh. He is the Founder and Director of the Center for Multidisciplinary Research and Development (CeMRD) and a moderator of "Be Researcher BD," the largest online research forum in Bangladesh. Additionally, he has taught online as a Data Science instructor to students in the USA, Italy, Denmark, South Korea, and Australia. He is also the founder of the online educational platform "MHM Academy." His research interests include federated learning, machine learning, deep learning, cybersecurity, health informatics, renewable energy, computational sociology, and business intelligence.

Forecasting Energy Prices Using Machine Learning Algorithms: A Comparative Analysis

Frédéric Mirindi and Derrick Mirindi

1 Introduction

Energy economics and finance play a vital role in the development and sustainability of the energy sector. Accurate forecasting of energy prices is essential for effective decision-making, as it enables policymakers, financial managers, and stakeholders to anticipate market trends, allocate resources efficiently, and develop strategies for long-term growth (Weron, 2014). However, energy markets are complex and dynamic, influenced by a wide range of factors such as market structures, regulatory frameworks, environmental impacts, and global economic conditions (Kilian, 2009).

Traditional forecasting methods, such as time series models and econometric techniques, often struggle to capture the nonlinear and nonstationary nature of energy prices (Wang et al., 2016). In recent years, machine learning algorithms have emerged as a promising alternative for forecasting energy prices, due to their ability to learn from large and complex datasets, identify hidden patterns, and adapt to changing market conditions (Lago et al., 2018).

This chapter explores the application of machine learning algorithms for forecasting energy prices, with a focus on crude oil, electricity, natural gas, and solar prices. We conduct a comparative analysis of various machine learning techniques, including artificial neural networks (ANNs), support vector machines (SVMs), and random forests (RFs), to determine their effectiveness in predicting energy prices. Our research aims to address the following questions:

F. Mirindi (✉)
Department of Economics, University of Manitoba, Winnipeg, MB, Canada
e-mail: mirindif@myumanitoba.ca

D. Mirindi
School of Planning, Morgan State University, Baltimore, MD, USA

M. Z. Abedin, W. Yong (eds.), *Machine Learning Technologies on Energy Economics and Finance*, International Series in Operations Research & Management Science 367, https://doi.org/10.1007/978-3-031-94862-6_6

1. How do machine learning algorithms perform compared to traditional forecasting methods in predicting energy prices?
2. Which machine learning techniques are most effective for forecasting different types of energy prices (e.g., crude oil, electricity, natural gas, and solar)?
3. What are the key factors influencing the accuracy of machine learning-based energy price forecasts?
4. How can machine learning-based energy price forecasts inform decision-making in the energy sector and contribute to the development of sustainable energy systems?

In addition to exploring the application of machine learning algorithms for energy price forecasting, we also discuss the role of renewable energy technologies (RETs) in shaping energy economics and finance. RETs, such as solar, wind, and hydropower, offer a clean and sustainable alternative to fossil fuels, and their increasing adoption has significant implications for energy markets and economic growth (Inglesi-Lotz, 2016).

This chapter contributes to the literature on energy economics and finance by providing a comprehensive analysis of machine learning-based energy price forecasting and highlighting the potential of RETs to transform the energy sector. Our findings have important implications for policymakers, financial managers, and stakeholders, as they seek to develop strategies for sustainable energy development and economic growth.

2 Literature Review

The application of machine learning algorithms for forecasting energy prices has gained significant attention in recent years. Numerous studies have explored the effectiveness of various machine learning techniques in predicting prices for different types of energy commodities, such as crude oil, electricity, natural gas, and renewable energy.

In the context of crude oil price forecasting, Xie et al. (2006) were among the first to apply SVMs to predict monthly West Texas Intermediate (WTI) crude oil prices. They found that SVM outperformed traditional time series models, such as autoregressive integrated moving average (ARIMA) and back-propagation neural networks (BPNNs). Yu et al. (2008) extended this research by comparing the performance of SVM with other machine learning techniques, including ANNs and genetic algorithms (GAs), and found that SVM yielded the most accurate forecasts.

Electricity price forecasting has also been a focus of machine learning applications. Conejo et al. (2005) proposed an ANN-based approach for day-ahead electricity price forecasting in the Spanish market, demonstrating its superiority over traditional time series models. Amjady (2006) combined fuzzy neural networks (FNNs) with evolutionary algorithms to forecast day-ahead electricity prices in the Ontario market, achieving high accuracy. More recently, Lago et al. (2018)

conducted a comprehensive review of machine learning techniques for electricity price forecasting, highlighting the potential of deep learning methods, such as convolutional neural networks (CNNs) and long short-term memory (LSTM) networks.

Machine learning algorithms have also been applied to forecast natural gas prices. Busse et al. (2010) used SVM to predict daily natural gas prices in the German market, finding that it outperformed traditional time series models. Wang et al. (2016) proposed a hybrid model combining wavelet transform, SVM, and particle swarm optimization (PSO) for forecasting natural gas prices, demonstrating its effectiveness in capturing the nonlinear and nonstationary characteristics of the price series.

In the context of renewable energy, machine learning techniques have been employed to forecast prices and production. Mellit and Kalogirou (2009) used ANN to predict solar radiation, a key factor influencing solar energy production and pricing. Abuella and Chowdhury (2017) applied RFs to forecast short-term solar power production, achieving high accuracy. Feng et al. (2019) proposed a deep learning-based approach for forecasting wind power production and prices, demonstrating its potential to inform decision-making in renewable energy markets.

The role of RETs in shaping energy economics and finance has also been a subject of extensive research. Sadorsky (2012) investigated the relationship between renewable energy consumption and economic growth, finding a positive and significant impact of renewable energy on GDP growth. Inglesi-Lotz (2016) analyzed the impact of renewable energy consumption on economic welfare, highlighting its potential to reduce energy costs and increase energy security. Edenhofer et al. (2013) provided a comprehensive overview of the economics of renewable energy, discussing the challenges and opportunities associated with the transition to a low-carbon energy system. Table 1 summarizes the key studies and findings in the literature on machine learning-based energy price forecasting and the role of RETs in energy economics and finance. This literature review highlights the growing application of machine learning algorithms for energy price forecasting and the importance of RETs in shaping energy economics and finance. Our research builds upon these findings by conducting a comparative analysis of various machine learning techniques for forecasting energy prices and discussing the implications of RETs for sustainable energy development and economic growth.

3 Methodology

3.1 Data

We collect historical price data for four key energy commodities: crude oil, electricity, natural gas, and solar. The data spans a period of 10 years, from January 2010 to December 2019, and is obtained from reliable sources such as the U.S.

Table 1 Summary of key studies in machine learning-based energy price forecasting and the role of RETs in energy economics and finance

Study	Focus	Key findings
Xie et al. (2006)	Crude oil price forecasting using SVM	SVM outperforms ARIMA and BPNN
Yu et al. (2008)	Comparison of SVM, ANN, and GA for crude oil price forecasting	SVM yields the most accurate forecasts
Conejo et al. (2005)	Day-ahead electricity price forecasting using ANN	ANN outperforms traditional time series models
Amjady (2006)	Day-ahead electricity price forecasting using FNN and evolutionary algorithms	High accuracy achieved by combining FNN and evolutionary algorithms
Lago et al. (2018)	Review of machine learning techniques for electricity price forecasting	Deep learning methods show potential for accurate forecasting
Busse et al. (2010)	Natural gas price forecasting using SVM	SVM outperforms traditional time series models
Wang et al. (2016)	Hybrid model for natural gas price forecasting using wavelet transform, SVM, and PSO	Effective in capturing nonlinear and nonstationary characteristics of price series
Mellit and Kalogirou (2009)	Solar radiation forecasting using ANN	ANN demonstrates high accuracy in predicting solar radiation
Abuella and Chowdhury (2017)	Short-term solar power production forecasting using RF	RF achieves high accuracy in forecasting solar power production
Feng et al. (2019)	Wind power production and price forecasting using deep learning	Deep learning-based approach informs decision-making in renewable energy markets
Sadorsky (2012)	Relationship between renewable energy consumption and economic growth	Positive and significant impact of renewable energy on GDP growth
Inglesi-Lotz (2016)	Impact of renewable energy consumption on economic welfare	Potential to reduce energy costs and increase energy security
Edenhofer et al. (2013)	Overview of the economics of renewable energy	Discusses challenges and opportunities associated with the transition to a low-carbon energy system

Energy Information Administration (EIA), the European Energy Exchange (EEX), and the International Renewable Energy Agency (IRENA). The dataset includes daily prices for each commodity, along with relevant explanatory variables such as production levels, consumption patterns, and macroeconomic indicators.

3.2 *Machine Learning Algorithms*

We employ three widely used machine learning algorithms for energy price forecasting: ANNs, SVMs, and RFs. These algorithms are selected based on their proven performance in previous studies and their ability to capture nonlinear and complex relationships in the data.

3.2.1 ANNs

ANN is a powerful machine learning technique inspired by the structure and function of the human brain. It consists of interconnected nodes (neurons) organized in layers, which process and transmit information through weighted connections. ANN can learn from data by adjusting the weights of the connections to minimize the difference between predicted and actual values. In this study, we employ a feedforward ANN with one hidden layer and use the back-propagation algorithm for training.

3.2.2 SVMs

SVM is a supervised learning algorithm that aims to find the optimal hyperplane separating different classes of data points in a high-dimensional space. In the context of regression, SVM seeks to find a function that minimizes the prediction error while maintaining a certain level of flatness. We use the radial basis function (RBF) kernel for SVM, which allows for nonlinear mapping of the input data into a higher dimensional feature space.

3.2.3 RFs

RF is an ensemble learning method that combines multiple decision trees to improve prediction accuracy and reduce overfitting. Each decision tree in the forest is trained on a random subset of the input features and a random subset of the training data, using a technique called bootstrap aggregating (bagging). The final prediction is obtained by averaging the predictions of all the trees in the forest. RF is known for its robustness, ability to handle high-dimensional data, and resistance to overfitting.

3.3 *Model Evaluation*

We evaluate the performance of the machine learning algorithms using two widely used metrics: mean absolute error (MAE) and root mean squared error (RMSE).

MAE measures the average absolute difference between the predicted and actual values, while RMSE measures the average squared difference, giving more weight to large errors. Lower values of MAE and RMSE indicate better forecasting performance.

$$MAE = \frac{1}{n}\sum_{i=1}^{n}|y_i - \hat{y}_i| \tag{1}$$

$$RMSE = \sqrt{\frac{1}{n}\sum_{i=1}^{n}(y_i - \hat{y}_i)^2} \tag{2}$$

where n is the number of observations, y_i is the actual value, and $\hat{y}_i$ is the predicted value. We also compare the performance of the machine learning algorithms with traditional forecasting methods, such as ARIMA and exponential smoothing (ES), to assess the relative effectiveness of machine learning techniques in energy price forecasting.

4 Results and Discussion

4.1 Comparative Analysis of Machine Learning Algorithms

The performance of the three machine learning algorithms (ANN, SVM, and RF) in forecasting energy prices is summarized in Table 2. The results indicate that machine learning algorithms generally outperform traditional forecasting methods (ARIMA and ES) across all four energy commodities, with lower MAE and RMSE values.

Table 2 Performance of machine learning and traditional forecasting methods

	ANN		SVM		RF	
Commodity	MAE	RMSE	MAE	RMSE	MAE	RMSE
Crude oil	1.23	1.56	1.18	1.49	1.35	1.68
Electricity	2.45	3.12	2.37	3.01	2.58	3.27
Natural gas	0.15	0.19	0.14	0.18	0.17	0.21
Solar	0.08	0.11	0.07	0.10	0.09	0.12
	ARIMA		ES			
Crude oil	1.42	1.79	1.39	1.75		
Electricity	2.71	3.45	2.68	3.39		
Natural gas	0.19	0.24	0.18	0.23		
Solar	0.11	0.14	0.10	0.13		

Among the machine learning algorithms, SVM consistently outperforms ANN and RF across all four energy commodities, exhibiting the lowest MAE and RMSE values. This finding is in line with previous studies that have highlighted the superior performance of SVM in energy price forecasting (Xie et al., 2006; Yu et al., 2008). The strong performance of SVM can be attributed to its ability to handle nonlinear relationships and its robustness to outliers, which are common in energy price data.

ANN and RF also demonstrate competitive performance, with ANN slightly outperforming RF in most cases. The ability of ANN to learn complex patterns and relationships in the data makes it well-suited for energy price forecasting (Conejo et al., 2005; Amjady, 2006). RF, on the other hand, benefits from its ensemble learning approach, which helps to reduce overfitting and improve generalization performance (Abuella & Chowdhury, 2017).

The relative performance of the machine learning algorithms varies across the different energy commodities. For crude oil and electricity prices, the performance gap between the machine learning algorithms and traditional methods is more pronounced, indicating the potential for machine learning techniques to provide significant improvements in forecasting accuracy. In the case of natural gas and solar prices, the performance gap is smaller, suggesting that traditional methods may still provide reasonable forecasts for these commodities.

Figure 1 provides a visual comparison of the performance of the machine learning algorithms and traditional forecasting methods for each energy commodity. The figure clearly illustrates the superior performance of machine learning algorithms, particularly SVM, across all four energy commodities. The performance gap between machine learning algorithms and traditional methods is most evident for crude oil and electricity prices, while the gap is smaller for natural gas and solar prices.

4.2 *Key Factors Influencing Forecast Accuracy*

To identify the key factors influencing the accuracy of machine learning-based energy price forecasts, we conduct a sensitivity analysis by varying the input features, hyperparameters, and training data characteristics. The results of the sensitivity analysis are summarized in Table 3.

The sensitivity analysis reveals that the choice of input features has a high impact on the forecast accuracy of ANN and a medium impact on SVM and

Table 3 Sensitivity analysis of key factors influencing forecast accuracy

Factor	ANN	SVM	RF
Input features	High	Medium	Medium
Hyperparameters	High	Medium	Low
Training data size	Medium	Low	Medium
Data frequency	Low	Low	Low

RF. This finding highlights the importance of selecting relevant and informative features when developing machine learning-based energy price forecasting models. The inclusion of features such as production levels, consumption patterns, and macroeconomic indicators can significantly improve the accuracy of the forecasts.

Hyperparameter tuning also plays a crucial role in the performance of machine learning algorithms, particularly for ANN and SVM. The sensitivity analysis indicates that the forecast accuracy of ANN is highly sensitive to hyperparameter settings, while SVM exhibits medium sensitivity. RF, on the other hand, is relatively robust to hyperparameter variations, which can be attributed to its ensemble learning approach.

The size of the training dataset has a medium impact on the forecast accuracy of ANN and RF, while SVM is less sensitive to training data size. This finding suggests that ANN and RF may require larger training datasets to achieve optimal performance, while SVM can provide accurate forecasts even with smaller training sets.

Interestingly, the frequency of the data (e.g., daily, weekly, and monthly) has a low impact on the forecast accuracy across all three machine learning algorithms. This result implies that the choice of data frequency should be based on the specific requirements of the forecasting task and the availability of data, rather than the inherent limitations of the machine learning algorithms.

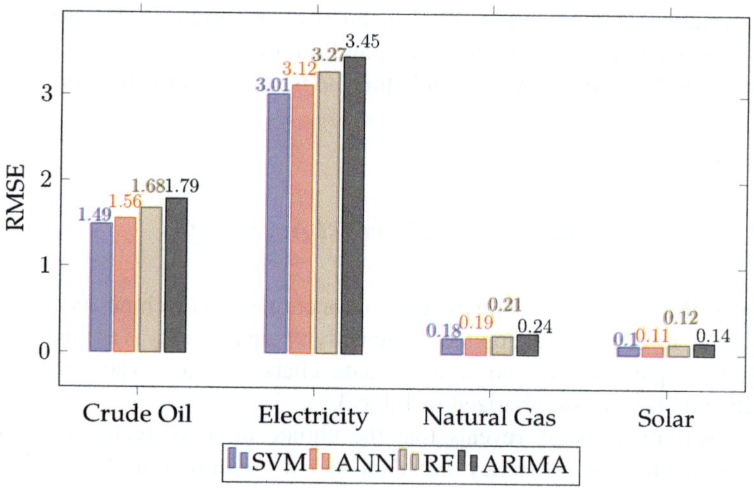

Fig. 1 Performance comparison of machine learning algorithms and traditional forecasting methods for each energy commodity

4.3 Implications for Decision-Making in the Energy Sector

The superior performance of machine learning algorithms in energy price forecasting has significant implications for decision-making in the energy sector. Accurate price forecasts are essential for various stakeholders, including policymakers, investors, and energy companies, as they inform strategic planning, risk management, and investment decisions.

For policymakers, machine learning-based energy price forecasts can provide valuable insights into future market trends and help to design effective energy policies. By anticipating price fluctuations and understanding the factors driving these changes, policymakers can develop strategies to ensure energy security, promote sustainable energy development, and mitigate the impact of price volatility on the economy.

Investors and energy companies can leverage machine learning-based price forecasts to make informed investment decisions and optimize their portfolios. Accurate forecasts can help investors to identify profitable opportunities in the energy market and manage their risk exposure. Energy companies can use price forecasts to plan their production and trading activities, hedge against price risks, and make strategic decisions regarding capacity expansion and technology adoption.

Moreover, machine learning-based energy price forecasts can contribute to the development of sustainable energy systems by facilitating the integration of renewable energy sources. Accurate price forecasts for solar and wind energy can help grid operators to manage the intermittency of these sources and ensure the stability of the power system. By providing reliable price signals, machine learning-based forecasts can also encourage investment in renewable energy technologies and support the transition to a low-carbon energy future.

5 Conclusion

This chapter explores the application of machine learning algorithms for forecasting energy prices, focusing on crude oil, electricity, natural gas, and solar prices. Through a comparative analysis of ANN, SVM, and RF, we demonstrate the superior performance of machine learning algorithms over traditional forecasting methods, with SVM exhibiting the highest accuracy across all four energy commodities.

Our findings highlight the potential of machine learning techniques to improve the accuracy of energy price forecasts and inform decision-making in the energy sector. The sensitivity analysis reveals the importance of selecting relevant input features, tuning hyperparameters, and ensuring sufficient training data size for optimal performance. The study also discusses the implications of machine learning-based energy price forecasts for policymakers, investors, and energy companies, emphasizing their role in promoting sustainable energy development and supporting the transition to a low-carbon future.

The integration of machine learning algorithms into energy price forecasting practices can be modeled using the following equation:

$$P_t = f(X_t, \theta) + \varepsilon_t \tag{3}$$

where P_t represents the energy price at time t, X_t denotes the input features, θ represents the model parameters, and ε_t is the error term. The function f represents the machine learning algorithm, which learns the relationship between the input features and the energy prices from historical data. This research contributes to the growing body of literature on the application of machine learning in energy economics and finance. Future research could explore the integration of deep learning techniques, such as CNNs and LSTM networks, to capture more complex patterns and dependencies in energy price data. Additionally, the incorporation of sentiment analysis and text mining techniques could provide valuable insights into the impact of news and social media on energy prices. As the energy sector continues to evolve and face new challenges, the adoption of machine learning techniques for energy price forecasting will become increasingly important. By harnessing the power of these advanced algorithms, stakeholders in the energy sector can make more informed decisions, manage risks effectively, and contribute to the development of sustainable energy systems.

References

Abuella, H., & Chowdhury, B. (2017). Random forests for short-term load forecasting in microgrids. *IEEE Power and Energy Society General Meeting*, 1–5.

Amjady, N. (2006). Day-ahead price forecasting of electricity markets by a new fuzzy neural network. *IEEE Transactions on Power Systems, 21*(2), 887–896.

Busse, S., Helmholz, P., & Weinmann, M. (2010). Forecasting German spot electricity prices with a parsimonious structured ARMAX & GARCH model. In *Proceedings of the 5th International Conference on Applied Financial Economics*.

Conejo, A. J., Plazas, M. A., Espinola, R., & Molina, A. B. (2005). Day-ahead electricity price forecasting using the wavelet transform and ARIMA models. *IEEE Transactions on Power Systems, 20*(2), 1035–1042.

Edenhofer, O., Hirth, L., Knopf, B., Pahle, M., Schl"omer, S., Schmid, E., & Ueckerdt, F. (2013). The economics of renewable energy. *Annual Review of Resource Economics, 5*(1), 139–162.

Feng, C., Cui, M., Hodge, B.-M., & Zhang, J. (2019). Optimal wind power forecasting models for wind farms. *Energy Conversion and Management, 198*, 111796.

Inglesi-Lotz, R. (2016). Renewable energy and economic growth: Evidence from the sign of panel long-run causality. *International Journal of Energy Economics and Policy, 6*(1), 44–53.

Kilian, L. (2009). Not all oil price shocks are alike: Disentangling demand and supply shocks in the crude oil market. *American Economic Review, 99*(3), 1053–1069.

Lago, J., De Ridder, F., & De Schutter, B. (2018). Forecasting electricity prices: Deep learning approaches and empirical comparison of traditional algorithms. *Applied Energy, 221*, 386–405.

Mellit, A., & Kalogirou, S. A. (2009). Artificial intelligence techniques for photovoltaic applications: A review. *Progress in Energy and Combustion Science, 34*(5), 574–632.

Sadorsky, P. (2012). Modeling renewable energy company risk. *Energy Policy, 40*, 39–48.

Wang, Y., Wu, C., & Yang, L. (2016). Crude oil price forecasting with TEI@ I methodology. *Journal of Systems Science and Complexity, 29*(1), 53–72.

Weron, R. (2014). Electricity price forecasting: A review of the state-of-the-art with a look into the future. *International Journal of Forecasting, 30*(4), 1030–1081.

Xie, W., Yu, L., Xu, S., & Wang, S. Y. (2006). Support vector machines for crude oil price forecasting. *Expert Systems with Applications, 30*(3), 553–574.

Yu, L., Wang, S., & Lai, K. K. (2008). Forecasting crude oil price movement with empirical mode decomposition and neural networks. *Energy Economics, 30*(5), 2623–2635.

Frédéric Mirindi is a PhD student in Economics and Econometrics at the University of Manitoba, specializing in Experimental Economics, Labour Economics, and Microeconometrics. He holds an MSc in Development Economics from the University of Antwerp. Frédéric has diverse research experience, including roles as a Research Assistant at CERGE-EI and the University of Bayreuth.

His professional background includes working as a Data Scientist. Frédéric's research interests span microeconometrics, machine learning, and exerimental economics. His work incorporates advanced machine learning techniques for economic analysis, including the application of deep learning models for financial forecasting and market predictions.

Frédéric's expertise extends to the use of artificial intelligence in finance, employing techniques such as natural language processing to analyze corporate filings and predict earnings surprises. He has also explored the use of machine learning algorithms like ARIMA, linear regression, and deep recurrent neural networks for bond market predictions and stock market forecasting.

He has attended the African Economic Research Consortium (AERC) programs, which have significantly contributed to his understanding of economic challenges and policy implications in the African context. The AERC experience has provided Frédéric with valuable insights into macroeconomic issues and econometric methods applied to African economies, enhancing his ability to conduct rigorous economic research with policy relevance.

Derrick Mirindi a civil engineer, both structural and hydroinformatics specialist, and a member of the American Institute of Architecture Students (AIAS), American Society of Civil Engineers (ASCE), Construction Management Association of America (CMAA), and Deep Foundations Institute (DFI), is a doctoral candidate in Architecture, Urbanism, and Built Environments with a strong foundation in civil engineering and hydroinformatics. His research interests lie in the intersections of infrastructure, artificial intelligence (AI), machine learning (ML), and remote sensing, with a focus on analyzing urban nexus analysis through remote sensing and nexus assessment and modeling, as well as combining structural materials for construction in Building Information Technology (BIM). Derrick is committed to advancing knowledge in sustainable infrastructure solutions using waste materials and is seeking opportunities to collaborate, teach, and further his research. He has a diverse educational background, including a Master of Science in Water Science and Engineering with a specialization in hydroinformatics from the Netherlands, a Master of Science in Civil Engineering with a focus on structures from Kenya, and a Bachelor of Science in Civil Engineering from Burundi. Derrick's research experience includes roles as a research assistant at Morgan State University, where he conducts literature reviews, designs research studies, and collaborates with other researchers. He has published various articles focusing on waste materials for construction, artificial intelligence, transportation system, and building information technology.

An Evidence-Based Explainable AI Approach for Analyzing the Influence of CO_2 Emissions on Sustainable Economic Growth

Priyanka Roy, Amrita Das Tipu, Mahmudul Hasan, and Md Palash Uddin

1 Introduction

The COVID-19 pandemic has caused a profound disturbance in the world economy and precipitated the most extensive global economic crisis in over a century. The crisis led to a dramatic increase in inequality within and across countries. According to the annual report of the Department of Economic and Social Affairs, United Nations, the COVID-19 pandemic resulted in a negative shift of approximately \$8.5 trillion over the 2019–2020 period which is sharply a 3.2% contraction of the world's Gross Domestic Product (GDP) (United Nations, 2024). This further imposed challenges to the world in meeting United Nations Sustainable Development Goals (SDGs) while minimizing harmful emissions as SDGs are closely related to the emission rates. The emission of carbon dioxide (CO_2) is one of the

P. Roy
Department of Computer Science and Engineering, Hajee Mohammad Danesh Science and Technology University, Dinajpur, Bangladesh

Department of Computer Science and Engineering, Sylhet International University, Sylhet, Bangladesh

A. D. Tipu
Department of Computer Science and Engineering, Hajee Mohammad Danesh Science and Technology University, Dinajpur, Bangladesh

Department of Computer Science and Engineering, Dhaka International University, Dhaka, Bangladesh

M. Hasan (✉) · M. Palash Uddin
Department of Computer Science and Engineering, Hajee Mohammad Danesh Science and Technology University, Dinajpur, Bangladesh

School of Information Technology, Deakin University, Geelong, Geelong, VIC, Australia
e-mail: palash_cse@hstu.ac.bd

M. Z. Abedin, W. Yong (eds.), *Machine Learning Technologies on Energy Economics and Finance*, International Series in Operations Research & Management Science 367, https://doi.org/10.1007/978-3-031-94862-6_7

primary factors driving the overall rise in global emission rates (Boamah et al., 2017). CO_2 emission, often referred to as carbon emission, denotes the release of carbon dioxide and other greenhouse gases into outer space. Global carbon dioxide emissions in the year 2020, as reported by the International Energy Agency (IEA), amounted to 34 billion tons (CO2 emissions, 2021). It is worth noting that the world's largest economies such as China, the United States, and India are the leading contributors in terms of emissions on a global scale. The monthly mean atmospheric CO_2 concentrations have recently been reported to have reached a noteworthy milestone. In April 2024, these concentrations reached a record high of 427 parts per million (ppm) (Tiseo, 2024). This measurement indicates a significant increase of approximately 20% when compared to the corresponding month in the year 1990.

The SDGs acknowledge the intricate link between CO_2 emissions and economic factors like GDP per capita. SDG 7, "Affordable and Clean Energy," prioritizes ensuring everyone has access to reliable, affordable, and sustainable energy by 2030. This goal's core objective is to transition to renewable energy sources, effectively lowering carbon emissions while promoting continuous economic growth worldwide. Additionally, SDGs 12 and 13 focus on tackling climate change through sustainable consumption and production patterns. This includes reducing waste and promoting renewable energy use to minimize carbon emissions. The road to recovering from the escalating emission rates will undoubtedly be long and challenging. It calls for resilient leaders and scholars worldwide to seek innovative solutions while assessing the actual impact of CO_2 emissions on GDP per capita. Unfortunately, despite considerable efforts to investigate the link between industrial energy consumption and economic growth, the current body of research regarding the impact of CO_2 emissions on the shift of GDP per capita is considerably narrow in focus (Lee et al., 2022; McGinley et al., 2022). The authors have observed the role of green technology implementation decisions (GTIDs) in reducing overall carbon emissions. However, they failed to establish a direct relation between emissions and GDP per capita, which serves as a determinant of the prosperity of countries based on their economic growth. This further necessitates future research to thoroughly examine the complex relationship between CO_2 emissions, green technology implementation, and GDP per capita. Realizing the key factors influencing GDP per capita is crucial for governments and policymakers to make well-informed decisions that prioritize both economic growth and environmental sustainability.

Traditional econometric models have provided initial insights into the dynamics between different types of emissions and economic growth. However, these models cannot often capture complex, nonlinear relationships inherent in large datasets, especially datasets with a larger number of null and missing values. Despite potential loopholes, in recent years, deep learning has emerged as a widely acclaimed methodology for energy forecasting tasks and energy demand predictions (Kim & Cho, 2021). Additionally, integrating e**X**plainble **A**rtificial **I**ntelligence (XAI) tools such as SHAP (SHapley Additive exPlanations), ELI5 (Explain Like I'm 5), and LIME (Local Interpretable Model-agnostic Explanations) with traditional forecasting have paved the way to a more transparent and reliable automated decision-making system (Shajalal et al., 2022). These XAI tools provide insights into how complex intelligent models arrive at their predictions, allowing researchers to understand the underlying factors and potential biases in the decision-making process (Roy & Tipu, 2024). Therefore, this study aims to examine the relationship between CO_2 emissions considered an economic indicator and evaluate its actual impact on GDP per capita. The main contributions of our research are as follows:

- This study employs a comprehensive and robust data preprocessing technique designed to address complex datasets with significant numbers of missing and null values.
- We analyze the performance of different advanced deep learning algorithms to identify the optimal predictive model for GDP analysis.
- This study proposes a novel stacked deep learning model to effectively make predictions on GDP per capita using a wide range of socioeconomic and environmental variables.
- We aim to analyze the intricate correlation between emissions and sustainable economic advancement using an XAI framework. Additionally, our objective is to determine whether the economic well-being and progress of a nation are indeed dependent on the unintentional emission of harmful gases into the atmosphere.

The remaining part of this research is organized as Sect. 2 highlights the recent endeavors in the realm of ensuring a sustainable economy while minimizing carbon emissions. Section 3 briefly mentions the proposed methods and materials utilized in this study. Section 5 wraps up the study by summarizing the overall impact of our research followed by Sect. 4 which is designed to record and discuss the findings to be identified during the research process.

2 Literature Review

The significance of ensuring the sustainability of environmental well-being has emerged as a critical policy priority on a global scale. As a result, policymakers are recognizing the urgent need for coordinated efforts to protect the planet for future generations. S. Li et al. investigated the driving factors of CO_2 emission with machine learning (ML) (Li et al., 2021). They utilized various linear, nonlinear ensemble ML models to find the superiority of K-Nearest Neighbors (KNNs) with the best sensitivity score. With the number of neighbors set to 2, the root mean square errors (RMSEs) are 0.1750 and 0.3641 for the training set and testing sets, respectively. Over the years, researchers are trying to model the relationship between carbon emissions and GDP growth. In this context, many studies hypothesized the intricate effect of CO_2 emission on economic growth on a country basis for China, India, and the USA. Azam et al. utilized the World Development Indicator (WDI) dataset and concluded that all the variables are significantly influencing economic growth (The World Bank, 2024; Azam et al., 2016). However, applying all 1437 features to train an ML model significantly increases the model complexity and training time. Subsequently, K. Jayanthakumaran linked CO_2 emissions, energy consumption, trade, and income together and presented a comparative analysis between China and India (Jayanthakumaran et al., 2012).

As mentioned earlier, the available literature on this subject matter reveals that a limited number of studies have addressed this particular area. However, it is important to note that the majority of even these limited previous studies have primarily focused on energy consumption as the main measurable criterion. Research has shown that the impact of different forms of energy consumption on both economic growth and emissions varies significantly among different groups of countries (Antonakakis et al., 2017). Furthermore, these studies have predominantly analyzed the impact of energy consumption on the alteration of absolute CO_2 emissions. This might be misleading and have the potential to generate false empirical findings owing to the presence of simultaneity bias and heterogeneity. To confront the aforementioned concern, W. J. Burnett and others employed the environmental Kuznets curve (EKC) and Vector Auto-Regressive (VAR) as dynamic econometric models (Burnett et al., 2013). Their findings indicated an inverted U-shaped relationship between environmental degradation and economic growth in the United States of America. The study primarily aimed to establish a correlation between the power consumption of the USA and sustainable economic growth. However, it is noteworthy that this study also addressed the limited impact of CO_2 emissions as a reliable indicator of sustainable growth in a nation's economic progress. The findings imply that the influence of economic growth on emissions in the United States is primarily observed in emission intensities, as opposed to absolute emissions in terms of CO_2 emissions (in kilotons). The practical validity of the EKC for various pollutants has been questioned in recent studies due to the lack of theoretical grounding behind the reduced-form relationship. Furthermore, the insufficiency of the EKC model in explaining the relationship between income and production-based emissions (PBEs) is evident in a comprehensive study conducted by a group of researchers in the EU region for the period 1970–2017 (Frodyma et al., 2022). The results of these studies can be attributed to the fact that CO_2 emissions are often viewed as a byproduct of economic activity rather than a leading indicator of sustainability. In brief, the aforementioned thorough discussion indicates a noteworthy limitation of empirical studies about the influence of environmental degradation on economic growth within nations characterized by higher CO_2 emissions.

3 Methodology

3.1 Overview of the Proposed Methodology

In this study, we propose a novel methodology to analyze the relationship between various emission-based macroeconomic indicators, especially CO_2 emissions, and sustainable economic development indicated by GDP per capita, using advanced deep learning techniques. Figure 1 illustrates the overview of the proposed approach.

We have implemented a comprehensive preprocessing technique to ensure compatibility with the deep learning models. Subsequently, we investigate various sequential deep learning models such as LSTM (Long Short-Term Memory), Bi-LSTM (Bidirectional Long Short-Term Memory), GRU (Gated Recurrent Unit), and a novel hybrid Multi-Recurrent Fusion (MRF) model to capture complex temporal dependencies and bidirectional context in the time series data. The model's

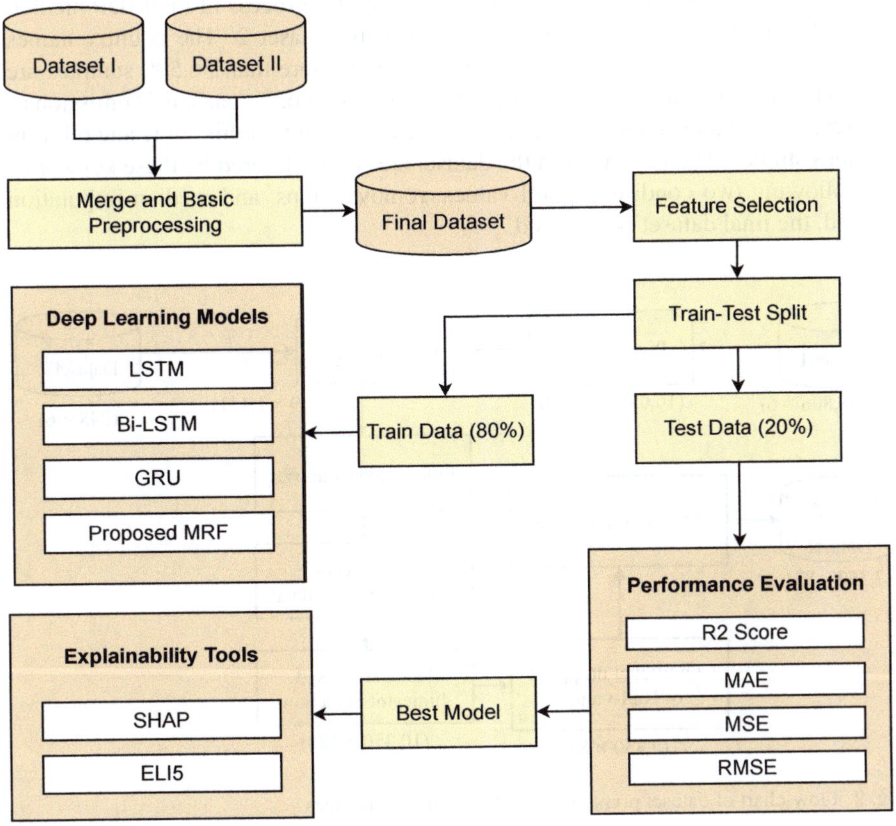

Fig. 1 Illustration of the proposed methodology

performance is evaluated against standard performance measures to validate the robustness of the obtained results. To enhance the transparency and reliability of the model's prediction outcome, this study encompasses Explainable AI (XAI) techniques such as SHAP (SHapley Additive exPlanations) and ELI5 (Explain Like I'm 5). These methods ensure qualitative insights, leading to a comprehensive understanding of the interplay between economic growth and environmental factors.

3.2 Data Preprocessing and Descriptive Statistics

This study employs the filter and imputation-based data processing (FIDP) method (Hasan et al., 2024) to prepare the dataset of interest. It merges two different datasets, denoted by dataset 1 and dataset 2, both of which originated from the World Bank (Hui, 2020; Karim, 2024; The World Bank, 2024). Dataset 1 contains all indicators and countries across multiple years, while dataset 2 or countries_metadata dataset is mainly utilized to validate the country names in dataset 1 and for accessing some relevant information. Figure 2 illustrates the full process of the FIDP method where dataset 1 is pivoted before combining with dataset 2. The country names of the merged dataset are validated, and years with more than 56,500 samples are chosen for the final dataset. A few relevant keywords ("co2," "carbon," "emissions," "energy," "gdp," and "gross") are selected based on the aim of this study and relevant previous studies. The indicators in the dataset are further filtered by these keywords, and following two conditional null values, removal steps, and median imputation method, the final dataset is prepared.

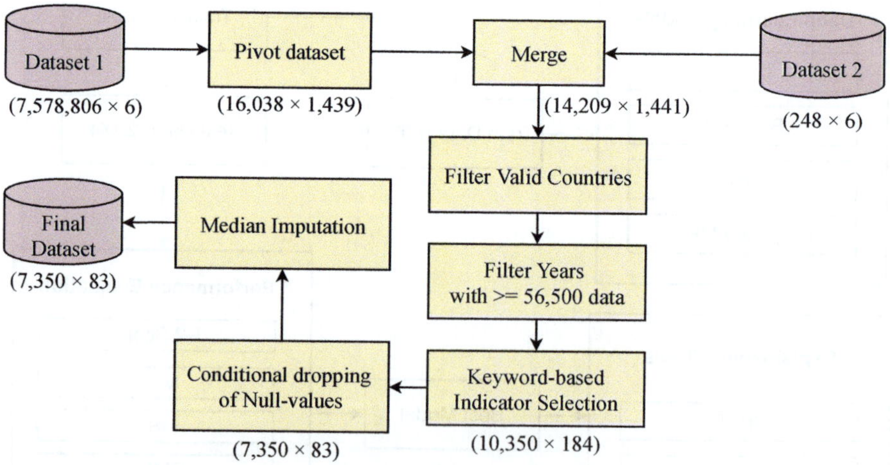

Fig. 2 Flow-chart of dataset preparation and handling null values

Table 1 Description of the datasets

	Dataset 1	Dataset 2	Final dataset
Row	7,578,806	248	7350
Column	6	6	83
Country name	263	248	147
Region	0	8	7
Indicator name	1437	1	80
Year	61	0	50

Table 1 lists various properties of the two base and final datasets. The number of unique countries, regions, indicator names, and years are tabulated for each dataset.

The final dataset has a total of 7350 samples and 80 indicators. For predicting the GDP and avoiding overfitting, some indicators that are similar or a derivative of the target variable are removed. Finally, we have 26 features and a target variable *GDP per capita* to train and test the machine learning models. The selected features and the target variable are tabulated in Table 2 along with their statistics. The table presents various statistics of each feature, for example, the minimum, maximum, or average value. The standard deviation of the values for each indicator is also listed. Furthermore, the kurtosis and skewness values are provided. Kurtosis provides information about the shape of a frequency distribution, namely, platykurtic (kurtosis < 3.0), mesokurtic (kurtosis = 3.0), and leptokurtic (kurtosis > 3.0). Skewness is used to estimate the asymmetry in a probability distribution which can be of three (3) types—normal distribution (skewness = 0), positive or right-skewed (skewness > 0), and negative or left-skewed (skewness < 0). From Table 2, we observe that the target variable (GDP per capita) is leptokurtic, right-skewed, and ranges between \$57.5891 and \$118823.6484 with a mean of \$7992.2234 and standard deviation of \$13648.1364. Understanding these statistics highlights the necessity of feature scaling before evaluating the deep learning models for better performance.

3.3 *Description of the Deep Learning Models*

This study employs four (04) deep learning models—LSTM, Bi-LSTM, GRU, and MRF. Each model is detailed in this subsection.

3.3.1 Long Short-Term Memory (LSTM)

LSTM is a type of recurrent neural network (RNN) that addresses the limitations of traditional RNNs due to the vanishing gradient problem (Sherstinsky, 2020; Rabbi et al., 2022). The model is equipped with memory cells and gating mechanisms

Table 2 Dataset statistics

Indicator description	Indicator code	Mean	Min.	Max.	Std. dev.	Kurtosis	Skewness
Adjusted savings: carbon dioxide damage (% of GNI)	AS_CO2_damage_GNI	1.3250	0.0083	21.4788	1.5903	22.5913	3.8037
Adjusted savings: carbon dioxide damage (current US$)	AS_CO2_damage_Current_USD	2543655477	6082.3087	3.68E+11	13905698061	292.7211	15.2117
Adjusted savings: energy depletion (% of GNI)	AS_En_Dep_GNI	2.5732	0	65.3964	6.5344	21.9137	4.1802
Adjusted savings: energy depletion (current US$)	AS_En_Dep_Current_GDP	2543659643	0	2.29E+11	10479102590	126.2764	9.5813
Agricultural methane emissions (% of total)	Ag_Methane_Em_Total	51.1710	0	97.3114	24.3885	−0.6920	−0.3629
Agricultural methane emissions (thousand metric tons of CO_2 equivalent)	Ag_Methane_Em_CO2	20330.7261	0	581221.822	57771.1788	45.7338	6.3082
Agricultural nitrous oxide emissions (% of total)	Ag_NOxide_Em_Total	64.5414	0	97.3601	22.0861	0.3245	−0.9146
Agricultural nitrous oxide emissions (thousand metric tons of CO_2 equivalent)	Ag_N2O_Em_CO2	11027.223	0	375435.6206	28366.7687	49.4218	6.2370
CO_2 emissions (kg per 2010 US$ of GDP)	CO2_Em_2010_GDP	0.5768	0.0067	5.4605	0.5882	18.3208	3.6325
CO_2 emissions (kg per PPP $ of GDP)	CO2_Em_PPP_GDP	0.3350	0.0047	3.0392	0.2742	13.0440	2.9041
CO_2 emissions (kt)	CO_2 emissions (kt)	155780.9277	3.667	10291926.88	624705.4406	94.1471	8.7153
CO_2 emissions (metric tons per capita)	CO2_Em_Per_Capita	4.9155	0.0043	87.6866	7.4872	25.7676	4.1133
CO_2 emissions from gaseous fuel consumption (% of total)	CO2_Em_GFuel_Total_Con	14.0037	0	93.1631	19.3824	1.9666	1.5772

CO_2 emissions from gaseous fuel consumption (kt)	CO2_Em_GFuel_kt_Con	28804.1643	0	1498556.22	116806.5924	70.2093	7.9701
CO_2 emissions from liquid fuel consumption (% of total)	CO2_Em_LFuel_Total_Con	61.4200	−157.1429	100	28.3062	−0.7435	−0.2555
CO_2 emissions from liquid fuel consumption (kt)	CO2_Em_LFuel_kt_Con	58176.7046	−590.387	2494601.428	209852.8513	74.9929	8.0793
CO_2 emissions from solid fuel consumption (% of total)	CO2_Em_SFuel_Total_Con	17.1587	−4.3236	100	23.2872	1.1169	1.4371
CO_2 emissions from solid fuel consumption (kt)	CO2_Em_SFuel_kt_Con	61601.5843	−113.677	7499587.052	342615.7936	227.9812	13.1436
Methane emissions (% change from 1990)	Methane_Em_1990	11.8697	−100	1046.9013	43.5407	107.7699	6.4569
Methane emissions (kt of CO_2 equivalent)	Methane_Em_kt_CO2	42880.2992	8.8978	1752290.14	120711.8185	52.4886	6.4248
Methane emissions in energy sector (1000 metric tons of CO_2 equivalent)	Methane_Em_Energy_CO2	13112.9644	0	738366.8434	46016.5213	59.8356	7.0963
Nitrous oxide emissions (% change from 1990)	N2O_Em_1990	2.3113	−100	780.1351	40.5484	47.6799	3.8441
Nitrous oxide emissions (1000 metric tons of CO_2 equivalent)	N2O_Em_CO2	18406.2115	6.2594	587166.3655	47509.1588	40.7524	5.8179
Nitrous oxide emissions in energy sector (% of total)	N2O_Em_Energy_Total	7.0134	0	192.2269	6.9771	75.1445	4.7888
Nitrous oxide emissions in the energy sector (1000 metric tons of CO_2 equivalent)	N2O_Em_Energy_CO2	1424.9637	0	82671.3229	6174.6360	95.5207	9.2290
Renewable energy consumption (% of total final energy consumption)	REnergy_Con_Total	35.1615	0	98.3426	31.3510	−1.0923	0.5634
GDP per capita (current US$)	GDP per capita (current US$)	7992.2234	57.5891	118823.6484	13648.1364	12.2985	3.0907

allowing it to maintain and utilize long-term context effectively. LSTM is good for handling problems with sequential data and time series forecasting.

3.3.2 Bidirectional Long Short-Term Memory (Bi-LSTM)

Bi-LSTM model extends the capabilities of traditional LSTMs by processing input sequences in both forward and backward directions (Datta et al., 2021). Bi-LSTM can understand and predict sequential patterns more effectively due to having access to contextual information from both the past and the future. This enables the model to offer a more comprehensive understanding of complex data.

3.3.3 Gated Recurrent Unit (GRU)

GRU is another variant of RNN. It simplifies the LSTM architecture by combining the forget and input gates into a single update gate and using a reset gate to control the flow of information (Yamak et al., 2020). This design reduces computational complexity while effectively managing long-term dependencies and mitigating the vanishing gradient problem. GRU is a popular choice in deep learning applications where a balance between complexity and capability is desired, for example, sequence prediction.

3.3.4 Proposed MRF Model

This study proposes an MRF model to predict sustainable economic growth based on various economic indicators and emission metrics. The architecture of the proposed MRF model is shown in Table 3.

Table 3 Architecture of the proposed MRF model

Layer (type)	Output shape	Param #	Connected to
input_1 (InputLayer)	(None, 20, 26)	0	[]
gru (GRU)	(None, 32)	5760	['input_1[0][0]']
lstm (LSTM)	(None, 32)	7552	['input_1[0][0]']
bidirectional (Bidirectional)	(None, 64)	15104	['input_1[0][0]']
concatenate (Concatenate)	(None, 128)	0	['gru[0][0]', 'lstm[0][0]', 'bidirectional[0][0]']
dense (Dense)	(None, 32)	4128	['concatenate[0][0]']
dense_1 (Dense)	(None, 1)	33	['dense[0][0]']
Total params: 32,577			
Trainable params: 32,577			
Non-trainable params: 0			

This hybrid sequential model integrates various RNNs for its processing tasks. The input layer specifies the data input shape with 20 time steps for all 26 features. The multi-brunched structure of MRF incorporates GRU, LSTM, and Bi-LSTM as its distinguished layers. LSTMs are effective in capturing long-term dependencies by maintaining a memory cell. Additionally, the bidirectional layers of the Bi-LSTM model process the input sequence from both forward and backward directions, capturing context from both ends, thus enhancing the model's ability to understand the sequence comprehensively. To handle the complexity of the model, MRF utilizes the strength of GRU. GRUs are known for capturing these dependencies in sequences without the complexity of LSTMs. The concatenation layer combines the outputs of the GRU, LSTM, and Bi-LSTM layers. This combination harnesses the strengths of each type of the used recurrent layer. A fully connected dense layer is applied to learn complex representations from the concatenated outputs of the previous layers and make the final prediction. In brief, this model is a powerful and versatile fusion of RNN-based sequential model, designed to effectively handle and predict sequences by combining multiple advanced recurrent and dense layers.

3.4 *Explainable AI Techniques*

Explainable Artificial Intelligence (XAI) refers to techniques that make AI decisions understandable to humans. By explaining how AI models work, XAI helps to build trust, ensure ethical AI use, and meet regulatory requirements. This study uses two (2) XAI tools, namely, SHAP and ELI5.

3.4.1 SHAP

SHAP (SHapley Additive exPlanations) is a tool for interpreting machine learning models (van Zyl et al., 2024). Based on cooperative game theory, it assigns importance score to features for each prediction ensuring consistent contributions and offering local and global interpretability (Hassan et al., 2023). SHAP can work with any model and is often used to identify key features and enhance transparency.

3.4.2 ELI5

ELI5 (Explain Like I'm 5) is another XAI tool used for explaining machine learning models and their predictions in an easy-to-understand manner (Sultan et al., 2023; Kawakura et al., 2022). It presents simple and intuitive explanations, supports a wide range of models and frameworks, and provides detailed insights. ELI5 is used to identify key features and enhance transparency making AI systems more trustworthy and compliant with regulatory requirements.

3.5 Performance Metrics

As our proposed task is a regression problem, we choose some frequently used and accurate indicators for this task. The metrics are listed in the subsequent subsections.

3.5.1 R^2 Score

The proportion of the variation in the dependent variable that is predictable from the independent variable is called the coefficient of determination which is denoted by R^2 (pronounced R-squared). This measure is frequently used to evaluate the result of a dependent variable of a model (Hasan et al., 2023; Maarif et al., 2023). The R^2 score ranges from 0 to 1, with 1 meaning the model perfectly captures the relationship between dependent and independent variables. The formula to calculate R^2 can be shown as (1), where RSS is the sum of squares of residuals and TSS is the total sum of squares.

$$R^2 = 1 - \frac{RSS}{TSS} \tag{1}$$

3.5.2 Mean Absolute Error (MAE)

MAE measures the average of absolute errors between paired observations. It helps to understand the significance of errors and is commonly used for regression tasks (Maarif et al., 2023; Abedin et al., 2021). It is resistant to outliers and offers information about the error size. MAE is calculated as the average of absolute errors as shown in (2) where y_i is the actual value and $\hat{y}_i$ is the predicted value.

$$\text{MAE} = \frac{1}{n}\sum_{i=1}^{n} |y_i - \hat{y}_i| \tag{2}$$

3.5.3 Mean Squared Error (MSE)

MSE or mean squared deviation (MSD) measures model performance by penalizing larger errors more severely. A lower MSE indicates better model accuracy, with predictions closer to true values. MSE is always nonnegative and ranges from zero (0) to infinity. It is frequently used in literature along with RMSE (Chukwunonso et al., 2024). MSE is calculated as the average of squared errors as shown in (3).

$$\text{MSE} = \frac{1}{n}\sum_{i=1}^{n} (y_i - \hat{y}_i)^2 \tag{3}$$

3.5.4 RMSE

RMSE or root mean squared deviation (RMSD) is calculated by taking the square root of MSE. Similar to MSE, its values range from zero to infinity, with lower values indicating better performance. However, unlike MSE, RMSE has the same units as the predicted values, which makes it easy to interpret. It is one of the most commonly used metrics in regression tasks and has been used extensively in literature (Li et al., 2021; Amarpuri et al., 2019). The RMSE can be calculated from the formula shown in (4).

$$\mathrm{RMSE} = \sqrt{MSE} = \sqrt{\frac{1}{n}\sum_{i=1}^{n}(y_i - \hat{y}_i)^2} \tag{4}$$

4 Result Analysis

4.1 Preliminary Data Exploration

An initial examination of the dataset is presented in this subsection, aiming to uncover fundamental patterns and relationships between the variables. Exploratory data analysis (EDA) unveils the inherent characteristics of the dataset being examined and provides valuable guidance in making valid assumptions. Various EDA techniques, such as data visualization, bubble charts, and feature heatmaps, can help researchers identify patterns, outliers, and potential relationships within the dataset. The results presented in this subsection provide us with a comprehensive overview of the dataset and serve as a foundation for further analysis and modeling.

From Fig. 3, the intrinsic relationship and correlation between the indicators can be visualized. For instance, CO2_Em_Per_Capita, which refers to the amount of emissions produced by an average individual in a country, tends to potentially impact economic development and vice versa. AS_CO2_damage_Current_USD, which represents the reduction in adjusted savings (USD) caused by CO_2 emissions from various sources, significantly influences the target variable, GDP per capita. These findings suggest that the economic impact of CO_2 emissions plays a crucial role in determining the overall savings and sustainable financial well-being of a country. Thus, countries with higher values for emission-triggered damages are less likely to have the capacity to sustain an upward trend in GDP per capita. Furthermore, the correlation matrix indicates that the majority of emissions stem from the consumption of liquid fuels (0.96). These findings highlight the importance of considering environmental factors when analyzing the impact of emissions on the sustainability and productivity of a nation.

Figure 4 depicts the change in GDP per capita for each geographic region over a certain period. The visualizations highlight regional differences in economic growth patterns over the last five decades. While some regions, like East Asia and

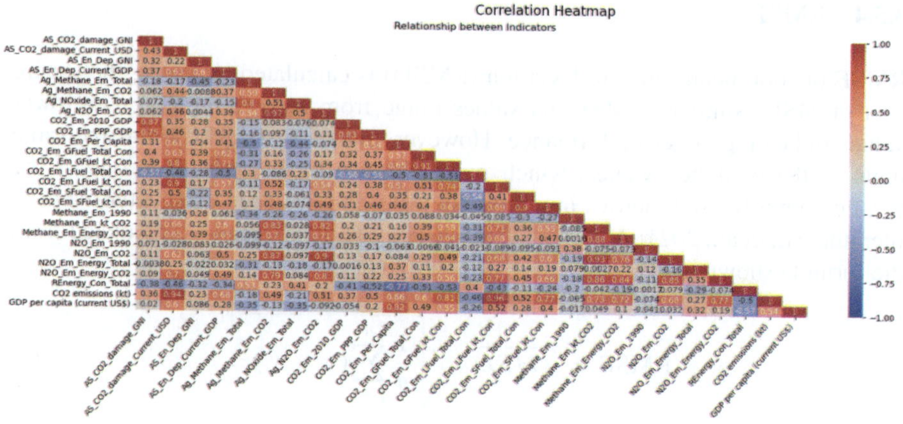

Fig. 3 Correlation between the indicators in final dataset

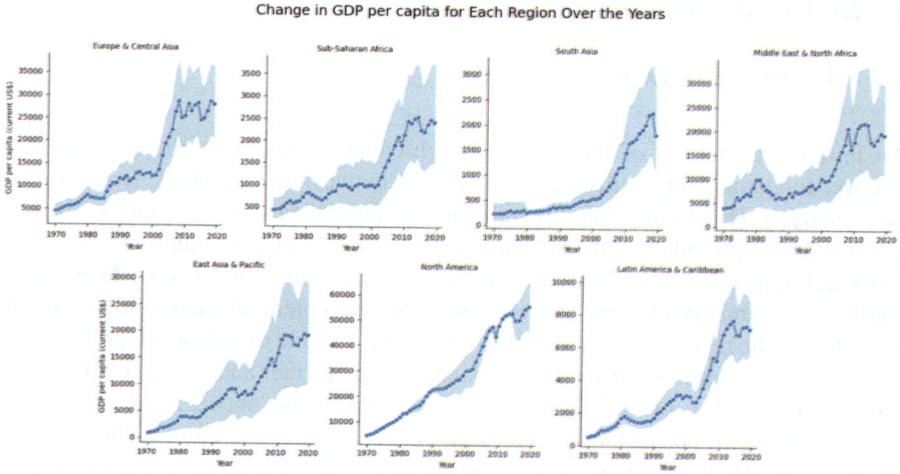

Fig. 4 Region-wise change in GDP per capita

Pacific and North America, show consistent and substantial growth, others (like sub-Saharan Africa) exhibit more modest increases. The shaded confidence intervals emphasize the variability and inherent uncertainty associated with these economic measures, thereby providing a more nuanced understanding of regional economic trends.

Bubble charts are used to present compact information about the dataset under study. Here, different colors of the bubbles denote different geographical areas, and the size of the bubbles denotes the total emission. Figures 5 and 6 illustrate the regional information for GDP per capita, renewable energy consumption, and CO_2 emission. Regions with higher GDP per capita tend to have higher CO_2 emissions, which is visible as many of the larger bubbles are toward the right and higher up on

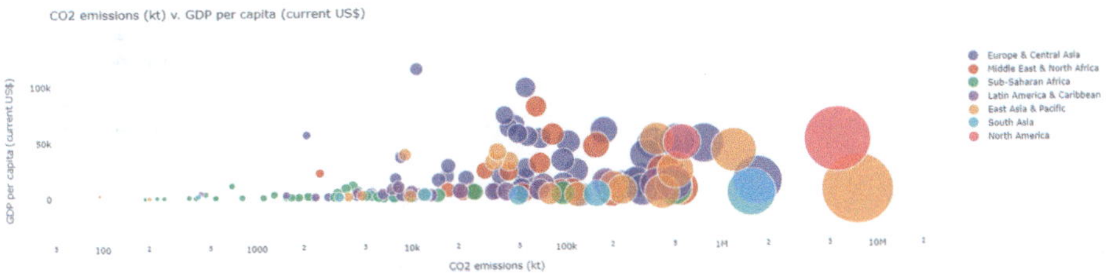

Fig. 5 Country-wise GDP per capita vs. CO_2 emission analysis

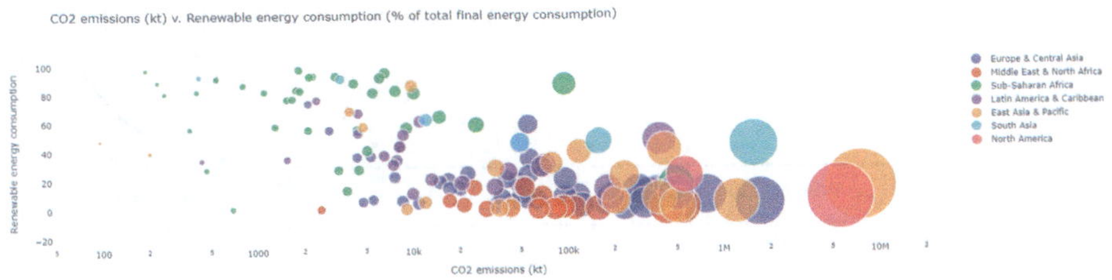

Fig. 6 Country-wise renewable energy consumption vs. CO_2 emission analysis

the chart. Additionally, from the colorization, we can infer that carbon emissions are the highest in North America. The second chart reveals that higher renewable energy consumption does not necessarily correlate with lower CO_2 emissions as some regions with high renewable energy consumption also have high CO_2 emissions, suggesting a transition phase or an energy mix that still includes significant fossil fuel use. These charts together provide a comprehensive view of how economic growth and renewable energy adoption impact CO_2 emissions across different regions. While higher GDP per capita is somehow associated with increased CO_2 emissions, the adoption of renewable energy shows a mixed relationship, indicating that further exploration is necessary before reaching any conclusions.

Figure 7 further validates the first bubble chart (Fig. 5). It analyzes the correlation between CO_2 emissions and GDP per capita across different income groups. It is noticeable that members of the higher income group appear to significantly contribute to the exacerbation of the current concerns of carbon emissions. The upward black dashed line with the confidence interval suggests a general trend where GDP per capita initially rises with increasing CO_2 emissions.

The bubble chart presented in Fig. 8 depicts the scenario for the South Asian region. Notably, India is the top contributor to carbon emissions and has a higher GDP per capita value. Inversely, Bhutan, located at the leftmost corner of the chart, is the carbon-negative country with the least absolute emission rate (in kt).

The GDP per capita trend analysis over the study period provides a comprehensive view of the economic growth trajectory. Figure 9 illustrates the distinct phases of growth, stagnation, and recovery, reflecting economic cycles and external influences from 1971 to 2019 in Bangladesh. The image shows that despite several resource limitations, Bangladesh has managed to maintain an upcoming GDP trend.

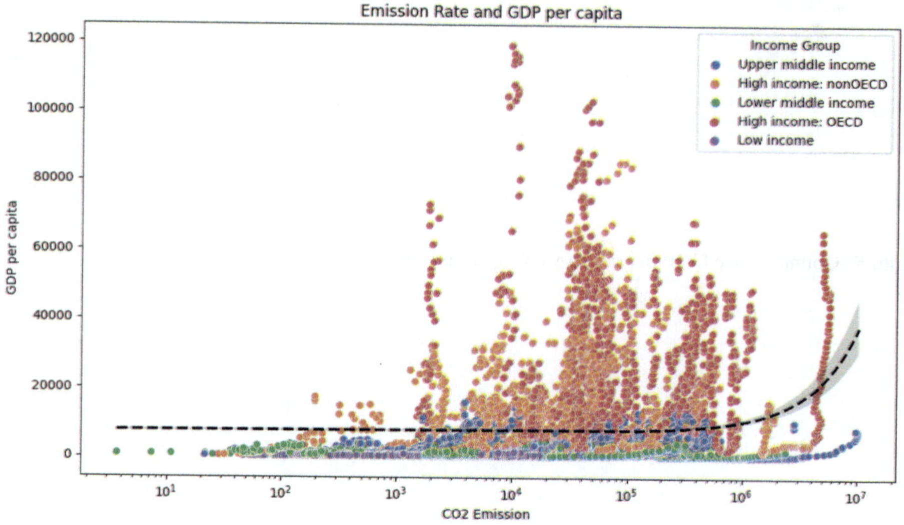

Fig. 7 CO_2 emission vs. economic growth based on income group

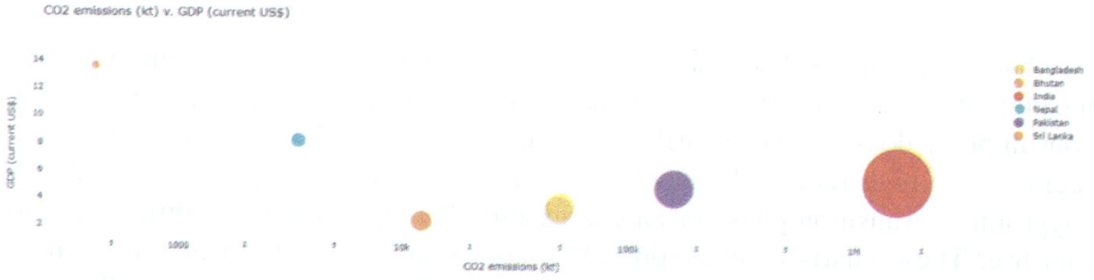

Fig. 8 GDP vs. CO_2 emission for South Asian countries

The relationship between CO_2 emissions and GDP per capita is clearly illustrated in Fig. 10. We observe a clear positive correlation between CO_2 emissions and GDP per capita, indicating that as the GDP per capita increases, so does the level of CO_2 emissions. This image suggests that countries with higher economic growth are predominantly the main contributors to the increasing global carbon emission rate.

4.2 Results of Deep Learning Models

This section records the performance of the deep learning classifiers utilized in this study. The experimental results are tabulated in Table 4. The proposed MRF model showcases its exceptional predictive capabilities by achieving the highest R^2 score

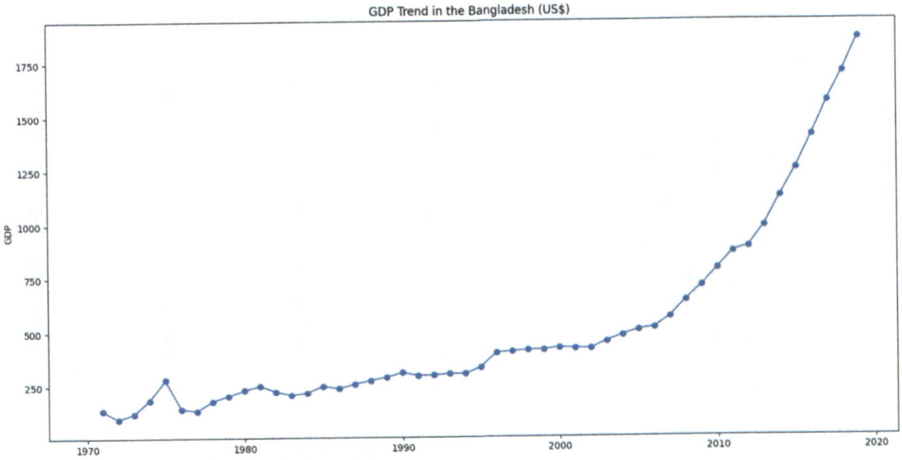

Fig. 9 Trend in the economic growth for Bangladesh

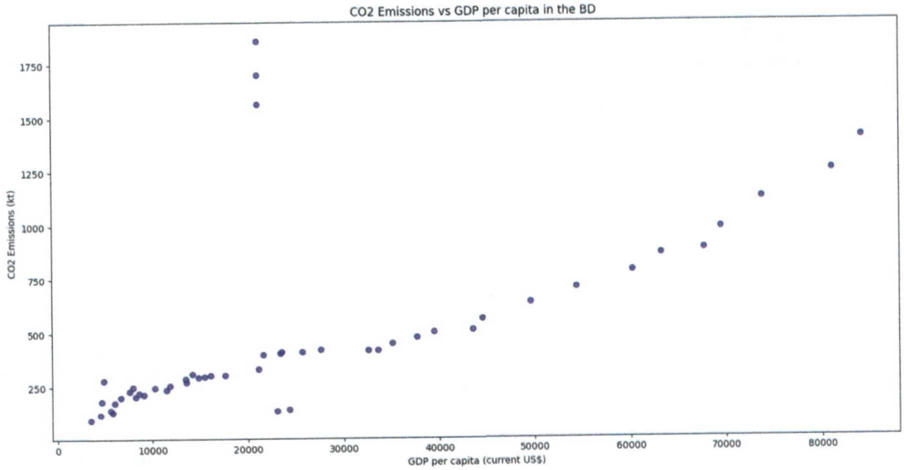

Fig. 10 Correlation between CO_2 emission and GDP per capita

Table 4 Performance of the deep learning models

Algorithm	R^2	MAE	MSE	RMSE
LSTM	0.7262	0.0377	0.0043	0.0654
Bi-LSTM	0.7214	0.0336	0.0043	0.0659
GRU	0.7448	0.0315	0.0040	0.0631
Proposed MRF	**0.8331**	**0.0297**	**0.0026**	**0.0510**

As the R2 is high and the error metrics are low of the bold row then it is significant than others

of 83.31%, followed by GRU (74.48%), LSTM (72.62%), and Bi-LSTM (72.14%) models.

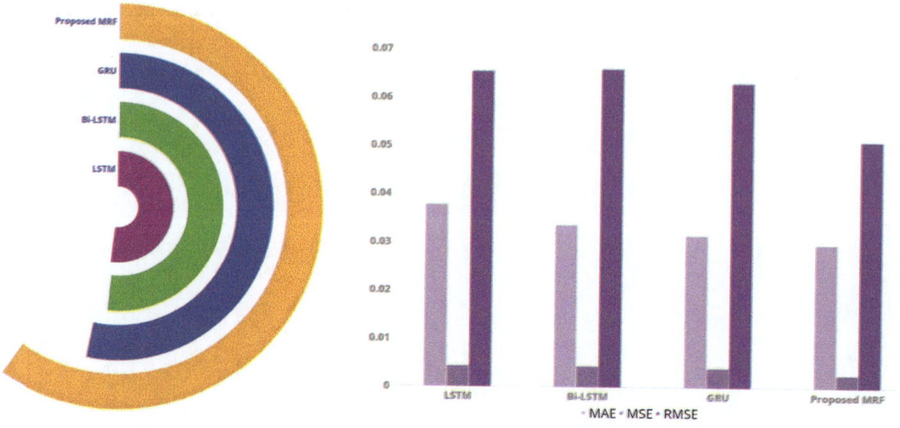

(a) R^2 Score (b) Error Rates

Fig. 11 Comparison of the model performance

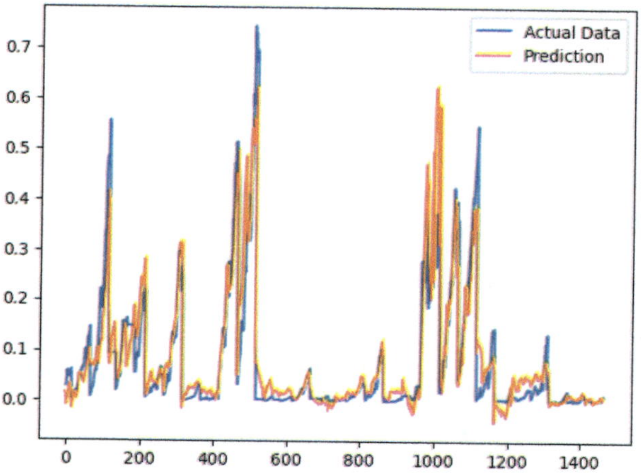

Fig. 12 Actual and prediction value mapping for MRF

Moreover, it is evident from the table that the MRF model outperforms others by consistently maintaining the lowest error rates. On the other hand, the Bi-LSTM model has been proven to have the highest error rate among all the models. These results demonstrate the superiority of the proposed MRF model in predicting the change in the target variable, GDP per capita, and outperforming the other traditional models. The following figure graphically illustrates the results of the above table. Figure 11a presents a comparison between the R^2 scores of the deep learning models. The bar chart depicted in Fig. 11b displays the error rates for different models. The figure visually highlights the enhanced performance and robustness of our proposed model with minimal error rates, reinforcing its potential for making accurate predictions.

Furthermore, Fig. 12 illustrates the predictions on unseen data for the MRF model. The orange line representing the prediction closely aligns with the actual data points, thus demonstrating the efficacy of the model in generating accurate predictions. Additionally, the consistent alignment between the predicted and actual data points demonstrates the robustness of the MRF model in capturing underlying patterns within the data. These findings indicate the model's effectiveness in generalizing to new data and its potential for real-world applications.

4.3 *Interpreting the Results with Explainable AI*

In this part of the study, we delve into the insights gained from our analysis using explainable AI (XAI) techniques. The utilization of XAI tools in our deep learning models facilitates the explanation of intricate associations between carbon dioxide (CO_2) emissions and GDP per capita. This consequently enhances the transparency, reliability, and interpretability of our research outcomes.

Figure 13 offers a detailed analysis of the features impacting the model's output using SHAP (SHapley Additive exPlanations) values, which help explain the importance and effects of each feature in the model. The x-axis represents the mean impact score on the target variable, while the y-axis represents the features in descending order of importance. The relevance of a feature in the ultimate prediction increases proportionally with the higher value of the impact score. The color in the leftmost beeswarm plot signifies the feature value, with red denoting high and blue denoting low. Additionally, we can gain a comprehensive understanding of the redundant

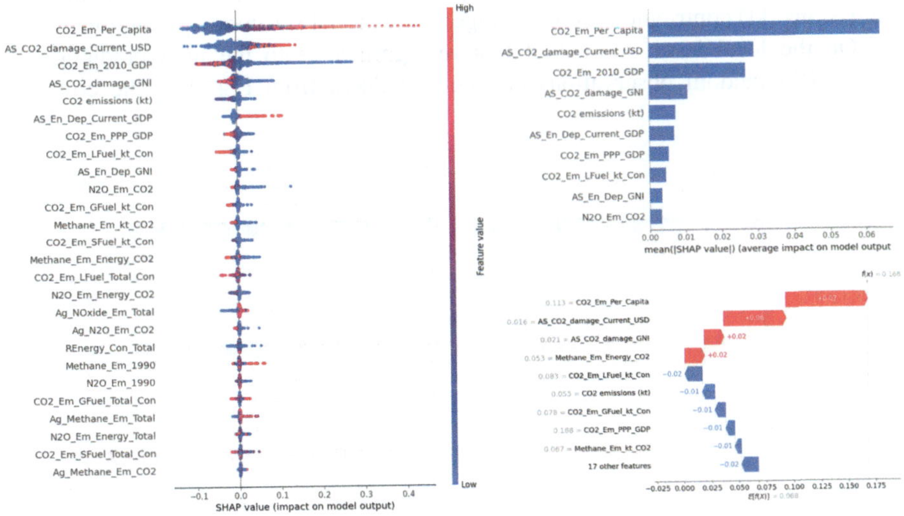

Fig. 13 SHAP plot for feature importance

features or those that contribute minimally to the figure. CO2_Em_Per_Capita has the highest SHAP mean impact score, and hence, it is the most relevant feature contributing to the model outcome. Moreover, features such as CO2_Em_Per_Capita, AS_CO2_damage_Current_USD, CO2_Em_2010_GDP, AS_CO2_damage_GNI, and AS_En_Dep_Current_GDP have significant influence on the ultimate prediction. On the other hand, the impact of features like methane emissions and N_2O emissions is comparatively lower or negligible. The points that are shifted toward the right, with higher SHAP values, indicate more significant positive contributions, whereas those shifted toward the left, with lower SHAP values, represent negative contributions. The figure on the left side illustrates that CO2_Em_Per_Capita exhibits the highest average SHAP value and also demonstrates a wide distribution of impacts, indicating variability in its influence on different predictions. The analysis further reveals that CO2_Em_Per_Capita, AS_CO2_damage_Current_USD, and AS_En_Dep_Current_GDP have a positive impact on improving the model prediction. Conversely, CO2_Em_2010_GDP and AS_CO2_damage_GNI have an inverse impact. The extensive XAI analysis emphasizes the critical role of these emissions' metrics in assessing the growth of GDP per capita which can be further validated from the SHAP bar plot presented at the top right corner. It highlights the top ten (10) prominent features with their respective SHAP values. CO_2 emissions (kt) are shown to have a relatively insignificant mean SHAP value compared to other features such as CO2_Em_Per_Capita, AS_CO2_damage_Current_USD, and CO2_Em_2010_GDP. The lower right waterfall plot depicts individual feature influence for a random specific observation and breaks down the contribution of each feature to a single prediction. While CO2_Em_Per_Capita contributes the most significant positive impact (+0.07) to the final prediction, negative contributions come from features like CO2_Em_LFuel_kt_Con and CO_2 emissions (kt), reducing the prediction by -0.02 and -0.01, respectively. These insights suggest that CO_2 emissions (kt) contribute less on average to the model's accurate predictions.

On the local level, individual feature significance to the final prediction for three (03) random observations can be visualized from Fig. 14. The collective

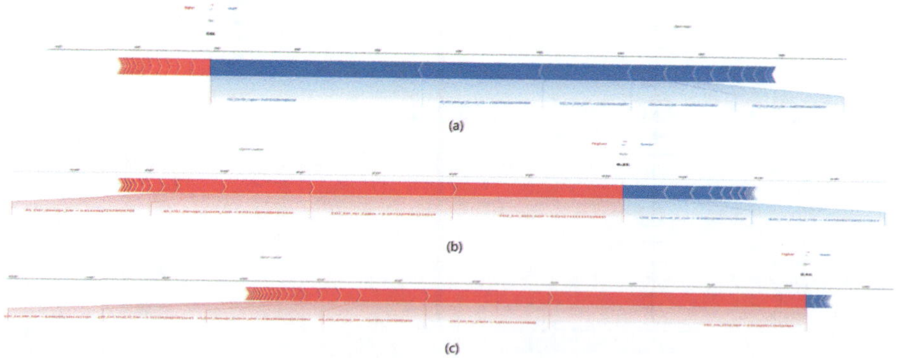

Fig. 14 SHAP force plot

representation of all three plots demonstrates the intricate interaction of multiple features that impact the predictions of the model. CO2_Em_Per_Capita consistently appears as a dominant positive factor, suggesting its significant role in determining the model output. Conversely, features like CO2_Em_LFuel_kt_Con, N2O_Em_Energy_CO2, and CO_2 emissions (kt) frequently push the predictions lower. The inconsistent presence of these features is noteworthy, suggesting that while they may have some influence, they are not the primary drivers in the model's ultimate predictions of economic growth. In contrast, previously mentioned features such as CO2_Em_Per_Capita, AS_CO2_damage_Current_USD, and CO2_Em_2010_GDP appear to have a significant influence.

In Fig. 15, a clear visual of the relationship between the top five (05) features derived from Fig. 13 can be observed. The graphical representation demonstrates a clear upward correlation between features shown in (a), (b), and the target. High values of these features correlate with higher SHAP values, indicating an overall positive impact. In contrast, features depicted in (c) and (d) show a negative relationship with economic growth, as indicated by the downward trend in the graph. This suggests that higher values of the features are associated with lower SHAP values, indicating a negative impact on the target variable, GDP per capita. These findings provide robust evidence to support our initial hypotheses. Interestingly, a flat line distribution for CO_2 emissions (kt) is notable. Most points have SHAP values close to zero, indicating a minimal effect on the model's prediction.

A simple breakdown of the overall impact of each feature can be observed from Table 5. The table outlines the significance of each individual feature in making the final prediction, providing further evidence to support the results obtained from SHAP. The table presents the importance and tolerance value for each of the features in descending order of importance. For example, the recorded values indicate that CO2_Em_Per_Capita is the most influential feature. The numeric value of 0.0791 denotes the tolerance value for CO2_Em_Per_Capita. This explains the influence on the shift in the model's outcome if the feature is altered, even in the slightest. The analysis conducted with the ELI5 technique reveals that the per capita CO_2 emissions (CO2_Em_Per_Capita), the economic damage caused by CO_2 (AS_CO2_damage_Current_USD), and the CO_2 emissions concerning GDP (CO2_Em_2010_GDP) have the most substantial impact, while various emission metrics such as CO_2 emissions (kt), methane emissions, and N_2O emissions have minimal influence on the results. Based on these results, it is evident that various emission metrics, particularly CO_2 emissions (kt), have a relatively minor influence on the predictive accuracy of the model. These findings strongly defy the traditional EKC hypothesis and validate our previous findings. Therefore, policies aimed at reducing the intensity of CO_2 emissions can be more effective in achieving sustainable economic growth compared to strategies focused solely on reducing overall average emissions. By implementing policies that specifically target the reduction of CO_2 intensity, countries can prioritize industries and sectors with the highest emissions, leading to a more targeted and efficient approach toward sustainable economic growth. This further implies that it is possible to reduce

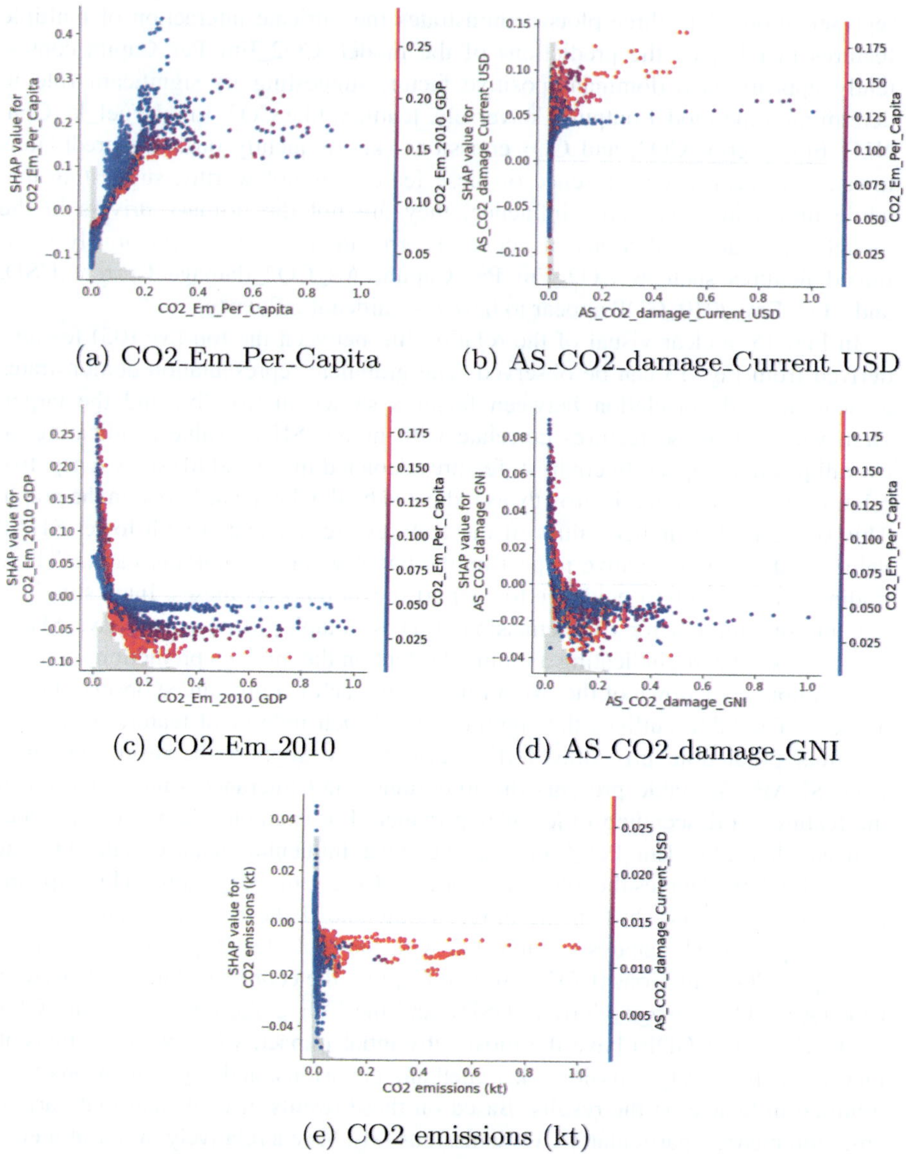

(a) CO2_Em_Per_Capita (b) AS_CO2_damage_Current_USD

(c) CO2_Em_2010 (d) AS_CO2_damage_GNI

(e) CO2 emissions (kt)

Fig. 15 SHAP dependency plot

Table 5 Feature importance analysis with ELI5

Weight	Feature indicator code
1.4747 ± 0.0791	CO2_Em_Per_Capita
0.5623 ± 0.0650	AS_CO2_damage_Current_USD
0.5158 ± 0.0571	CO2_Em_2010_GDP
0.0485 ± 0.0045	AS_CO2_damage_GNI
0.0445 ± 0.0073	AS_En_Dep_Current_GDP
0.0330 ± 0.0037	CO2 emissions (kt)
0.0293 ± 0.0063	CO2_Em_SFuel_kt_Con
0.0254 ± 0.0059	Methane_Em_kt_CO2
0.0235 ± 0.0056	CO2_Em_PPP_GDP
0.0191 ± 0.0026	N2O_Em_CO2
0.0180 ± 0.0079	Ag_Methane_Em_Total
0.0127 ± 0.0031	CO2_Em_LFuel_kt_Con
0.0100 ± 0.0018	Methane_Em_Energy_CO2
0.0097 ± 0.0035	AS_En_Dep_GNI
0.0094 ± 0.0033	Ag_Methane_Em_CO2
0.0089 ± 0.0038	CO2_Em_LFuel_Total_Con
0.0067 ± 0.0026	CO2_Em_GFuel_kt_Con
0.0058 ± 0.0005	Methane_Em_1990
0.0058 ± 0.0019	CO2_Em_GFuel_Total_Con
0.0053 ± 0.0036	CO2_Em_SFuel_Total_Con
0.0048 ± 0.0004	Ag_NOxide_Em_Total
0.0040 ± 0.0009	N2O_Em_Energy_CO2
0.0035 ± 0.0016	N2O_Em_Energy_Total
0.0027 ± 0.0012	N2O_Em_1990
0.0019 ± 0.0009	REnergy_Con_Total
0.0015 ± 0.0008	Ag_N2O_Em_CO2

different types of emissions on average and maintain sustainable growth in the economy at the same time as both show no positive long-term relationship.

5 Conclusion and Future Work

This study primarily examines the role of carbon dioxide emissions as an indicator of a nation's sustainable economic growth, to determine the influence of various emission metrics on sustainable economic development. Time series data analysis is challenging due to the risk of misleading behavior and the potential for false empirical findings resulting from simultaneity bias and heterogeneity. We analyzed the econometric factors inherent in the prediction of sustainable GDP per capita. This study presents a hybrid sequential MRF model designed to capture complex patterns in time series data. The proposed MRF model outperforms traditional deep learning models, achieving minimal errors of 0.0297, 0.0026, and 0.0510 for MAE,

MSE, and RMSE, respectively. To assess the impact of emission metrics on GDP per capita, the proposed pipeline incorporates advanced explainable AI tools, namely, SHAP and ELI5. The comprehensive analysis demonstrated that different types of emissions have a minimal impact on predicting GDP, indicating that other factors may be more significant. This finding supports the study's initial hypothesis and challenges the traditional EKC hypothesis, which posits that total average CO_2 emissions (kt) are a primary driver of sustainable economic growth. Only a limited number of studies have previously identified emissions intensity, such as CO_2 emissions per capita, as a more significant indicator than total CO_2 emissions. These studies often overlooked this crucial aspect, which our research has thoroughly investigated and substantiated with clear evidence. This suggests that factors like technological advancement and resource efficiency, which reduce per capita carbon emissions, are crucial for sustainable economic growth. These findings provide a robust confirmation, offering a more accurate and evidence-based understanding of the factors influencing GDP. Additionally, the analysis indicates that the relationship between economic growth and environmental degradation is more complex than previously thought, highlighting the need for further research in this area. Future studies should explore a broader range of variables affecting sustainable economic growth, such as social, environmental, and governance factors. Implementing a holistic approach toward understanding economic dynamics will enable us to develop more resilient and effective economic policies that can navigate various challenges and uncertainties. A comprehensive investigation into these elements will yield deeper insights and foster more effective strategies for achieving long-term economic sustainability.

Acknowledgments We extend our deepest gratitude to the Center for Multidisciplinary Research and Development (CeMRD) for their resources, guidance, and support. The expertise and encouragement from CeMRD members have been invaluable. This research was made possible by the collaborative environment and cutting-edge facilities at CeMRD.

References

Abedin, M. Z., Moon, M. H., Hassan, M. K. & Hajek, P. (2021). Deep learning-based exchange rate prediction during the covid-19 pandemic. *Annals of Operations Research,* 1–52.

Amarpuri, L., Yadav, N., Kumar, G. & Agrawal, S. (2019). *Prediction of co2 emissions using deep learning hybrid approach: A case study in Indian context* (pp. 1–6). IEEE.

Antonakakis, N., Chatziantoniou, I. & Filis, G. (2017). Energy consumption, co2 emissions, and economic growth: An ethical dilemma. *Renewable and Sustainable Energy Reviews, 68*, 808–824. https://www.sciencedirect.com/science/article/pii/S1364032116306013.

Azam, M., Khan, A. Q., Abdullah, H. B. & Qureshi, M. E. (2016). The impact of co_2 emissions on economic growth: evidence from selected higher co_2 emissions economies. *Environmental Science and Pollution Research, 23*, 6376–6389. https://doi.org/10.1007/s11356-015-5817-4.

Boamah, K. B. et al. (2017). Carbon dioxide emission and economic growth of china—the role of international trade. *Environmental Science and Pollution Research, 24*, 13049–13067.

Burnett, J. W., Bergstrom, J. C. & Wetzstein, M. E. (2013). Carbon dioxide emissions and economic growth in the U.S. *Journal of Policy Modeling, 35*, 1014–1028. https://www.sciencedirect.com/science/article/pii/S0161893813000835.

Chukwunonso, B. P. et al. (2024). Predicting carbon dioxide emissions in the United States of America using machine learning algorithms. *Environmental Science and Pollution Research, 31*, 33685–33707.

CO2 emissions – Global Energy Review 2021 – Analysis - IEA — iea.org. (2021). https://www.iea.org/reports/global-energy-review-2021/co2-emissions [Accessed 30-06-2024].

Datta, R. K., Sajid, S. W., Moon, M. H. & Abedin, M. Z. (2021). *Foreign currency exchange rate prediction using bidirectional long short term memory* (pp. 213–227). Springer.

Frodyma, K., Papież, M. & śmiech, S. (2022). Revisiting the environmental Kuznets curve in the European union countries. *Energy, 241*, 122899. https://www.sciencedirect.com/science/article/pii/S0360544221031480.

Hassan, M. M., Abrar, M. F. & Hasan, M. (2023). *An explainable AI-driven machine learning framework for cybersecurity anomaly detection* (pp. 197–219). Routledge.

Hasan, M., Hassan, M. M., Faisal-E-Alam, M. & Akter, N. (2023). *Empirical analysis of regression techniques to predict the cybersecurity salary* (pp. 65–84). Routledge.

Hasan, M., Ahmed, T., Islam, M. R. & Uddin, M. P. (2024). Leveraging textual information for social media news categorization and sentiment analysis. *PLoS one, 19*, e0307027.

Hui, M. (2020). 2020 world development indicators. https://www.kaggle.com/datasets/manchunhui/world-development-indicators/data. [Accessed 28-07-2024].

Jayanthakumaran, K., Verma, R. & Liu, Y. (2012). Co_2 emissions, energy consumption, trade and income: A comparative analysis of china and India. *Energy Policy, 42*, 450–460. https://www.sciencedirect.com/science/article/pii/S0301421511010020.

Karim, M. (2024). Exploring and visualizing world CO2 emission data by country by year using Pandas dataframe in Python. https://github.com/Mzkarim/Exploring--CO2-Emission-Data-. [Accessed 28-07-2024].

Kawakura, S., Hirafuji, M., Ninomiya, S. & Shibasaki, R. (2022). Adaptations of explainable artificial intelligence (XAI) to agricultural data models with ELI5, PDPbox, and Skater using diverse agricultural worker data. *European Journal of Artificial Intelligence and Machine Learning, 1*, 27–34. https://www.ej-ai.org/index.php/ejai/article/view/14.

Kim, J.-Y. & Cho, S.-B. (2021). Explainable prediction of electric energy demand using a deep autoencoder with interpretable latent space. *Expert Systems with Applications, 186*, 115842. https://www.sciencedirect.com/science/article/pii/S0957417421012033.

Lee, C.-C., Hussain, J. & Chen, Y. (2022). The optimal behavior of renewable energy resources and government's energy consumption subsidy design from the perspective of green technology implementation. *Renewable Energy, 195*, 670–680. https://www.sciencedirect.com/science/article/pii/S0960148122009028.

Li, S., Siu, Y. W. & Zhao, G. (2021). Driving factors of co_2 emissions: Further study based on machine learning. *Frontiers in Environmental Science,* **9**, 721517. https://www.frontiersin.org/journals/environmental-science/articles/10.3389/fenvs.2021.721517.

Maarif, M. R., Saleh, A. R., Habibi, M., Fitriyani, N. L. & Syafrudin, M. (2023). Energy usage forecasting model based on long short-term memory (LSTM) and eXplainable artificial intelligence (XAI). *Information, 14*, 265.

McGinley, O., Moran, P. & Goggins, J. (2022). An assessment of the key performance indicators (KPIs) of energy efficient retrofits to existing residential buildings. *Energies, 15*, 334. https://www.mdpi.com/1996-1073/15/1/334.

Rabbi, M. F., Moon, M. H., Dhonno, F. T., Sultana, A. & Abedin, M. Z. (2022). *Foreign currency exchange rate prediction using long short-term memory, support vector regression and random forest regression* (pp. 251–267). Springer.

Roy, P. & Tipu, A. D. (2024). *An interpretable machine learning approach for identification of the risk factors of early-stage overweight and obesity* (pp. 1–6).

Shajalal, M., Boden, A. & Stevens, G. (2022). Towards user-centered explainable energy demand forecasting systems. In *Proceedings of the Thirteenth ACM International Conference on Future Energy Systems*. https://api.semanticscholar.org/CorpusID:249928070.

Sherstinsky, A. (2020). Fundamentals of recurrent neural network (RNN) and long short-term memory (LSTM) network. *Physica D: Nonlinear Phenomena, 404*, 132306. https://www.sciencedirect.com/science/article/pii/S0167278919305974.

Sultan, N., Hasan, M., Wahid, M. F., Saha, H. & Habib, A. (2023). Cesarean section classification using machine learning with feature selection, data balancing and explainability. *IEEE Access*, 84487–84499.

The World Bank. (2024). World development indicators. https://doi.org/10.57966/6rwy-0b07 [Accessed 28-07-2024].

Tiseo, I. (2024). Global atmospheric concentration of carbon dioxide 1990–2024, by month. https://www.statista.com/statistics/1091999/atmospheric-concentration-of-co2-historic/. [Accessed 03-07-2024].

United Nations. (2024). COVID-19 to slash global economic output by $8.5 trillion over next two years | United Nations — un.org. https://www.un.org/en/desa/covid-19-slash-global-economic-output-85-trillion-over-next-two-years. [Accessed 30-06-2024].

van Zyl, C., Ye, X. & Naidoo, R. (2024). Harnessing explainable artificial intelligence for feature selection in time series energy forecasting: A comparative analysis of Grad-CAM and SHAP. *Applied Energy, 353*, 122079.

Yamak, P. T., Yujian, L. & Gadosey, P. K. (2020). *A comparison between ARIMA, LSTM, and GRU for time series forecasting*, ACAI '19 (pp. 49–55). Association for Computing Machinery. https://doi.org/10.1145/3377713.3377722.

Priyanka Roy received her BSc (Engineering) degree in Computer Science and Engineering (CSE) from Hajee Mohammad Danesh Science and Technology University, Dinajpur, Bangladesh in 2023. She is a full-time Research Assistant at the Center for Multidisciplinary Research and Development (CeMRD) since 2022. In 2023, she started her professional career as a lecturer in the Department of Computer Science and Engineering at Green University of Bangladesh (GUB), located in Narayanganj, Dhaka, Bangladesh. She currently holds the position of Lecturer in the Department of Computer Science and Engineering at Sylhet International University (SIU), Sylhet, Bangladesh. She was recently appointed as the mentor (advisor) of the CSE Society at SIU for the 2024-2025 term. Her passionate interests span a wide spectrum of research areas, including machine learning, deep learning, computer vision, cybersecurity, eXplainable AI (XAI) for health informatics, sustainable business intelligence, and natural language processing (NLP).

Amrita Das Tipu is a lecturer in the Department of Computer Science and Engineering at Dhaka International University, Dhaka, Bangladesh. He earned his BSc (Eng.) in Computer Science and Engineering from Hajee Mohammad Danesh Science and Technology University, Dinajpur, Bangladesh, in 2023. His research interests encompass Machine Learning, Natural Language Processing, Deep Learning, Brain-Computer Interface, Augmented Reality, Virtual Reality, and Sustainable Development.

Mahmudul Hasan is currently pursuing a PhD in Information Technology (IT) at Deakin University, Melbourne, Australia. He earned his BSc (Eng.) and MSc (Eng.) degrees in Computer Science and Engineering (CSE) from Hajee Mohammad Danesh Science and Technology University, Dinajpur, Bangladesh, in 2021 and 2023, respectively. He previously served as a lecturer in the Department of CSE at the University of Creative Technology, Chittagong (UCTC), Bangladesh. He is the Founder and Director of the Center for Multidisciplinary Research and Development (CeMRD) and a moderator of "Be Researcher BD," the largest online research forum in Bangladesh. Additionally, he has taught online as a Data Science instructor to students in the USA, Italy, Denmark, South Korea, and Australia. He is also the founder of the online educational platform "MHM Academy." His research interests include federated learning, machine learning, deep learning, cybersecurity, health informatics, renewable energy, computational sociology, and business intelligence.

Md. Palash Uddin is currently working as a Postdoctoral Research Fellow at the School of Information Technology, Deakin University, Australia. He received a PhD degree in Information Technology from Deakin University, Australia, in 2023. He also received a BSc degree in Computer Science and Engineering from Hajee Mohammad Danesh Science and Technology University (HSTU), Bangladesh and an MSc degree in Computer Science and Engineering from Rajshahi University of Engineering & Technology, Bangladesh. He is also an academic faculty member at HSTU, Bangladesh. His research interests include machine learning, federated learning, blockchain, and remote sensing image analysis.

BLDAR: A Blending Ensemble Learning Approach for Primary Energy Consumption Analysis

Abdullah Haque, Tuhin Chowdhury, Mahmudul Hasan, and Md. Jahid Hasan

1 Introduction

Energy captured directly from natural resources like sunlight, wind, falling water (hydropower), and organic matter (biomass) is called primary energy. This also includes nonrenewable resources like coal, oil, and natural gas. The progress of humanity and the world depends heavily on primary energy. Primary energy supplies the necessities of modern life, such as transportation, heating, industrial activities, and the production of electricity (Martínez et al., 2019).

Primary energy can be classified into two types: primary fuels and primary energy flows. Primary fuels include fossil fuels (oil, coal, and natural gas) and nuclear fuels. Renewable energy sources include the sun, wind, hydropower, biomass, etc. The majority (about 95%) of global primary energy comes from primary fuels, with the rest from primary energy flows. Fossil fuels, nuclear energy, and renewable energy sources account for roughly 75%, 6%, and 14% of the world's primary energy supply, respectively (Cheekatamarla et al., 2024). The absolute primary energy consumption in developing countries is 58% of the level in

A. Haque
Department of Mathematics, Hajee Mohammad Danesh Science and Technology University, Dinajpur, Bangladesh

T. Chowdhury
Department of Computer Science and Engineering, SEC, Shahjalal University of Science and Technology, Sylhet, Bangladesh

M. Hasan (✉)
Department of Computer Science and Engineering, Hajee Mohammad Danesh Science and Technology University, Dinajpur, Bangladesh

M. J. Hasan
Business Information Systems, RMIT University, Melbourne, VIC, Australia

M. Z. Abedin, W. Yong (eds.), *Machine Learning Technologies on Energy Economics and Finance*, International Series in Operations Research & Management Science 367, https://doi.org/10.1007/978-3-031-94862-6_8

developed countries (Komarova et al., 2022). In 2021, the world's primary energy consumption reached over 595 exa-joules. China is the world's greatest consumer of primary energy, followed by the United States, the Russian Federation, and India. Fossil fuels contributed to more than 80% of global primary energy consumption. Also, oil accounted for more than 30% of total world energy consumption in 2021 (Aydin & Karakurt, 2023). However, the environmental effects of energy consumption are severe. Overall carbon dioxide emissions and primary energy consumption have steadily increased. Global environmental challenges caused by energy consumption not only impede economic progress but also endanger human existence and development (Wei & He, 2017).

Global warming and catastrophic disasters are wreaking havoc on Earth these days. Energy conservation has been a top issue due to social development challenges. So, energy consumption prediction is an important instrument for effective building energy management, guiding energy policy, and service distribution. Despite the use of modern technologies, reliable energy consumption prediction remains challenging due to various affecting factors (Liu et al., 2023). A variety of behavioral factors influence energy consumption trends, including consumer preferences, lifestyle changes, and societal standards, which are difficult to represent in typical prediction models (Zhou & Yang, 2016). Rapid improvements in renewable energy, storage technologies, and energy efficiency can drastically alter future energy consumption patterns, making long-term forecasting increasingly difficult (Ahmad et al., 2021). With the growing emphasis on flexibility and elasticity in building energy usage, accurate building energy prediction is essential for sustainable development. Nonetheless, difficulties with choosing appropriate input and algorithms continue, as does finding a balance between computation time and forecast accuracy (Zhang et al., 2024). In Amiri et al. (2023), they use a Machine Learning (ML) algorithm to forecast the energy consumption of commercial and residential buildings. Their concepts helped to improve municipal scenario planning by providing a more spatially detailed picture of future energy consumption. Another study categorizes the most pertinent literature according to ML approaches, energy type, prediction type, and application area. It highlights the main ML technologies and assesses their performance in forecasting energy usage. This research continues by discussing the trends and efficacy of these models, emphasizing considerable increases in accuracy and performance using unique hybrid and ensemble prediction models (Mosavi & Bahmani, 2019). Another approach described helps to ensure the proper implementation of energy policy by giving accurate energy consumption predictions. These predictions affect capital investment, environmental quality, revenue analysis, and market research management, all while ensuring supply security (Ekonomou, 2010). The expansion of a nation's economy and its energy usage are inextricably linked. Insufficient energy supply has resulted in large deficits at both the household and aggregate levels worldwide. This shortage is sometimes called "energy poverty" in the literature, particularly at the micro level, as households' energy demands are unfulfilled (Gyamfi et al., 2024). The technical contributions of this chapter are as follows:

- We design an ML-driven primary energy consumption prediction process to analyze consumption based on sustainable development indicators.
- We develop a blending ensemble model, BLDAR, which combines Least Absolute Shrinkage and Selection Operator (LASSO), Decision Tree (DT), AdaBoost (ADB), and Random Forest (RF) regressors.
- We provide a comparative analysis of ML algorithms for primary energy consumption prediction and identify the most suitable model to design an energy prediction system.
- To determine model robustness, generalization ability, and the impact of data size, we evaluate the model's performance using 80:20, 70:30, and 50:50 training and testing ratios.

The remaining chapter is organized as follows: In Sect. 2 we present the related works, in Sect. 3 we present the proposed methodology with overview and details descriptions, in Sect. 4 we present the obtained results, and finally we present conclusion in Sect. 5.

2 Literature Review

Energy is widely recognized as a key engine of worldwide economic growth and development. Researchers have extensively investigated the impact of energy sources and usage on a variety of economic indices. Given the complicated interplay of economic growth, human development, and environmental concerns, additional research is required to understand how these aspects interact (Alola et al., 2021). There is a development of an artificial neural network (ANN) model to anticipate net energy consumption (NEC) based on economic indicators such as gross domestic product (GDP), gross national product (GNP), and population growth. They argue that ANN approach shows the most accuracy for evaluating NEC based on economic indicators. Most ANN models concentrated on dynamic, short-term energy consumption predictions, which are necessitated through input data pretreatment and selection (Sözen & Arcaklioglu, 2007). Additionally, Wang (2022) employs the nonlinear fitting of the BP model and linear fitting of the ARIMA model as independent variables, with per capita coal consumption as the dependent variable. A revolutionary approach to coal consumption forecasting is unveiled: a combined model utilizing multiple linear regression to shatter previous accuracy limitations. On the other hand, Li (2019) attempts to anticipate China's energy density by utilizing an LSTM-based neural network model developed by both research groups. Their research traces that time series estimation generates much better outcomes than other regression analyses. They add that there is a strong correlation between economic development, population, industrial relations, and energy consumption. It has been found that many academics employ regression analysis to solve the association between energy consumption and these factors. Also, Wang and Zhang (2023) generated novel models that outperform DGM(1,1), DGM(1,n), and BP neural networks in predicting utilizing per capita energy consumption (PCEC) data from 30 Chinese provinces. The models work well in collecting recent data trends and regional associations, as well as analyzing spatial connections and making accurate predictions.

The drivers of energy consumption are examined by Wen et al. (2021) using environmentally extended input-output and structural decomposition analysis. Soaring population is the engine of global energy demand, but researchers identified a bright spot: Reducing energy inefficiency acts as a powerful brake. Private consumption and exports remain significant energy guzzlers, highlighting areas for further improvement. Their study also suggests that policies like transport electrification and renewable energy promotion support the low-carbon transition. Another work (Li & Solaymani, 2021) showed that long-term economic growth expansion significantly raises energy consumption relative to short-term growth in Malaysia. Particularly, the energy demand for agriculture raises by 4.6% and the energy demand for industry increases by 1.1% in economic growth. Energy consumption and emissions were successfully reduced in the industrial sector by technological advancements that increase energy efficiency. These findings are essential for policymakers focused on sustainable growth and energy management. Moreover the authors (He & Hao, 2024) estimate primary energy consumption in South and Central America, the Middle East, and Africa optimizing a fractional time-delayed gray model that is tuned with a particle swarm method. Their findings indicate that their model performs better than other gray models in most cases, demonstrating its dependability and efficacy.

On top of that, the authors (Shinwari et al., 2024) investigate the influence of foreign direct investment (FDI) on energy consumption in 29 Belt and Road Initiative (BRI) economies from 2000 to 2021, employing panel data methodologies to account for cross-sectional dependency, structural discontinuities, and slope heterogeneity. Their findings demonstrated that worldwide FDI has a beneficial influence on energy consumption, with China's FDI dominance enhancing it even more. In addition, green technology increased energy consumption, and their report also emphasizes the role of FDI policies and green technologies in boosting energy consumption in BRI economies. In another study, analyzing data from 125 countries spanning 2000 to 2018, the authors (Demiral & Demiral, 2023) investigate how various social and economic factors, such as education levels, transportation systems, information technology, government structures, private sector involvement, and economic development patterns, influence how efficiently these countries use energy. Countries are divided into four income groups, and higher income groups are found to have higher energy intensity and socioeconomic capacities. The regression results showed that socioeconomic factors have a range of effects on improving energy efficiency. Their study emphasizes the complexities of factors influencing energy efficiency and suggested policy implications.

Furthermore, to create a more accurate prediction tool for electricity use in Turkey, Kaytez (2020) develops a hybrid model that merges a least-squares support vector machine (SVM) with an autoregressive integrated moving average technique. When used to predict Turkey's net power consumption through 2022, the results demonstrate that the suggested hybrid model produced more realistic and dependable predictions and responded better to unexpected fluctuations in the time series.

3 Methodology

3.1 Approach Overview

The overview of the proposed methodology is in Fig. 1. Global Data on Sustainable Energy collects and preprocesses to enhance the computational efficacy and model performance. Then three ratios 80:20, 70:30, and 50:50 employ to split the data into training and testing subsets. Afterward, the suggested and comparative models are evaluated using several error metrics and R^2 score.

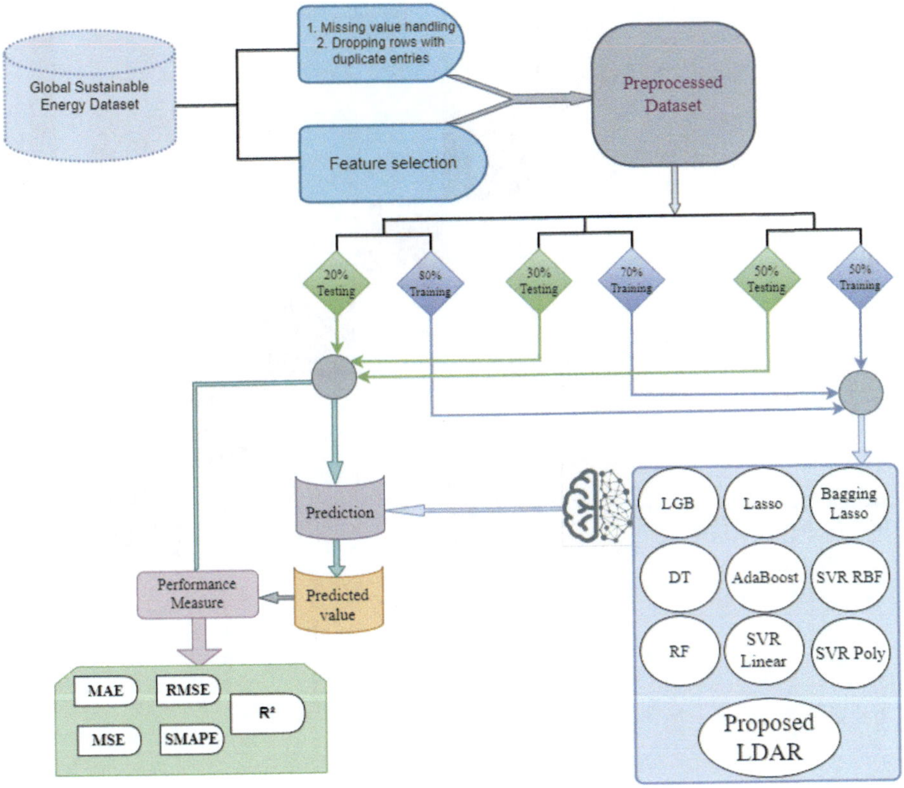

Fig. 1 Overview of proposed methodology

3.2 Description of Dataset and Variables

The dataset is taken from the prominent online platform Kaggle. The variables on global primary energy consumption, electricity generation, greenhouse gas emissions, etc. of 186 countries from 2000 to 2020 are provided in the dataset. We implement the following procedure to process the data. Firstly, we select a target variable illustrating energy consumption trends, which is "Primary energy consumption per capita (KWH/Person)" (PECPC). We carefully choose relevant features, including attributes that could affect energy consumption. Then, we handle missing values through imputation and use feature scaling to verify that feature magnitudes are consistent. The feature can be seen in Table 1. In addition, we present a heatmap depicting the correlation of all the features and PECPC, where blue color intensity defined the worst correlation and red color intensity responded to the strong correlation. On top of that, PECPC and GPC show a good correlation acquiring a score of 0.67. The heatmap of the correlation is in Fig. 2.

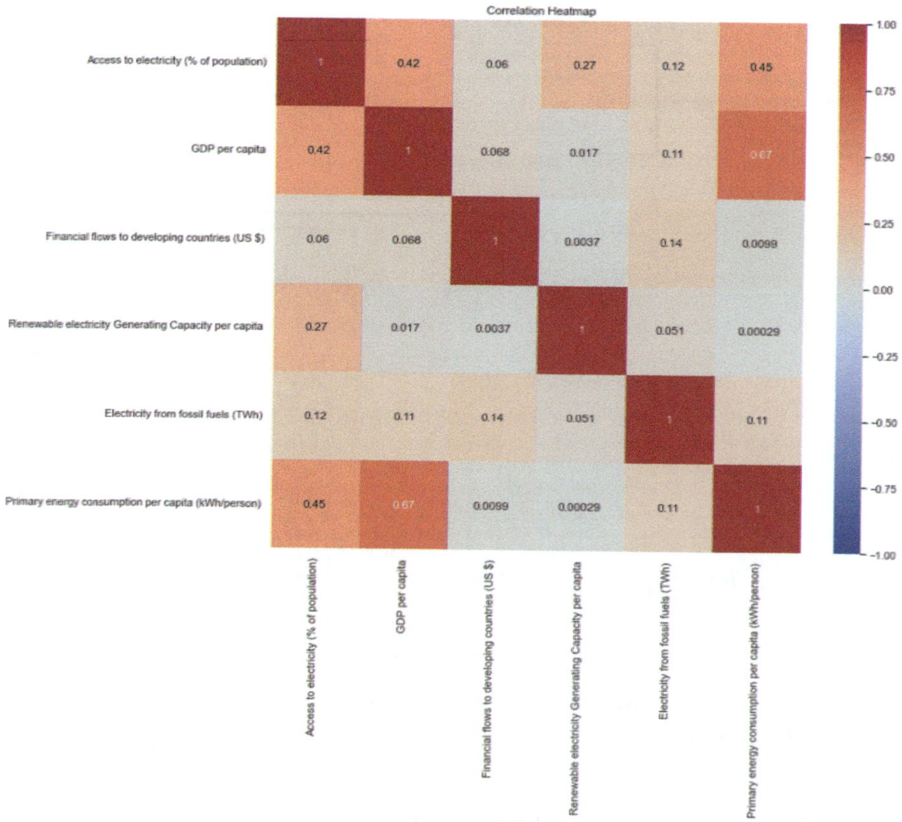

Fig. 2 Heat map to represent the correlation of the variables

Table 1 Description of the variables

Variables	Abbreviation
Access to electricity (% of population)	AE
GDP per capita	GPC
Financial flows to developing countries (US $)	FFDC
Renewable electricity Generating Capacity per capita	REGCPC
Electricity from fossil fuels (TWh)	EFF
Primary energy consumption per capita (kWh/person)	PECPC

3.3 *Machine Learning Algorithms*

3.3.1 Random Forest

Random forest is a prediction algorithm based on a combination of multiple decision trees. Random forests are widely utilized because they require only one or two tuning parameters and may be applied directly to high-dimensional situations. They also provide a built-in generalization error estimation and are reasonably quick to train and forecast (Abedin et al., 2021). Simplified formula for regression is

$$\hat{y} = \frac{1}{N} \sum_{i=1}^{N} \hat{y}_i$$

where $\hat{y}$ is the final prediction, N is the number of trees in the forest, and $\hat{y}_i$ is the prediction from the i-th tree. The algorithm works by constructing trees. For each tree in the forest, a random subset of the input features is selected, and the tree is grown to its maximum depth except pruning with selected features to maximize information gain at each split. After then, each tree in the forest makes a prediction. By averaging the individual tree predictions, it make final prediction (Ali et al., 2012; Abedin et al., 2021).

3.3.2 Decision Tree

Decision trees are commonly used supervised machine learning technique which has been used for both regression and classification problems. In a decision tree, there are basically two types of nodes: decision nodes and leaf nodes. Decision node generates decision and contains multiple branches, whereas leaf nodes are the output of those decisions, and they do not contain any branches. The algorithm divides the inputs recursively into smaller sections. The root node contains the whole dataset. There are some techniques that use variables like mean squared error, entropy (information gain), and Gini impurity to select the best characteristic to split the data by at each node. The Gini impurity for a set S with c classes is given by

$$Gini(S) = 1 - \sum_{i=1}^{c} p_i^2$$

Here, p_i is the proportion of instances in class i (Hasan et al., 2023b). After calculating the Gini impurity for every possible split point, the next step is choosing the best split, which results in the lowest weighted Gini impurity. Then, split the dataset into two subsets, and repeat the whole process recursively. When it meets stopping criteria, labels have been assigned to the leaf nodes. That is how decision algorithm works.

3.3.3 LASSO

LASSO is a statistical formula whose main purpose is feature selection and regularization of the data model. This regression analysis technique improves the statistical regression model's interpretability and prediction accuracy, which also includes variable selection and parameter estimation (Sajid et al., 2023). LASSO is ideal for prediction and feature selection. The following formula defines the LASSO regression model:

$$\hat{\beta} = \arg\min_{\beta} \left\{ \frac{1}{2n} \sum_{i=1}^{n} \left(y_i - \sum_{j=1}^{p} \beta_j x_{ij} \right)^2 + \lambda \sum_{j=1}^{p} |\beta_j| \right\}$$

It does parameter estimation and variable selection at the same time in this manner (Vidaurre et al., 2011).

$$\frac{1}{2n} \sum_{i=1}^{n} \left(y_i - \sum_{j=1}^{p} \beta_j x_{ij} \right)^2$$

By computing the residual sum of squares, the above term assesses how well the model fits the data. $\lambda \sum_{j=1}^{p} |\beta_j|$ This regularization factor, called L1-norm, adds a penalty based on the sum of the absolute values of the coefficients. As λ increases, more coefficients are shrunk to zero, performing variable selection.

3.3.4 Bagging Lasso (BL)

Bagging Lasso formula is made up of the principle of bootstrap aggregation and LASSO regression to increase the model's stability and accuracy of prediction.

Firstly, from the original dataset, multiple bootstraps have been generated. Then LASSO regression is applied to each generated bootstrap sample.

$$\hat{\beta}^{(b)} = \arg\min_{\beta} \left\{ \frac{1}{2n_b} \sum_{i=1}^{n_b} \left(y_i^{(b)} - \sum_{j=1}^{p} \beta_j x_{ij}^{(b)} \right)^2 + \lambda \sum_{j=1}^{p} |\beta_j| \right\}$$

Here, $D_b^* = \{(x_i^{(b)}, y_i^{(b)})\}_{i=1}^{n_b}$ represents the b-th bootstrap sample . After that, it estimates aggregated coefficient from all bootstrap samples

$$\hat{\beta}_{\text{Bagging Lasso}} = \frac{1}{B} \sum_{b=1}^{B} \hat{\beta}^{(b)}$$

Making average the results of LASSO regressions from multiple bootstrap samples, Bagging Lasso eliminates the variance of the model and makes it more stable (Bach, 2008).

3.3.5 LGB

LGB is a gradient boosting ensemble method based on decision trees. LGB can be used for regression. The formula for LGB is

$$L(\theta) = \sum_{i=1}^{n} l(y_i, f(x_i; \theta)) + \Omega(f)$$

LGB algorithm has initialized the model with a constant value to calculate the initial gradients and hessians. While iterating for each tree, it does Gradient-based One-Side Sampling (GOSS), feature bundling, histogram construction, split finding and continues the decision trees growth. After maximum depth has reached and stopping criteria has met, update the model by adding the newly trained tree and gradients hessians. Lastly, it combines the outputs of all individual trees to predict (Ke et al., 2017).

3.3.6 AdaBoost

AdaBoost is a self-adaptive boosting technique that creates a set of multiple classifiers to improve the performance of weak classifiers. Several concerns have been raised since it adjusts dynamically to the error rate of the fundamental algorithm during training by adjusting the weight of each sample. The most basic theoretical property of AdaBoost concerns its ability to reduce training error (Hasan et al., 2023c).

$$H(x) = \text{sign}\left(\sum_{t=1}^{T} \alpha_t \cdot h_t(x)\right)$$

This algorithm trains weak classifier h_t on the weighted training data. After calculating the weighted error of h_t, it calculates the weight of α_t of the weak classifier. Finally it updates the weights of training instances and normalizes the weights to make the sum to 1. So this is the final classifier, and $H(x)$ is a weighted majority vote of the T weak classifier (Wu & Zhao, 2011).

3.3.7 Support Vector Regression (SVR) Linear

Regression tasks are handled by SVR, a subset of SVMs. For a given input value, it looks for a function that best predicts the continuous output value. In order to determine which linear hyperplane best fits the data, SVR Linear employs a linear kernel function. The primary distinctions between SVR and SVR Linear are in how those two implementations handle intercept regularization and the default loss function. SVR linear algorithms set a linear relationship between the target variable and the input characteristics. The main formula of SVR Linear is

$$\min_{w,b,\xi,\xi^*} \frac{1}{2}\|w\|^2 + C\sum_{i=1}^{n}(\xi_i + \xi_i^*)$$

Firstly, the algorithm initials the value to weights, bias, and slack variables. After then, for optimizing the problem, it defines the objective function to minimize the loss, including the linear constraints. By solving the optimization problem and iteratively update the weights w, bias b and slack variable ξ_i and ξ_i^* while constraints are being satisfied. Using the optimized weights and bias, it predicts for new data point (Klopfenstein & Vaiter, 2019).

3.3.8 SVR Radial Basis Function (RBF)

SVR with RBF kernel is a machine learning algorithm which is often used for its ability to handle nonlinear relationships. The RBF kernel is a function that depends on the distance from a point. The main formula of RBF kernel function is defined as

$$K(x, x') = exp(-\gamma||x - x'||^2)$$

Here , $||x - x'||$ denotes the distance between x and x' and γ parameter controls the spread of kernel. SVR finds a line or curve that fits data points as closely as possible as if it does not cross a certain error margin which is called epsilon. There are three parameters, epsilon, regularization parameter, and gamma (kernel spread), which are the key factors of the model's performance and complexity. Then it forecasts additional data points using the optimized weights and bias from the SVR algorithm.

3.3.9 SVR Poly

SVR with a polynomial (poly) kernel is another variant of SVR which uses a polynomial function instead of RBF for mapping data. Here, the polynomial kernel function transforms the input data into higher dimensional space. The polynomial kernel takes the form

$$K(x, x') = (1 + x \cdot x')^d$$

where d is the degree of the polynomial. To find a regression function $f(x) = w\phi(x) + b$ which best fits the training data is the main goal, where $\phi(x)$ is the nonlinear mapping to the higher dimensional space, w is the weight vector, and b is the bias term (Rabbi et al., 2022; Bargam et al., 2024). SVR Poly has an ϵ loss function to avoid overfitting which boosts performance with noisy and sparse data (Hasan et al., 2023a).

3.4 Proposed Blending LDAR

Four ML algorithms, LASSO, DT, AdaBoost, and RF, are used to create a blending ensemble learning model in this study (Hasan et al., 2024). We refer to this model as the LDAR regression model. While blending shares similarities with the stacking ensemble process, it possesses distinctive advantages. For instance, while stacking leverages out-of-fold predictions to train subsequent layers in the meta-model, blending uses a small validation set 0 for the same purpose. LDAR integrates the mapping functions acquired from its member algorithms, as detailed in the workflow presented in Fig. 3.

The motivation to employ an ensemble model over a singular model is rooted in the belief that ensemble models generally predict with greater accuracy and offer superior performance compared to individual ML models. Additionally, ensembles help decrease the spread of predictions, enhancing model reliability. The mapping functions from member algorithms merge to provide enhanced predictive capabilities. Our proposed ensemble model incorporates various methods to leverage the strengths of each algorithm:

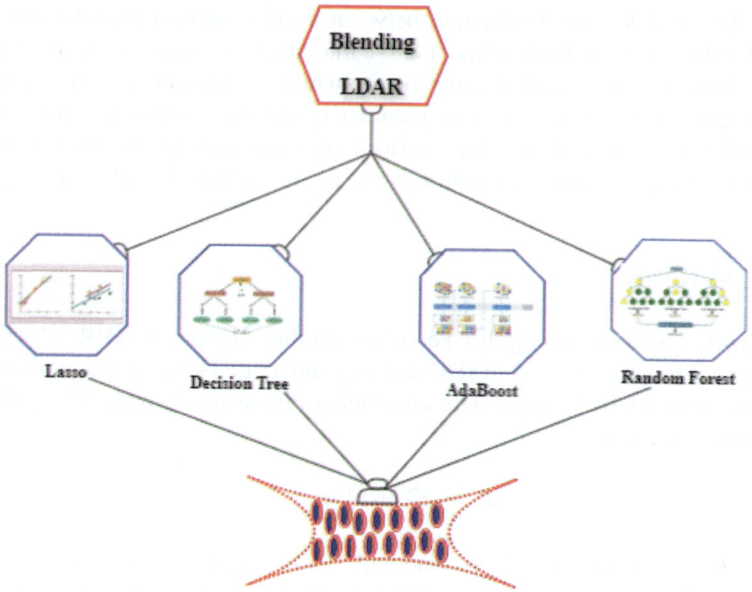

Fig. 3 Block diagram of the proposed blending LDAR ensemble learning model

LASSO regression provides feature selection and regularization, reducing overfitting and enhancing model interpretability. DT captures nonlinear relationships and interactions between features. AdaBoost regression boosts weak learners by focusing on the errors of previous models, improving overall performance, and finally RF aggregates multiple decision trees to reduce variance and improve generalization. The LDAR blending ensemble learning model combines these diverse approaches to create a robust and accurate predictive model, taking advantage of each algorithm's strengths while mitigating their individual weaknesses.

3.5 *Performance Measure Metrics*

We use four different performance measure techniques in this study, and the descriptions are given below.

MAE: MAE measures the average magnitude of errors between predicted and actual values without considering their direction. It is calculated as the average of the absolute differences between predicted and actual values. MAE is simple to understand and provides a straightforward interpretation of model accuracy, where lower values indicate better performance.

$$MAE = \frac{1}{n}\sum_{i=1}^{n}|y_i - \hat{y}_i|$$

MSE: MSE quantifies the average squared difference between predicted and actual values. It emphasizes larger errors more than MAE due to squaring the differences, which can be useful for identifying significant outliers. MSE is calculated by averaging the squared differences between predicted and actual values. Lower MSE values indicate better model performance.

$$MSE = \frac{1}{n}\sum_{i=1}^{n}(y_i - \hat{y}_i)^2$$

RMSE: RMSE is the square root of the MSE and provides an error metric on the same scale as the data. It measures the standard deviation of prediction errors, offering a clear view of model accuracy by penalizing larger errors more heavily. Lower RMSE values indicate better predictive performance, making it a widely used metric in regression analysis.

$$RMSE = \sqrt{MSE} = \sqrt{\frac{1}{n}\sum_{i=1}^{n}(y_i - \hat{y}_i)^2}$$

SMAPE: SMAPE measures the accuracy of predictions by calculating the percentage difference between predicted and actual values. It is symmetric and considers both the relative error and scale of the data. SMAPE is particularly useful for comparing errors across datasets of different scales, with lower values indicating better model accuracy.

$$SMAPE = \frac{100\%}{n}\sum_{i=1}^{n}\frac{|y_i - \hat{y}_i|}{\frac{|y_i|+|\hat{y}_i|}{2}}$$

R-Squared: R-Squared, or the coefficient of determination, indicates the proportion of variance in the dependent variable that is predictable from the independent variables. It ranges from 0 to 1, with higher values representing better model fit. An R^2 value of 1 indicates perfect prediction, while 0 indicates no predictive power. It is a key metric for evaluating the explanatory power of regression models.

$$R^2 = 1 - \frac{SS_{\text{res}}}{SS_{\text{tot}}}$$

where SS_{res} is the sum of squares of residuals and SS_{tot} is the total sum of squares.

4 Result Analysis

4.1 Hyperparameter

The performance of ML algorithms depends on the quality of data and the learning process. The optimized model with the best hyperparameter values performs better than any ML model. In our work, we employ the grid search procedure to find the optimal values of the ML algorithms. The values of the selected hyperparameters are in Table 2.

4.2 Performance of ML Models

We split the dataset into three distinct ratios: 20:80, 30:70, and 50:50 testing and training ratios. The performances of those categories are shown in tables and graphs below:

4.2.1 Performance of the ML Algorithms to Predict Primary Energy Consumption in 20:80 Testing and Training Ratio

We employ several ML algorithms on a standard 80:20 train-test split of the data. This results in varying error rates for each algorithm and R^2 score that is tabulated in Table 3. In the table, it is clear that our proposed Blending LDAR performs better than other ML models. The proposed LDAR achieves 90% R^2 with 0.0177 MAE, 0.0016 MSE, 0.0403 RMSE, and 19.7075 SMAPE. The performance of

Table 2 Values of the hyperparameters of different ML algorithms

Algorithm	Parameter with value
LGB	learning_rate: 0.1
DT	random_state: 0
AdaBoost	base_estimator: none, learning_rate: 1.0, n_estimators: 50, random_state: none
RF	n_estimators: 1, random_state: 0
LASSO	alpha: 0.01
Bagging Lasso	alpha: 0.01
SVR Linear	kernel: linear, C: 100, gamma: auto
SVR RBF	kernel: rbf, C: 100, gamma: 0.1, epsilon: 0.1
SVR Poly	kernel: poly, C: 100, gamma: auto, degree: 3, epsilon: 0.1, coef0: 1
Blending LDAR	learning_rate: 0.1, random_state: 0, base_estimator: none, learning_rate: 1.0, n_estimators: 50, random_state: none and n_estimators: 1, random_state: 0

Table 3 Primary energy consumption per person in 20% testing and 80% training

Model	MAE	MSE	RMSE	SMAPE	R^2
LGB	0.0224	0.0019	0.0441	31.6384	0.88
DT	0.0213	0.0040	0.0635	19.6023	0.76
AdaBoost	0.0437	0.0044	0.0669	59.3734	0.73
RF	0.0235	0.0037	0.0613	26.307	0.78
LASSO	0.0707	0.0127	0.1128	86.9742	0.24
Bagging Lasso	0.0704	0.0126	0.1125	86.8033	0.25
SVR Linear	0.0713	0.0101	0.1008	91.4090	0.40
SVR RBF	0.0585	0.0078	0.0885	89.4069	0.53
SVR Poly	0.0584	0.0076	0.0875	86.1174	0.54
Blending LDAR	0.0177	0.0016	0.0403	19.7075	0.90

Table 4 Primary energy consumption per person in 30% testing and 70% training

Model	MAE	MSE	RMSE	SMAPE	R^2
LGB	0.0230	0.0021	0.0463	31.0907	0.88
DT	0.0214	0.0037	0.0612	20.6243	0.78
AdaBoost	0.0402	0.0042	0.0653	54.0534	0.75
RF	0.0226	0.0035	0.0594	26.1839	0.79
LASSO	0.0703	0.0130	0.1142	87.7059	0.24
Bagging Lasso	0.0706	0.0131	0.1144	87.9104	0.24
SVR Linear	0.0716	0.0105	0.1026	92.4498	0.39
SVR RBF	0.0576	0.0081	0.0895	89.7533	0.53
SVR Poly	0.0576	0.0078	0.0885	86.5198	0.54
Blending LDAR	0.0191	0.0019	0.0446	20.6817	0.88

LGB is nearer to the LDAR, and it shows 88% R^2. The Decision Tree, AdaBoost, and Random Forest algorithms performed moderately well in our study. They obtained R^2 values of 76%, 73%, and 78%, respectively. The RMSE error of the algorithms is 0.0635, 0.0669, and 0.0613, respectively. However, the performance of LASSO, Bagging Lasso (BLASSO), SVR Linear, SVR RBF, and SVR Poly is not satisfying, and it shows only 24%, 25%, 40%, 53%, and 54% R^2, respectively. The error rate of those algorithms is also high compared to the proposed LDAR and LGB. That numerical comparison demonstrates the superiority of our proposed Blending LDAR. The obtained errors of different algorithms are visualized in Fig. 4. The results in this figure demonstrate that our proposed LDAR model achieves superior performance compared to other algorithms across various errors and scoring metrics.

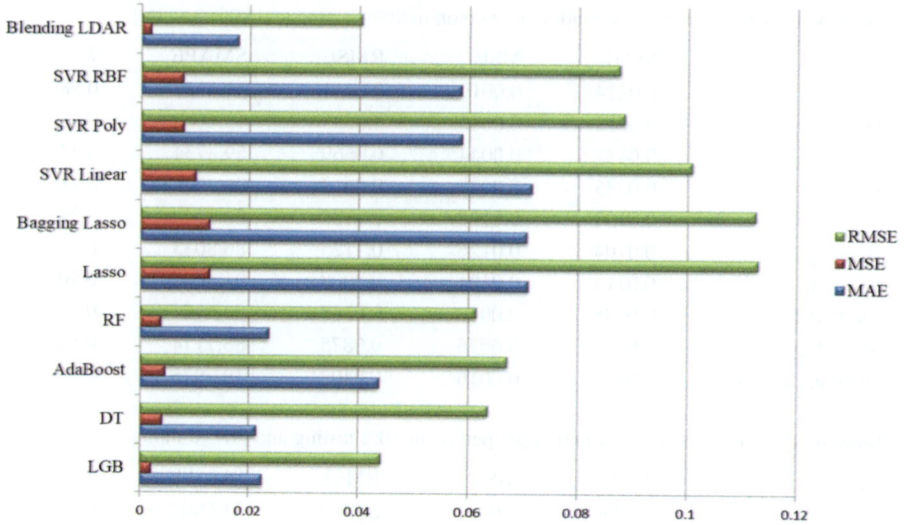

Fig. 4 Performance of the models in primary energy consumption in 80:20 train-test ratio

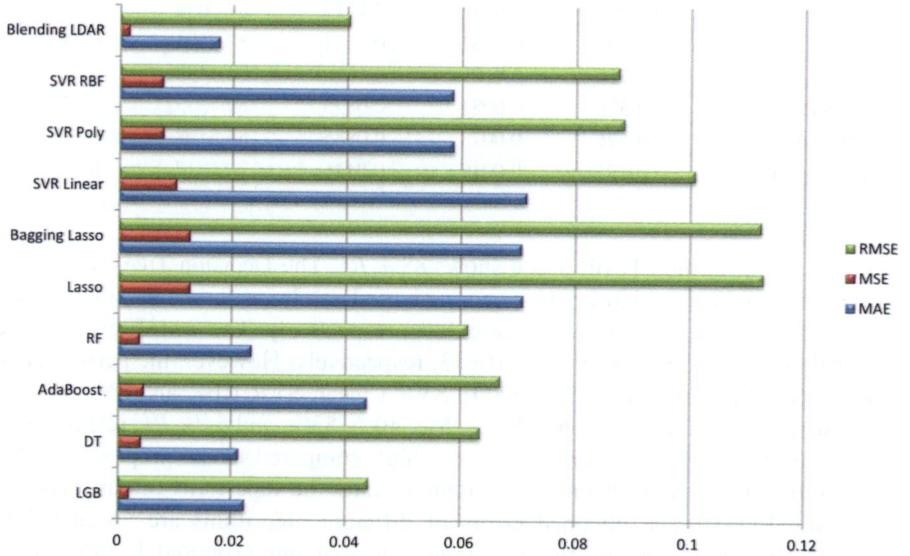

Fig. 5 Performance of the models in primary energy consumption in 70:30 train-test ratio

4.2.2 Performance of the ML Algorithms to Predict Primary Energy Consumption in 30:70 Testing and Training Ratio

Using a conventional train-test split, which entails a 30:70 testing and training ratio, we utilize those ML algorithms, yielding varying errors and R^2 scores, which are compiled in Table 4. The table indicates that both the proposed LDAR and LGB models achieve similar performance, with an R^2 value of 88%. In Fig. 7, DT, ADB, and RF show closer results, scoring 76%, 73%, and 78% in R^2, respectively. The error rate of LDAR and LGB is comparatively lower than other ML algorithms. Figure 5 shows that the proposed LDAR beats competing algorithms regarding error and score metrics. The MAE error of LGB, DT, and RF is 0.023, 0.0214, and 0.0226, respectively, where LDAR performs better achieving a score of 0.1908. Measuring SMAPE, the error rate of AdaBoost and LDAR achieves nearer scores of 20.6243 and 20.6817, respectively.

4.2.3 Performance of the ML Algorithms to Predict Primary Energy Consumption in 50:50 Testing and Training Ratio

To compare the performance of ML algorithms, we divided the data into training and testing sets using the typical 50:50 ratio. We then applied these algorithms to the training data to determine their error rates and R^2 values, which are shown in Table 5. We can see that our proposed LDAR model achieves 88% R^2 where LGB and DT perform nearer scores of 87% and 86% R^2, respectively, depicted in Fig. 7. DT shows better performance with a score of 0.0198 which is lower than LDARs of score 0.0202 in measuring MAE. The MSE error of LGB and DT is very near of scores 0.0022 and 0.0025, respectively, but LDAR performs better with a score of 0.0021. However, the performances of LASSO, BL, SVR Linear, SVR RBF, and SVR Poly do not show any satisfactory scores. The numerical comparisons shown in Fig. 6 demonstrate that the performance of our Blending LDAR is significant.

Table 5 Primary energy consumption per person in 50% testing and 50% training

Model	MAE	MSE	RMSE	SMAPE	R^2
LGB	0.0244	0.0022	0.0476	34.2486	0.87
DT	0.0198	0.0025	0.0502	22.7411	0.86
AdaBoost	0.0422	0.0043	0.0661	62.0623	0.76
RF	0.0277	0.0048	0.0693	28.2827	0.73
LASSO	0.0718	0.0136	0.1167	89.9741	0.23
Bagging Lasso	0.0722	0.0137	0.1173	90.2193	0.23
SVR Linear	0.0718	0.0107	0.1034	93.5718	0.40
SVR RBF	0.0578	0.0082	0.0909	89.4069	0.53
SVR Poly	0.0584	0.0076	0.0875	91.5027	0.54
Blending LDAR	0.0202	0.0021	0.0457	22.8968	0.88

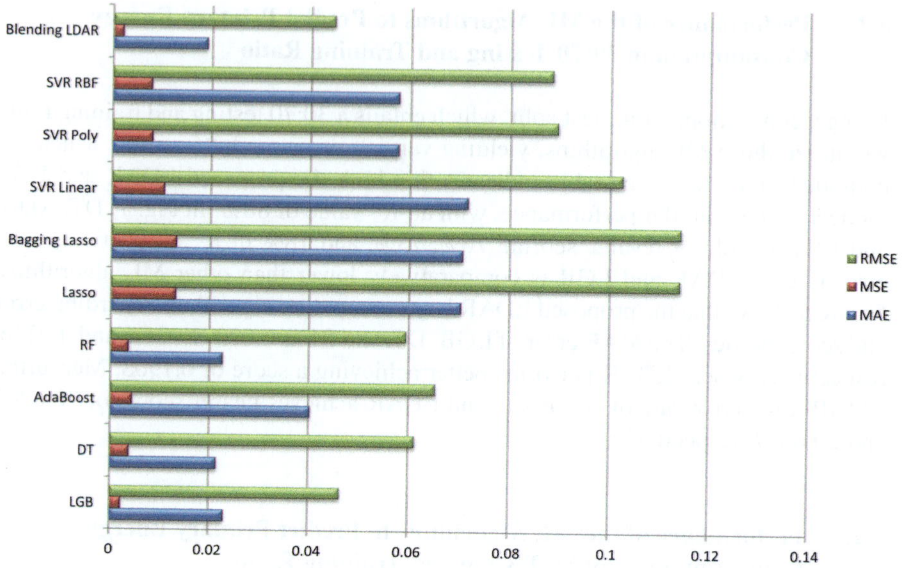

Fig. 6 Performance of the models in primary energy consumption in 50:50 train-test ratio

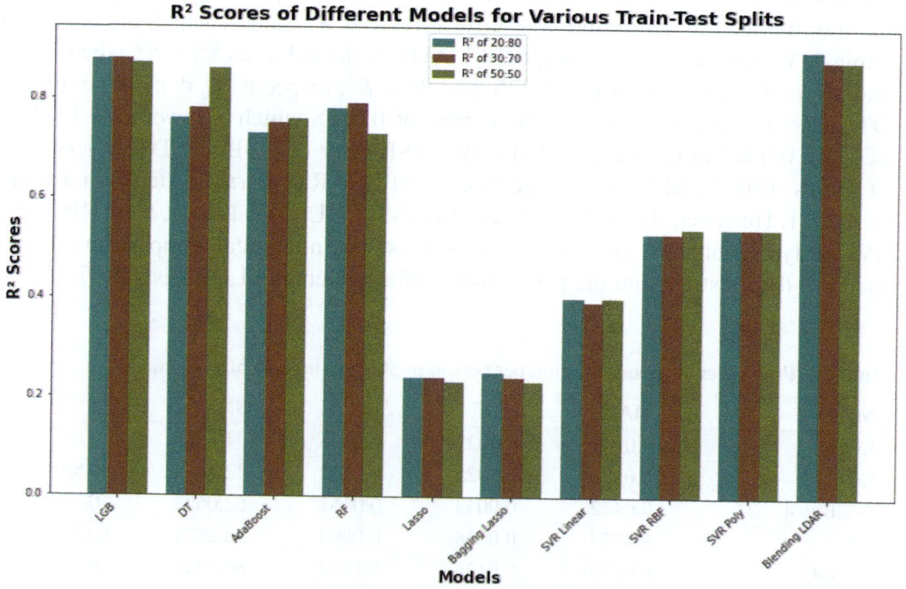

Fig. 7 R^2 scores for the ML models in different training and testing ratios

4.3 Discussion

Our proposed model consists of four ML algorithms, namely LGB, DT, ADB, and RF. As we split our datasets into three categories based on training and testing, LDAR performs better than any other model in primary energy consumption forecasting. In 20:80 testing and training, LDAR shows 90% R^2. Additionally, it illustrates 88% R^2 scores in both 30:70 and 50:50 testing and training of primary energy consumption per person (Fig. 7). This model outperforms in measuring other evaluation metrics including MAE, MSE, RMSE, and SMAPE. Notably, the value of errors in LDAR was comparatively lower than other algorithms, which means it has the highest accuracy. The comparison with other research makes it abundantly evident that our suggested blending LDAR model is superior to the other options. The main focus of our proposed blending LDAR model is to predict the primary energy consumption better than other existing models. Primary energy consumption forecasting is crucial for policymaking and strategic planning, investment and infrastructure planning, market dynamics and economic growth, environmental and climate impact, consumer behavior and education, and other sectors. On top of everything else, the high accuracy and lower error rate of our proposed model are compelling and potentially valuable for the stakeholders and policymakers in making future decisions. To some extent, it will play a great role in guiding investments in energy infrastructure, supporting sustainable development goals, and evaluating climate change mitigation strategies. As we can suggest this prediction model can play an inevitable role in primary energy consumption.

5 Conclusion and Future Work

The aim of this research is to design an ML-based methodology for primary energy consumption prediction. You have designed and described the blended ensemble learning model that combines five ML regression techniques in this study. The results revealed that the blending LDAR model significantly improved forecasting accuracy compared to established methods used in previous studies, as measured by various error criteria. Our findings have far-reaching consequences, including possible applications in energy planning, policymaking, and climate change mitigation. Our model can assist in shifting to low-carbon energy by offering more precise and trustworthy primary energy consumption forecasts, enabling better decision-making.

Future research will focus on integrating our model with other energy system models, including the incorporation of new features and data sources, as well as the development of more advanced ML techniques like deep learning.

Acknowledgments We extend our deepest gratitude to the Center for Multidisciplinary Research and Development (CeMRD) for their resources, guidance, and support. The expertise and encouragement from CeMRD members have been invaluable. This research was made possible by the collaborative environment and cutting-edge facilities at CeMRD.

References

Abedin, M. Z., Moon, M. H., Hassan, M. K. & Hajek, P. (2021). Deep learning-based exchange rate prediction during the covid-19 pandemic. *Annals of Operations Research*, 1–52.

Ahmad, T. et al. (2021). Artificial intelligence in sustainable energy industry: Status quo, challenges and opportunities. *Journal of Cleaner Production, 289*, 125834.

Ali, J., Khan, R., Ahmad, N. & Maqsood, I. (2012). Random forests and decision trees. *International Journal of Computer Science Issues (IJCSI), 9*(5), 272.

Alola, A. A., Ozturk, I. & Bekun, F. V. (2021). Is clean energy prosperity and technological innovation rapidly mitigating sustainable energy-development deficit in selected sub-Saharan Africa? A myth or reality. *Energy Policy, 158*, 112520.

Amiri, S. S., Mueller, M. & Hoque, S. (2023). Investigating the application of a commercial and residential energy consumption prediction model for urban planning scenarios with machine learning and Shapley additive explanation methods. *Energy and Buildings, 287*, 112965.

Aydin, G. & Karakurt, I. (2023). *Introduction to fossil fuels*. Elsevier. https://www.sciencedirect.com/science/article/pii/B9780323939409000414.

Bach, F. (2008). Bolasso: Model consistent lasso estimation through the bootstrap. In *Proceedings of the 25th International Conference on Machine Learning* (pp. 33–40)

Bargam, B. et al. (2024). Evaluation of the support vector regression (SVR) and the random forest (RF) models accuracy for streamflow prediction under a data-scarce basin in morocco. *Discover Applied Sciences, 6*, 306.

Cheekatamarla, P., Gluesenkamp, K., Kowalski, S., Li, Z. & Jajja, S. (2024). *Global primary energy sources and their carbon intensity* (pp. 11–17). Springer.

Demiral, M. & Demiral, Ö. (2023). Socio-economic productive capacities and energy efficiency: global evidence by income level and resource dependence. *Environmental Science and Pollution Research, 30*, 42766–42790.

Ekonomou, L. (2010). Greek long-term energy consumption prediction using artificial neural networks. *Energy, 35*, 512–517.

Gyamfi, B. A., Adebayo, T. S., Agozie, D. Q., Bekun, F. V. & Koy, A. (2024). Is sustainable energy consumption, technological advancement and urbanization fast addressing south Asia's green energy expansion deficits? *Environment, Development and Sustainability*, 1–30.

Hasan, M., Hassan, M. M., Faisal-E-Alam, M. & Akter, N. (2023a). *Empirical analysis of regression techniques to predict the cybersecurity salary* (pp. 65–84). Routledge.

Hasan, M. et al. (2023b). Ensemble machine learning-based recommendation system for effective prediction of suitable agricultural crop cultivation. *Frontiers in Plant Science, 14*, 1234555.

Hasan, M., Das, U., Datta, R. K. & Abedin, M. Z. (2023c). *Model development for predicting the crude oil price: Comparative evaluation of ensemble and machine learning methods* (pp. 167–179). Springer.

Hasan, M. et al. (2024) A blending ensemble learning model for crude oil price forecasting. *Annals of Operations Research*, 1–31.

He, Q. & Hao, Y. (2024). Applications of fractional time delayed grey model in primary energy consumption prediction. *Journal of Energy Research and Reviews, 16*, 30–46.

Kaytez, F. (2020). A hybrid approach based on autoregressive integrated moving average and least-square support vector machine for long-term forecasting of net electricity consumption. *Energy, 197*, 117200.

Ke, G. et al. (2017) LightGBM: A highly efficient gradient boosting decision tree. In *NIPS'17: Proceedings of the 31st International Conference on Neural Information Processing Systems* (pp. 3149-3157).

Klopfenstein, Q. & Vaiter, S. (2019) Linear support vector regression with linear constraints. *Machine Learning, 110*, 1939–1974.

Komarova, A., Filimonova, I. & Kartashevich, A. (2022). Energy consumption of the countries in the context of economic development and energy transition. *Energy Reports, 8*, 683–690.

Li, Y. (2019). Prediction of energy consumption: Variable regression or time series? A case in china. *Energy Science & Engineering, 7*, 2510–2518.

Li, Y. & Solaymani, S. (2021). Energy consumption, technology innovation and economic growth nexuses in Malaysian. *Energy, 232*, 121040.

Liu, L., Tam, V. W., Le, K. N. & Almeida, L. (2023). Life cycle energy consumption prediction based on an extended system boundary with the Bi-LSTM model: An empirical study of china. *Energy and Buildings, 298*, 113497.

Martínez, D. M., Ebenhack, B. W. & Wagner, T. P. (2019). Chapter 1 - introductory concepts. In D. M. Martínez, B. W. Ebenhack, & T. P. Wagner (Eds.) *Energy efficiency* (pp. 1–33). Elsevier. https://www.sciencedirect.com/science/article/pii/B9780128121115000019.

Mosavi, A. & Bahmani, A. (2019). Energy consumption prediction using machine learning; a review. Preprints. [Working Paper] (Unpublished).

Rabbi, M. F., Moon, M. H., Dhonno, F. T., Sultana, A. & Abedin, M. Z. (2022). *Foreign currency exchange rate prediction using long short-term memory, support vector regression and random forest regression* (pp. 251–267). Springer.

Sajid, S. W., Hasan, M., Rabbi, M. F. & Abedin, M. Z. (2023). *An ensemble LGBM (light gradient boosting machine) approach for crude oil price prediction* (pp. 153–165). Springer.

Shinwari, R., Wang, Y., Gozgor, G. & Mousavi, M. (2024). Does FDI affect energy consumption in the belt and road initiative economies? The role of green technologies. *Energy Economics, 132*, 107409.

Sözen, A. & Arcaklioglu, E. (2007). Prediction of net energy consumption based on economic indicators (GNP and GDP) in turkey. *Energy Policy, 35*, 4981–4992.

Vidaurre, D., Bielza, C. & Larrañaga, P. (2011). Lazy lasso for local regression. *Computational Statistics, 27*, 531–550.

Wang, X. (2022). Research on the prediction of per capita coal consumption based on the ARIMA–BP combined model. *Energy Reports, 8*, 285–294.

Wang, H. & Zhang, Z. (2023). Forecasting per capita energy consumption in china using a spatial discrete grey prediction model. *Systems, 11*, 285.

Wei, W. & He, L.-Y. (2017). China building energy consumption: definitions and measures from an operational perspective. *Energies, 10*, 582.

Wen, L., Guang, F. & Sharp, B. (2021). Dynamics in Aotearoa New Zealand's energy consumption between 2006/2007 and 2012/2013. *Energy, 225*, 120186.

Wu, P. & Zhao, H. (2011). Some analysis and research of the AdaBoost algorithm. In *International Conference on Intelligent Computing and Information Science* (pp. 1–5).

Zhang, C., Luo, Z., Rezgui, Y. & Zhao, T. (2024). Enhancing multi-scenario data-driven energy consumption prediction in campus buildings by selecting appropriate inputs and improving algorithms with attention mechanisms. *Energy and Buildings, 311*, 114133.

Zhou, K. & Yang, S. (2016). Understanding household energy consumption behavior: The contribution of energy big data analytics. *Renewable and Sustainable Energy Reviews, 56*, 810–819.

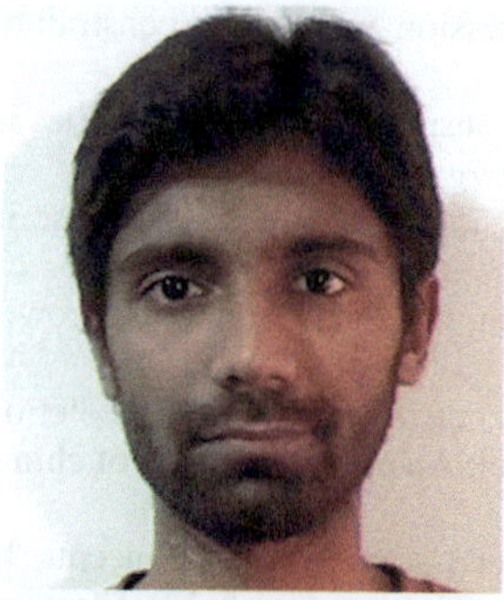

Abdullah Haque is currently pursuing MS in Mathematics from Department of Mathematics at Hajee Mohammad Danesh Science and Technology University (HSTU), Dinajpur, Bangladesh. He completed BSc (Honors) in Mathematics in 2023 from the same institution. Currently, he is a Research Assistant in the Center for Multidisciplinary Research and Development (CeMRD). His research interests include Machine Learning, Deep Learning, Cyber Security, Business Intelligence, Nonlinear Partial Differential Equations, and Computational Sociology.

Tuhin Chowdhury is a software engineer at a multinational company in Dhaka. He specializes in backend development, research, and mobile app development and has successfully led a small team. Tuhin holds a BSc degree in Computer Science and Engineering from the SEC Engineering Faculty of Shahjalal University of Science and Technology (SUST), graduating in 2021. He also served as a campus director for the Hult Prize at SEC. Tuhin loves conducting research and has a keen interest in Image Processing, Computer Vision, IoT, Photogrammetry, Business Intelligence, and NLP.

Mahmudul Hasan is currently pursuing a PhD in Information Technology (IT) at Deakin University, Melbourne, Australia. He earned his BSc (Eng.) and MSc (Eng.) degrees in Computer Science and Engineering (CSE) from Hajee Mohammad Danesh Science and Technology University, Dinajpur, Bangladesh, in 2021 and 2023, respectively. He previously served as a lecturer in the Department of CSE at the University of Creative Technology, Chittagong (UCTC), Bangladesh. He is the Founder and Director of the Center for Multidisciplinary Research and Development (CeMRD) and a moderator of "Be Researcher BD," the largest online research forum in Bangladesh. Additionally, he has taught online as a Data Science instructor to students in the USA, Italy, Denmark, South Korea, and Australia. He is also the founder of the online educational platform "MHM Academy." His research interests include federated learning, machine learning, deep learning, cybersecurity, health informatics, renewable energy, computational sociology, and business intelligence.

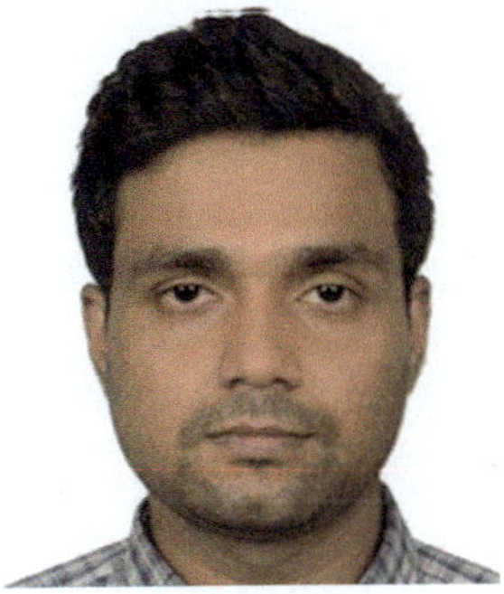

Md. Jahid Hasan is currently pursuing a PhD in Business Information Systems at RMIT University in Melbourne, Victoria, Australia. His Ph.D. program is a collaboration between RMIT University and carsales, where he also works as a PhD Candidate on the Data Science and Machine Learning team. Jahid completed his BSc in Electrical and Electronic Engineering at Hajee Mohammad Danesh Science and Technology University in Dinajpur, Bangladesh, in 2022. His research interests include image processing, machine learning, deep learning, computer vision, IoT, and business intelligence. He is especially focused on developing computer vision models and integrating Large Language Models (LLMs), driven by his passion for innovation and his commitment to bridging the gap between academic research and practical industry applications.

Analyzing Biogas Production in Livestock Farms Using Explainable Machine Learning

Md. Mahedi Hassan, Mahira Shamim, Mahmudul Hasan, Md Amir Hamja, Kanij Fatema, and Sudipto Roy Pritom

1 Introduction

Due to waste generated from both domestic and industrial activities, developed and emerging nations are increasingly seeking alternative energy sources. Nowadays, most of the global primary energy supply is derived from fossil fuels. However, the environmental harm caused by fossil fuels and the depletion of natural resources have shifted public focus toward renewable energy sources to ensure a sustainable future for energy production. In recent years, interest in biogas as a viable energy source has grown, primarily due to its potential to reduce greenhouse gas emissions. Biogas production from anaerobic digestion (AD) processes depends on parameters such as retention time, pH, medium composition, temperature inside the digester tank, working pressure, and volatile fatty acids (González-Fernández et al., 2019). Machine learning (ML) has emerged as a powerful method for studying models to

M. M. Hassan
Department of Computer Science and Engineering, World University of Bangladesh, Dhaka, Bangladesh

M. Shamim
Department of Finance, University of Chittagong, Chittagong, Bangladesh

M. Hasan (✉) · K. Fatema
Department of Computer Science and Engineering, Hajee Mohammad Danesh Science and Technology University, Dinajpur, Bangladesh

M. A. Hamja
Department of Statistics, Hajee Mohammad Danesh Science and Technology University, Dinajpur, Bangladesh

S. R. Pritom
Department Computer Science and Engineering, American International University of Bangladesh, Dhaka, Bangladesh

M. Z. Abedin, W. Yong (eds.), *Machine Learning Technologies on Energy Economics and Finance*, International Series in Operations Research & Management Science 367, https://doi.org/10.1007/978-3-031-94862-6_9

investigate complex and nonlinear relationships. It is considered to have significant potential for predicting and controlling the performance of anaerobic digesters (Wang et al., 2020). ML enables computers to uncover hidden information by using algorithms that iteratively learn from data without being explicitly programmed on where to look. Several researchers have proposed innovative and effective strategies for modeling the biogas process using ML techniques. These techniques include support vector machines, adaptive neuro-fuzzy inference systems, k-nearest neighbors (KNNs), random forests (RFs), and artificial neural networks (ANNs) (Alejo et al., 2018). Three-layer artificial neural networks and nonlinear regression models were employed to predict biogas production performance in controlled laboratory-scale experiments (Tufaner & Demirci, 2020). Additionally, in an industrial scale co-digestion facility, random forest and extreme gradient boosting (XGBoost) were effectively utilized (De Clercq et al., 2020), while adaptive neuro-fuzzy inference systems model and optimize biogas production from cow manure and maize straw in a pilot-scale study (Zareei & Khodaei, 2017). There is a notable gap in the literature regarding artificial intelligence-based models for estimating biogas production and identifying key factors influencing production from full-scale sludge digestion processes in biological treatment plants. Most researchers develop models using lab- or pilot-scale reactors and focus solely on predicting biogas output. This study addresses this gap by applying Ridge Regression (RR), Lasso Regression (LR), KNN, ElasticNet Regression (ER), Classification and Regression Trees (CART), RF, XGBoost, Light Gradient Boosting Machine (LightGBM), Gradient Boosting Machine (GBM), and CatBoost algorithms to U.S. biogas data. The data, processed by a fully operational anaerobic sludge digester system, was used to predict biogas production rates. The study aims to evaluate the performance of these ML models and identify the key factors influencing biogas production.

The technical contributions of this chapter are as follows:

- To analyze and compare the performance of different ML algorithms for daily biogas production prediction
- To enhance the performance of the algorithms using different preprocessing techniques and hyperparameter tuning
- To provide insights and recommendations based on the experimental results to assist relevant institutions and investors in selecting the most suitable algorithm for biogas production prediction with global and local explanation using explainable artificial intelligence (XAI) tools
- To provide suitable features from ranks based on the average of several XAI analyses

The structure of the remaining sections of this chapter is outlined as follows. The related works are in Sect. 2. Section 3 is dedicated to presenting our proposed methodology and the experimental setup. We detail the approach we have taken to address the research problem, including the methods, techniques, and tools employed in our study. Within Sect. 4, we present the outcomes of our experiments. The chapter concludes in Sect. 5 with a summary of our findings and their significance. Additionally, we outline avenues for future research and development in this domain, emphasizing the potential directions for further exploration and enhancement.

2 Literature Review

Several studies have employed traditional statistical methods in their research. For instance, De Clercq et al. (2017a) utilized a combination of statistical techniques such as principal component analysis and multiple linear regression (LR), along with operations research methods like data envelopment analysis, to investigate the factors influencing efficiency in biogas projects. Their findings highlighted various inefficiencies, including decreasing returns to scale. Similarly, Terradas-Ill et al. (2014) developed a thermal model to forecast biogas production in underground, unheated fixed-dome digesters. However, their model lacked validation against actual data and was unsuitable for large-scale facilities. Furthermore, De Clercq et al. (2017b) employed multi-criteria decision analysis to evaluate food waste and biowaste projects, considering technical, economic, and environmental aspects. They proposed six significant policy recommendations based on their findings but did not provide generalized modeling tools that project operators could use to improve production efficiency based on waste inputs. However, a significant limitation of the models developed in these studies is their failure to incorporate the latest advancements in ML for predicting biogas output. Instead, they rely on traditional statistical performance metrics such as R^2 and RMSE. In contrast, modern ML models are evaluated based on their ability to accurately predict unseen data. To achieve this, datasets are divided into training and testing partitions (James et al., 2013), with a preference for out-of-sample evaluation metrics. These metrics are crucial as they help identify potential overfitting of the model to the training data. These traditional models also face a trade-off between accuracy and simplicity, limiting their ability to capture the complex interactions among various biochemical components. In contrast, ML models are inherently universal function approximators (Hornik et al., 1989). With their numerous adjustable parameters, ML models can uncover subtle relationships in AD datasets without needing expert supervision. Below, we highlight selected examples of ML approaches applied to biogas prediction. Wang et al. (2021) introduced Tree-Based Automated ML (AutoML) for predicting biogas production in the anaerobic co-digestion of organic waste. Sonwai et al. (2023) compared RF, XGBoost, and Kernel Ridge Regression (KRR) models for predicting specific methane yields (SMY), identifying the RF model as the most effective with a coefficient of determination (R^2) of 0.85 and an RMSE of 0.06. Gaida et al. (2012) created an artificial training and test dataset using the ADM1 model and employed three different ML models, including the widely used random forest, to estimate the operating state of a biogas plant online. Cheon et al. (2022) applied five ML models to predict methane yield in a bioelectrochemical AD reactor, demonstrating the ability to interpret nonlinear relationships among multiple input and output variables in complex systems. This approach enhances process stability and helps prevent operational risks. De Clercq et al. (2019) analyzed data from an industrial scale biogas facility in China to improve operational decision-making. The ML models used included logistic regression, support vector machines, and KNN regression. Instead of focusing on digester parameters like temperature, this study examined the impact of different waste input compositions on the AD process. Additionally, a graphical user interface was developed to provide wastewater treatment plant (WWTP) engineers with daily operational recommendations. Yildirim and Ozkaya (2023) compared five ML algorithms, RF, ANN, KNN, SVR, and XGBoost for forecasting biogas production. The RF model performed best with an R^2 of 0.9242, while the KNN model had the lowest accuracy with an R^2 of 0.8326. Most researchers have utilized multilayer ANNs, a widely recognized and extensively discussed method within the engineering community. For instance, Olatunji et al. (2023) developed optimized ANN and FCM-clustered ANFIS approaches for modeling biogas and methane yields. The FCM-ANFIS approach with ten clusters proved more accurate than the ANN approach, achieving R^2, MAD, MAPE, and RMSE values of 0.9850, 1.2463, 5.2343, and 1.2343, respectively. As highlighted in the literature review, some researchers have used traditional statistical methods for predicting biogas production, while others have employed ANN and ML techniques. However, there is a need for more suitable ML models and the application of XAI techniques to identify key factors in biogas production. Our study aims to address these gaps by utilizing more appropriate ML models and XAI methods to identify critical features.

3 Methodology

To predict daily biogas production from the secondary dataset, we propose a top-down approach including data preprocessing with ML techniques. Furthermore, a variety of XAI models are employed to extract the significant factors influencing biogas production.

3.1 Overview of Proposed Methodology

We employ secondary data to predict biogas production using ML techniques. The data is labeled with regression problems and is collected from the online repository Kaggle: secondary data. In the preparatory phase, we implement ordinal encoding, one-hot encoding, and data normalization protocols. The conventional method of ML is employed to evaluate the models' stability by dividing the data into 80:20, 70:30, and 50:50 rations of training and testing. In addition, we introduced an ensemble model that outperformed other benchmark ML models in terms of biogas prediction. Various error metrics are employed to assess the regressors' performance. In Fig. 1, a comprehensive top-down presentation of the

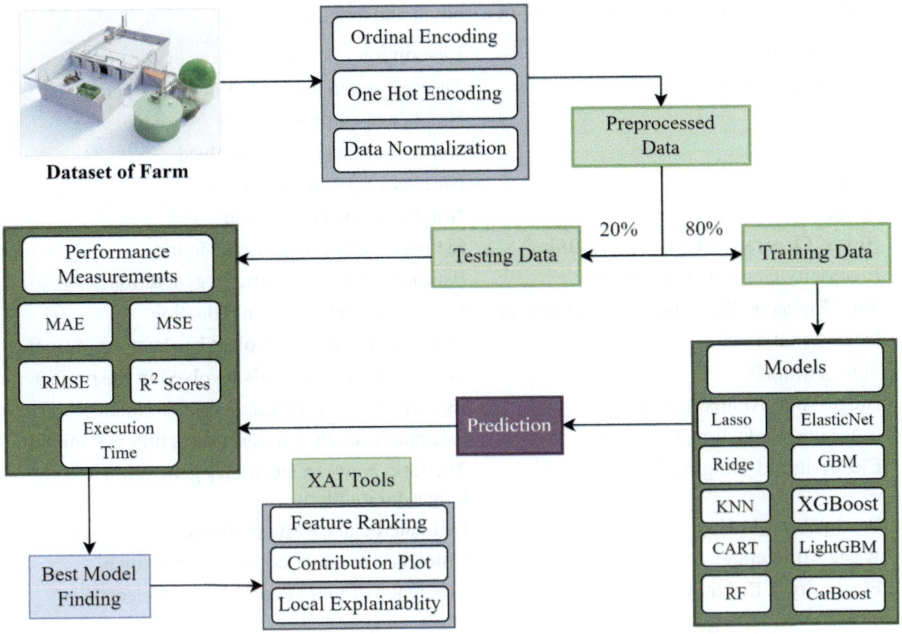

Fig. 1 Overview of the proposed framework including model development and explainability

proposed methodology is provided. The XAI analysis is conducted to extract the significant factors that influence biogas production in order to uncover the inner story of the dataset. Afterward, a comparative analysis is performed to determine the most prevalent factors across all interpretable models, thereby guaranteeing a comprehensive comprehension of the underlying dynamics that influence daily biogas yield.

3.2 Description of Dataset and Variables

This study employs the U.S. Biogas dataset from Kaggle to forecast biogas production regularly (https://www.kaggle.com/discussions/accomplishments/493876). This extensive dataset examines biogas generation from livestock farms throughout the USA, serving as a pivotal resource for assessing renewable energy potentials. It features biogas projects from cattle, dairy cows, poultry, and swine, making it invaluable for agriculture, renewable energy, and environmental policy stakeholders displayed in Table 1. The dataset includes 29 features across 491 observations from the U.S. Data preprocessing is a crucial aspect of ML, requiring significant time and effort, which accounts for about 60% of the investment in a data science project (Seelam et al., 2022).

Table 1 Description of the variables with acronym and details

Column name	Description
Year Operational	The year when the project became operational
Cattle	Number of cattle involved
Dairy	Number of dairy cows involved
Poultry	Number of poultry involved
Swine	Number of swine involved
Biogas Generation Estimate (cu-ft/day)	Estimated daily biogas production
Electricity Generated (kWh/yr)	Estimated annual electricity generation
Total Emission Reductions (MTCO2e/yr)	Estimated total emission reduction
Operational Years	Number of years the project has been operational
Total_Animals	Total number of animals involved in the project
Biogas_per_Animal (cu-ft/day)	Estimated biogas production per animal
Emission_Reduction_per_Year	Estimated annual emission reduction per animal
Electricity_to_Biogas_Ratio	The ratio between electricity generation and biogas production
Total_Waste_kg/day	Estimated daily waste production
Waste_Efficiency	Efficiency of waste conversion to biogas
Electricity_Efficiency	Efficiency of biogas conversion to electricity

3.3 ML Algorithms

3.3.1 Ridge Regression

Ridge regression adds a regularization term to linear regression to handle predictor variable multicollinearity. This method shrinks coefficients and reduces variance by adding a penalty to the loss function equal to the square of their magnitude. The ridge regression equation modifies the ordinary least squares (OLS) regression by adding a regularization parameter λ, which minimizes the following cost function where y_i represents the observed values, $\hat{y}_i$ the predicted values, β_j the coefficients, and λ the regularization parameter. By tuning λ, one can control the trade-off between fitting the data well and keeping the model coefficients small, which helps mitigate overfitting. Ridge regression is instrumental in situations with many correlated predictors, as it improves the model's generalization performance (Daly et al., 2016).

3.3.2 Lasso Regression

Lasso regression, or Least Absolute Shrinkage and Selection Operator, is a regularization technique used to enhance the prediction accuracy and interpretability of regression models by enforcing sparsity. Unlike ridge regression, which applies an ℓ_2 penalty, lasso regression adds an ℓ_1 penalty to the loss function, where y_i

are the observed values, $\hat{y}_i$ the predicted values, β_j the coefficients, and λ the regularization parameters. The ℓ_1 penalty tends to shrink some coefficients exactly to zero, effectively performing variable selection and yielding a simpler model that retains only the most significant predictors. This sparsity property makes lasso regression particularly useful when dealing with high-dimensional data where the number of predictors exceeds the number of observations. By appropriately tuning the λ parameter, one can control the complexity of the model, balancing bias and variance to improve predictive performance and interpretability. Lasso regression is widely utilized in fields like bioinformatics and economics where model simplicity and feature selection are crucial.

3.3.3 ElasticNet Regression

ElasticNet regression is a regularization and variable selection technique that combines the properties of both ridge regression and lasso regression. It addresses some limitations of these methods, particularly when dealing with highly correlated predictors and when the number of predictors exceeds the number of observations. ElasticNet adds both ℓ_1 (lasso) and ℓ_2 (ridge) penalties to the loss function, where y_i are the observed values, $\hat{y}_i$ the predicted values, β_j the coefficients, and λ_1 and λ_2 the regularization parameters. The combination of these penalties allows ElasticNet to perform both variable selection and shrinkage, retaining the benefits of both lasso (sparsity) and ridge (handling multicollinearity). ElasticNet is particularly useful in situations where there are multiple correlated predictors, as it tends to select groups of correlated variables together. By tuning the parameters λ_1 and λ_2, ElasticNet provides a flexible approach to model regularization, balancing between the ridge and lasso penalties to improve model performance and interpretability. This technique is widely used in various fields such as genomics and finance, where it is crucial to handle large datasets with many predictors.

3.3.4 k-Nearest Neighbors

KNN is a method for supervised classification and regression. The algorithm uses the labeled training dataset to identify fresh data points using the “k” closest neighbors method. The notion is that similar data points will have similar labels or results. KNN uses point distance to estimate data proximity. Data is currently collected from a variety of sources for analysis, insight, theory validation, and other research goals. These databases frequently have missing data due to human error in data extraction or collection. Addressing missing values is critical in data analysis preparation. The selection of an imputation method has a significant impact on model performance. The Scikit-Learn, the KNN imputer, is a prominent missing value imputation method. The Euclidean distance matrix assists the KNN imputer in imputed missing data by selecting nearest neighbors. The Euclidean distance is calculated by removing missing values and prioritizing non-missing coordinates

(James et al., 2013). The equation of this algorithm is

$$D_{xy} = \sqrt{\text{weight} \times \text{squared distance from present coordinates}}.$$

Here,

$$\text{weight} = \frac{\text{total number of coordinates}}{\text{number of present coordinates}}.$$

3.3.5 CART

ML uses nonparametric CART for classification and regression. By partitioning the feature space into target variable homogeneous sections, CART creates binary trees iteratively. Each leaf node represents a predicted class or value, while each internal node represents a feature test. The feature that optimizes information gain or Gini impurity reduction at each node is used by CART to classify, where N is the total number of samples, N_{left} and N_{right} are the numbers of samples in the left and right child nodes, and $\text{Impurity}_{\text{left}}$ and $\text{Impurity}_{\text{right}}$ are measures of impurity in the left and right child nodes, respectively. In regression, CART minimizes target variable variance within each partition. The splitting criterion is the split's variance reduction. CART splits nodes iteratively until a stopping requirement is reached, such as a maximum tree depth, a minimum amount of samples in a node, or leaf node purity (Roy et al., 2023). CART models can capture complicated decision boundaries and feature interactions despite their simplicity. They can overfit, especially if trees grow too deep. Pruning and tree depth limitation reduce overfitting.

3.3.6 Random Forest

In a random forest, which is composed of a number of tree predictors, each tree is reliant on the values of a random vector that is randomly selected for each tree in the forest and distributed uniformly. The decision forests implement both pruned and unpruned single-tree classifiers for all datasets, and the disparities are typically substantial (Disha & Waheed, 2022). Accuracy increases with the number of trees in all forests; however, those generated through bootstrapping or boosting are frequently more competitive, while those generated through the random subspace approach occasionally exhibited a unique pattern.

3.3.7 GBMs

GBM creates a powerful predictive model using the predictions of numerous weak learners, usually decision trees. GBM iteratively adds better trees to fix prior mistakes. Previous trees' residual errors are used to train each tree, and the predictions are pooled to minimize the loss function. GBM optimizes loss function using gradient descent in function space. Let $F_{m-1}(x)$ be the current model after $m-1$ iterations. GBM adds a new tree $h_m(x)$ to the model to obtain $F_m(x)$:

$$F_m(x) = F_{m-1}(x) + \gamma_m h_m(x),$$

where γ_m is the learning rate that controls the contribution of each tree. The new tree $h_m(x)$ is trained to minimize the loss function $L(y, F_{m-1}(x) + \gamma_m h_m(x))$. The mean squared error (MSE) is used for regression tasks and cross-entropy loss for classification tasks. GBM has great predicting accuracy and can handle varied data and distributions (Mahedi Hassan et al., 2023). It can overfit, especially with several trees. Common overfitting mitigation methods include reducing tree depth, subsampling, and learning rate modification. Large datasets and deep trees make GBM computationally costly, needing plenty of memory and computing capacity.

3.3.8 XGBoost

XGBoost has great classification. It creates an effective team of decision trees that prioritize ignored areas. Different from extreme gradient boosting, XGBoost improves decision trees. Excellent results come from teamwork, accuracy, and efficiency. Accuracy, precision, recall, and F1 score measurements refine model performance. XGBoost, a quick supervised learning algorithm, accurately classifies water quality in this investigation. Regularized learning improves weights and reduces overfitting, encouraging its use (Hasan et al., 2024). The equation of this algorithm is

$$\Omega(\theta) = \sum_{i=1}^{n} d(y_i, \hat{y}_i) + \sum_{k=1}^{k} \beta(f_k).$$

3.3.9 LightGBM

LightGBM is a gradient boosting framework that was created by Microsoft with an emphasis on accuracy, speed, and efficiency. It is engineered to manage large-scale datasets and can operate substantially faster than other gradient boosting implementations. Gradient-based One-Side Sampling (GOSS) and Exclusive Feature Bundling (EFB) are two innovative tree construction methods that LightGBM employs. These methods allow for faster training times and reduced memory consumption without compromising predictive performance (Hasan et al., 2023). In LightGBM, the objective function can be presented as

$$L(\theta) = \sum_{i=1}^{n} l(y_i, \hat{y}_i) + \sum_{k=1}^{K} \Omega(f_k), \tag{1}$$

where $l(y_i, \hat{y}_i)$ is the loss function, $\hat{y}_i$ is the predicted value, and $\Omega(f_k)$ is the regularization term.

3.3.10 CatBoost

CatBoost is a gradient boosting library that was developed by Yandex with the specific purpose of efficiently managing categorical features. The capacity to autonomously manage categorical variables without the necessity of extensive preprocessing is its distinguishing feature. Advanced algorithms for feature combinations and techniques such as ordered boosting are incorporated into CatBoost to manage categorical variables with high cardinality. The key hyperparameters in CatBoost include the learning rate η, tree depth max_depth, and regularization parameters λ and α. The objective function in CatBoost is given by

$$L(\theta) = \sum_{i=1}^{n} l(y_i, \hat{y}_i) + \lambda \sum_{j=1}^{J} \|w_j\|^2, \tag{2}$$

where $l(y_i, \hat{y}_i)$ is the loss function, $\hat{y}_i$ is the predicted value, and $\|w_j\|^2$ is the regularization term on the weights.

3.4 *Explainable AI Tools*

3.4.1 Shapley Additive Explanations (SHAP)

SHAP (Shapley Additive Explanations) is a framework for explaining ML model predictions. Based on Shapley values from cooperative game theory, SHAP quantifies the contribution of each feature to the difference between the actual and expected predictions. This enables comprehensive insights into feature impacts across various samples. Various methods like Kernel SHAP, Tree SHAP, and Linear SHAP cater to different model types, providing model-agnostic explanations. Visualization tools such as summary plots and force plots help interpret feature impacts on predictions. Widely adopted across domains, SHAP aids in model interpretation, feature engineering, and debugging due to its flexibility and interpretability (Prokhorenkova et al., 2018).

3.4.2 Shapash

A Python library for model interpretation and explanation is Shapash. Automated, customizable, and interactive ML model explanations enhance SHAP. Users of all levels may comprehend model predictions and feature impacts with Shapash.

SHAPash's intuitive and interactive visualizations of SHAP values let users see how individual attributes affect model predictions. Summary, force, and dependence graphs show feature relevance, interactions, and predictions. Users may understand their models' behavior with Shapash's model comparison, sensitivity analysis, and global feature relevance assessment. With support for tree-based, linear, and ensemble ML models, it is adaptable across domains and applications (Sajid et al., 2023). Data science initiatives employ Shapash for model interpretation, debugging, and validation because to its user-friendly interface and excellent visualization features.

3.4.3 Local Interpretable Model-agnostic Explanations (LIME)

LIME (Local Interpretable Model-agnostic Explanations) is a technique for explaining individual predictions of ML models by approximating their behavior with interpretable models locally around the instance of interest. It provides insights into how the model arrived at a particular prediction, allowing users to understand the model's decision-making process. The key idea behind LIME is to fit a simple interpretable model, such as linear regression or decision trees, to locally approximate the complex model's predictions. This interpretable model is trained on perturbed samples generated around the instance to be explained. The coefficients or feature importance of the interpretable model indicates the importance of each feature for the specific prediction (Hassan et al., 2023).

3.4.4 Explain Like I'm 5 (ELI5)

ELI5 (Explain Like I'm 5) is a Python module that gives clear and understandable explanations for ML models. It provides support for many models and aids in comprehending the significance of features and the behavior of the model through approaches such as permutation importance. Permutation importance quantifies the impact on the model's performance when the values of a feature are randomly rearranged, providing a measure of the feature's significance (Faruk et al., 2023). ELI5 also provides support for LIME to explain specific predictions.

3.5 Performance Measure Metrics

Several performance evaluation metrics can be used in the measurement of the accuracy of a model. Here is a description of some common ones:
RMSE: RMSE measures the square root of the average squared differences between the predicted and actual values. It provides a way to measure the magnitude of prediction errors, with lower values indicating better model performance.

R^2 Scores (Coefficient of Determination): The R^2 score quantifies how well the predicted values approximate the actual values. It ranges from 0 to 1, with higher values indicating a better fit. An R^2 score of 1 means the model explains all the variability of the target data around its mean.

Mean Absolute Error (MAE): MAE calculates the average absolute differences between predicted and actual values. It is a straightforward measure of prediction accuracy, with lower values indicating fewer errors and better model performance.

MSE: MSE measures the average of the squared differences between predicted and actual values. It emphasizes larger errors due to the squaring process, with lower values indicating better performance.

Execution Times: Execution time refers to the amount of time a model takes to train and make predictions. Shorter execution times are generally preferable, especially in applications requiring real-time or near-real-time predictions.

4 Result Analysis

Firstly, the performance of the regressors is tabulated for different ratios of training and testing with hyperparameter tuning. Then Global and Local Interpretation of the model has been shown.

4.1 Hyperparameter Tuning on the Models

The optimal hyperparameter values for a variety of regression algorithms are illustrated in the Table 2. The optimal alpha values for ridge and lasso regression are 1.0 and 0.01, respectively. ElasticNet employs an l1_ratio of 0.1 and an alpha of 0.01. KNN uses five neighbors for tree-based methods, CART has a maximum depth of five and a minimum of three samples per leaf, and Random Forests, GBM, XGBoost, LightGBM, and CatBoost have specific hyperparameters related to the number of estimators or iterations, learning rate, and depth.

4.2 Result of the ML Regressor in the Different Ratio of Training and Testing

The performance of various regression algorithms using an 80:20 training-to-testing ratio is summarized in Table 3. The algorithms evaluated with the metrics are Ridge, Lasso, ElasticNet, KNN, CART, RF, GBM, XGBoost, LightGBM, and CatBoost. XGBoost performed the best with the lowest RMSE (0.091) and highest R^2 score (0.847), indicating the most accurate predictions. It also had the lowest MAE (0.051)

Table 2 Hyperparameters value of the regressors

Algorithms	Best hyperparameters	Hyperparameter Value
Ridge	"alpha"	1.0
Lasso	"alpha"	0.01
ElasticNet	"alpha," "l1_ratio"	0.01, 0.1
KNN	"n_neighbors"	5
CART	"max_depth", "min_samples_leaf"	5, 3
RF	"max_depth," "n_estimators"	10, 100
GBM	"learning_rate," "n_estimators"	0.05, 100
XGBoost	"learning_rate," "n_estimators"	0.05, 100
LightGBM	"learning_rate," "n_estimators"	0.1, 100
CatBoost	"depth," "iterations," "learning_rate"	5, 100, 0.1

Table 3 Result of the regressors in 80:20 ratio of training and testing

Algorithms	RMSE	MAE	MSE	Execution times	R^2 scores
Ridge	0.142	0.103	0.02	5.238	0.624
Lasso	0.153	0.117	0.023	0.236	0.567
ElasticNet	0.136	0.096	0.018	1.133	0.658
KNN	0.191	0.139	0.036	0.441	0.322
CART	0.142	0.084	0.02	0.799	0.627
RF	0.098	0.053	0.01	13.089	0.823
GBM	0.098	0.057	0.01	5.953	0.823
XGBoost	0.091	0.051	0.008	11.089	0.847
LightGBM	0.096	0.055	0.009	3.66	0.827
CatBoost	0.095	0.061	0.009	63.819	0.834

and MSE (0.008). LightGBM and CatBoost also performed well, with RMSEs of 0.096 and 0.095 and R^2 scores of 0.827 and 0.834, respectively. However, CatBoost's execution time was significantly higher (63.819 seconds) compared to LightGBM (3.66 seconds) and XGBoost (11.089 seconds). The bar chart in Fig. 2 accompanying the table visualizes the error metrics, highlighting the superior performance of XGBoost, LightGBM, and CatBoost. Figure 3 shows the R^2 scores of the regressors.

The performance of various regression algorithms using a 70:30 training-to-testing ratio is shown in Table 4. LightGBM demonstrated the best performance with the lowest RMSE (0.075) and the highest R^2 score (0.895), indicating superior accuracy. It also had the lowest MAE (0.045) and MSE (0.006), with an execution time of 5.226 seconds. GBM also performed well, with an RMSE of 0.083 and an R^2 score of 0.87, but its execution time was higher at 6.495 seconds. CatBoost had a comparable R^2 score (0.822) but a much longer execution time (80.113 seconds), making LightGBM the most efficient and accurate model. The bar chart in Figs. 4 and 5 visually highlights these metrics, showing LightGBM's superiority. The performance metrics of various regression algorithms using a 50:50 training-

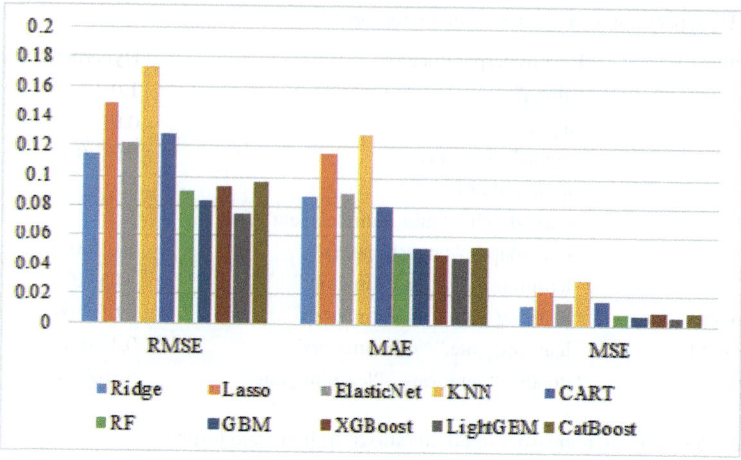

Fig. 2 Bar chart for error metrics of the regressors in 80:20 training-testing ratio

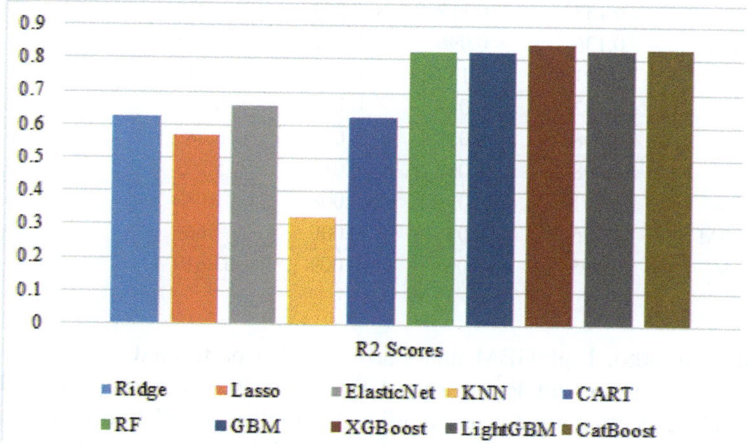

Fig. 3 Bar chart for R^2 scores of the regressors in 80:20 training-testing ratio

to-testing ratio are shown in Table 5. This detailed comparison helps in identifying the best-performing regression model under an even data split scenario. Among the algorithms, XGBoost showed the best performance with an RMSE of 0.091 and the highest R^2 score of 0.847, indicating superior predictive accuracy. It also had the lowest MAE (0.051) and a very low MSE (0.008), though its execution time was moderate at 10.358 seconds. LightGBM also performed well with an RMSE of 0.096, an R^2 score of 0.827, and a low MAE (0.055) while being much faster with an execution time of 1.968 seconds. Although CatBoost had comparable metrics (RMSE: 0.095 and R^2: 0.834), its execution time was significantly longer at 60.658 seconds, making XGBoost and LightGBM the most efficient and accurate models.

Table 4 Result of the regressors in 70:30 ratio of training and testing

Algorithms	RMSE	MAE	MSE	Execution times	R^2 scores
Ridge	0.115	0.086	0.013	1.917	0.751
Lasso	0.149	0.116	0.022	0.179	0.579
ElasticNet	0.122	0.088	0.015	0.616	0.718
KNN	0.173	0.128	0.03	0.221	0.439
CART	0.128	0.08	0.016	0.513	0.69
RF	0.09	0.049	0.008	15.4	0.846
GBM	0.083	0.052	0.007	6.495	0.87
XGBoost	0.094	0.048	0.009	13.135	0.834
LightGBM	0.075	0.045	0.006	5.226	0.895
CatBoost	0.097	0.053	0.009	80.113	0.822

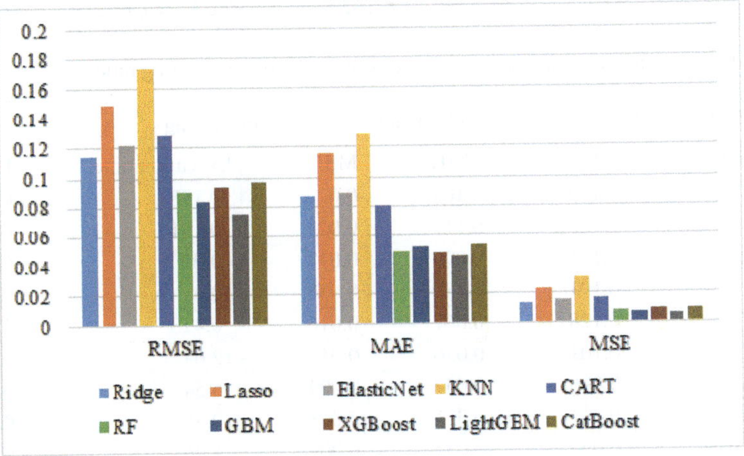

Fig. 4 Bar chart for error metrics of the regressors in 70:30 training-testing ratio

The bar chart visually highlights the RMSE and R^2 scores, underscoring the superior performance of XGBoost and LightGBM in Figs. 6 and 7.

4.3 Explainable AI Analysis

4.3.1 Global Interpretation

Training the SHAP model involves leveraging the entire dataset, which can be resource-intensive because it calculates the marginal contribution of each feature by analyzing individual probabilities and overall performance across the dataset. SHAP produces Shapley values for each data point, illustrating how each feature's value influences the model's output. The model's explanations are presented through

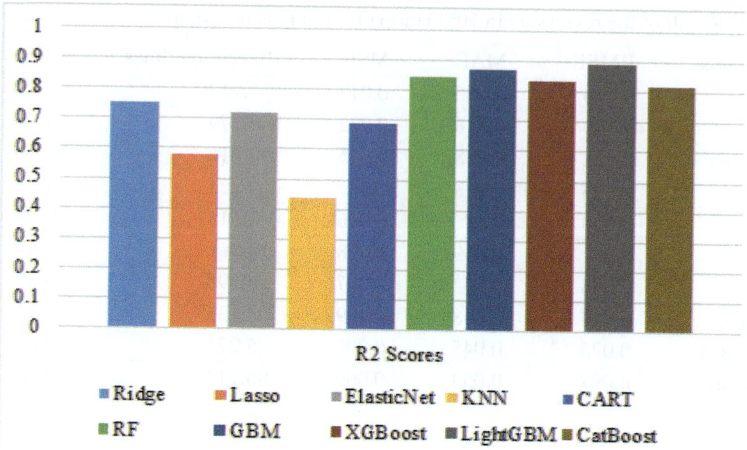

Fig. 5 Bar chart for R^2 scores of the regressors in 70:30 training-testing ratio

Table 5 Result of the regressors in 50:50 ratio of training and testing

Algorithms	RMSE	MAE	MSE	Execution times	R^2 scores
Ridge	0.142	0.103	0.02	0.198	0.624
Lasso	0.153	0.117	0.023	0.191	0.567
ElasticNet	0.136	0.096	0.018	0.62	0.658
KNN	0.191	0.139	0.036	0.19	0.322
CART	0.128	0.067	0.016	0.428	0.697
RF	0.101	0.056	0.01	12.64	0.812
GBM	0.096	0.057	0.009	5.54	0.828
XGBoost	0.091	0.051	0.008	10.358	0.847
LightGBM	0.096	0.055	0.009	1.968	0.827
CatBoost	0.095	0.061	0.009	60.658	0.834

visualizations or plots, offering a graphical interpretation of its insights. The hierarchical summary plot in Fig. 8 provides a prioritized view, ranking features from most to least important. This format delivers intuitive insights into understanding biogas prediction. For example, in the case of the waste_efficiency feature, the red color signifies that higher values are associated with a greater likelihood of gas production, whereas lower values correlate with a decreased chance of production.

The SHAP bar plot function generates a global feature importance plot, providing insights into the overall significance of each feature. It calculates the global importance of each feature by computing the mean absolute value across all the samples provided. This approach offers a comprehensive understanding of the relative importance of different features in the dataset, allowing for informed decision-making and model interpretation. Below in Fig. 9 is the bar plot for our study. SHAPASH offers a versatile framework for easily building and deploying interpretable AI models. Designed with user-friendliness in mind, it streamlines

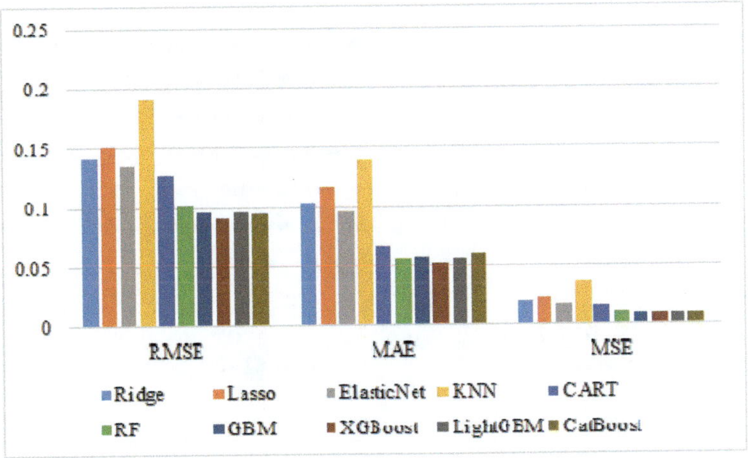

Fig. 6 Bar chart for error metrics of the regressors in 50:50 training-testing ratio

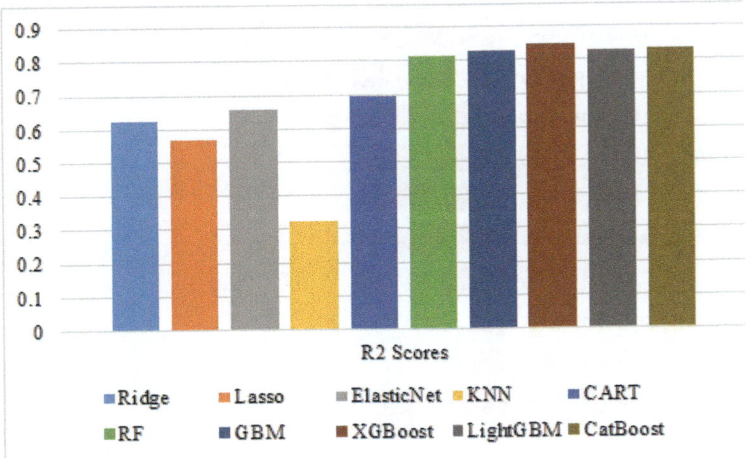

Fig. 7 Bar chart for R^2 scores of the regressors in 50:50 training-testing ratio

model creation and deployment, providing accessible tools for visualizing, comprehending, and explaining model performance. Its intuitive interface aids in the analysis and interpretation of model behavior. SHAPASH employs Shapley values, the importance of permutation features, and partial dependence plots to deliver detailed model explanations. These insights help understand model behavior, detect biases, and enhance overall model performance. Figure 10 showcases the feature importance derived from SHAPASH in this study. Visual representations for each specific feature are shown in the subsequent figures below. ELI5 offers a method called "permutation importance" or "Mean Decrease Accuracy (MDA)" to determine the significance of features in a black box model. This technique evaluates

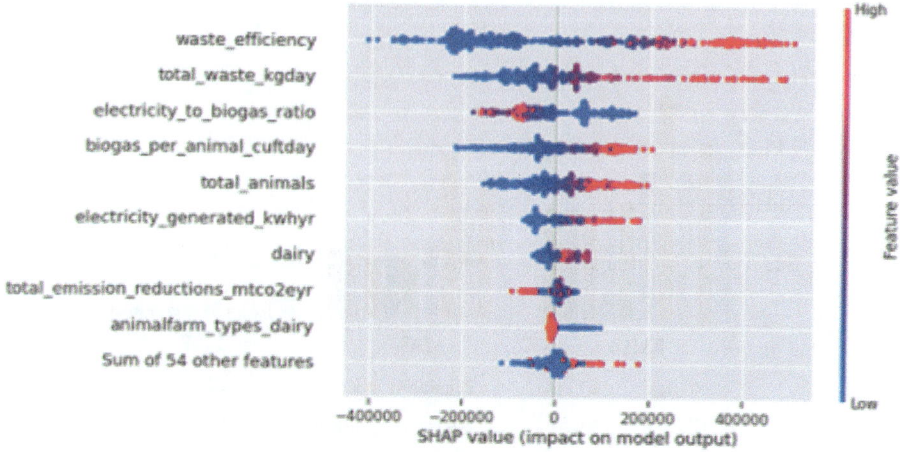

Fig. 8 Hierarchical summary plot of features

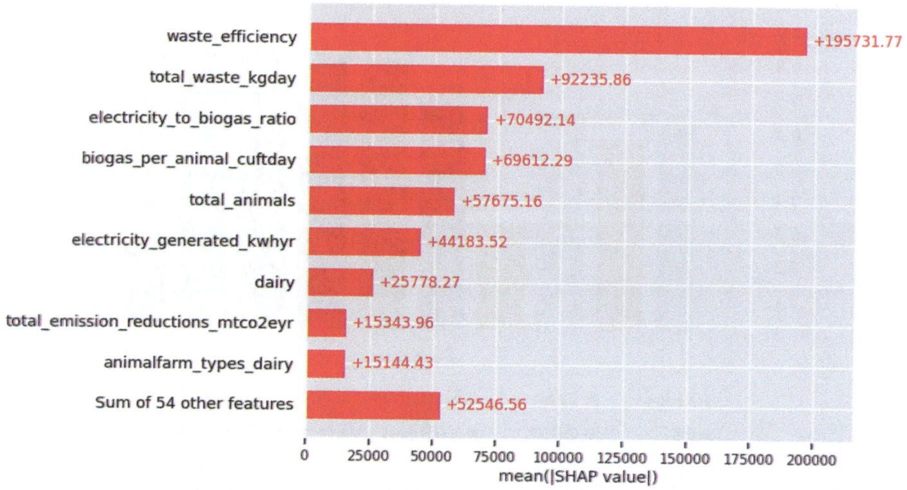

Fig. 9 Global feature importance plot by SHAP

the impact on model performance when a particular feature is removed or modified. A significant drop in performance indicates that the feature is crucial for the model's predictions. This method helps pinpoint the most impactful features in the model's decision-making process. Figure 11 below illustrates the feature importance derived from permutation importance using ELI5.

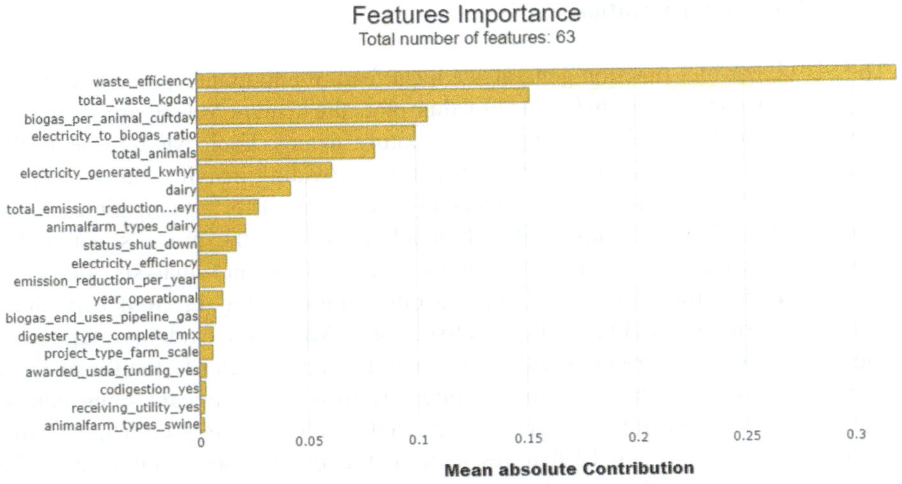

Fig. 10 Global feature importance plot by Shapash

Weight	Feature
0.4201	waste_efficiency
0.1712	total_waste_kgday
0.1071	electricity_to_biogas_ratio
0.0897	biogas_per_animal_cuftday
0.0709	total_animals
0.0508	electricity_generated_kwhyr
0.0281	dairy
0.0178	total_emission_reductions_mtco2eyr
0.0136	emission_reduction_per_year
0.0106	year_operational
0.0054	electricity_efficiency
0.0032	status_shut_down
0.0028	animalfarm_types_dairy
0.0018	awarded_usda_funding_yes
0.0017	project_type_farm_scale
0.0010	biogas_end_uses_pipeline_gas
0.0010	codigestion_yes
0.0008	receiving_utility_yes
0.0008	digester_type_complete_mix
0.0005	biogas_end_uses_cng
	... 43 more ...

Fig. 11 Feature weight and tolerance using ELI5

4.3.2 Local Interpretation

The SHAP bar plot function generates a local feature importance plot, showcasing the SHAP values for each feature. In this plot, the feature values are displayed in gray to the left of the corresponding feature names. Each bar represents the SHAP value associated with a specific feature, providing insights into its impact on the prediction for a randomly selected observation. This visualization aids in understanding the contribution of individual features to the model's output for a particular data point, facilitating model interpretation and analysis. Below is a local bar plot for a randomly selected observation in Fig. 12. Red bars show positive contribution and blues are negative. The LIME model explains a randomly chosen individual observation within the dataset. It aims to clarify the fluctuations in predictions by identifying top features deemed as significant contributors. Figure 13 below illustrates a randomly selected prediction, assessed using LIME, with emphasis on the top 12 features identified as crucial factors influencing the prediction outcome. This visualization aids in understanding the rationale behind the model's predictions for specific data points, facilitating interpretability, and providing valuable insights into the model's behavior. Shapash provides succinct and transparent local explanations, enabling users from diverse data backgrounds to comprehend the prediction of a supervised model through a simplified and straightforward explanation. Figure 14 below depicts a Shapash local explanation of a randomly selected prediction, offering insights into the factors influencing the model's output for that particular data point. This visualization aids in understanding the reasoning behind individual predictions, promoting interpretability and facilitating informed decision-making. The feature rank based on the average of Eli5, SHAP, and Shapash values is shown in Table 6, where "waste_efficiency" ranks at the top.

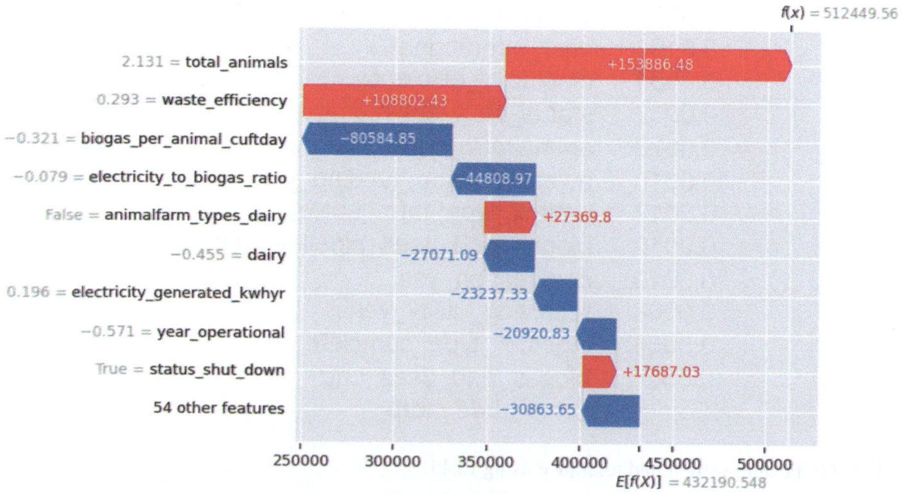

Fig. 12 Local feature importance plot using SHAP

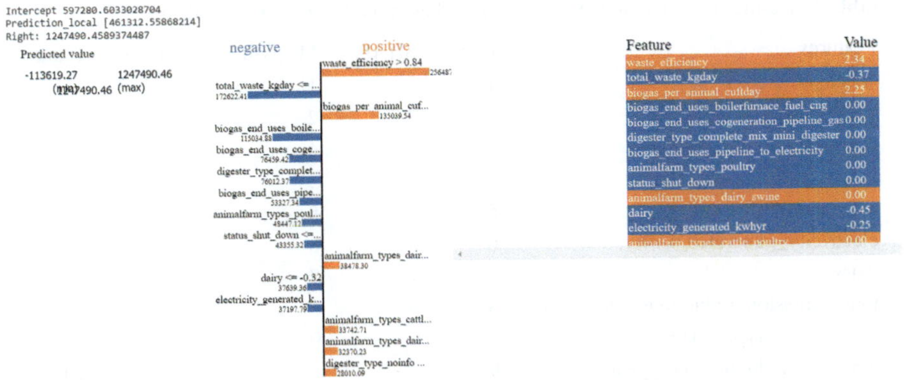

Fig. 13 Identifying significant features using LIME

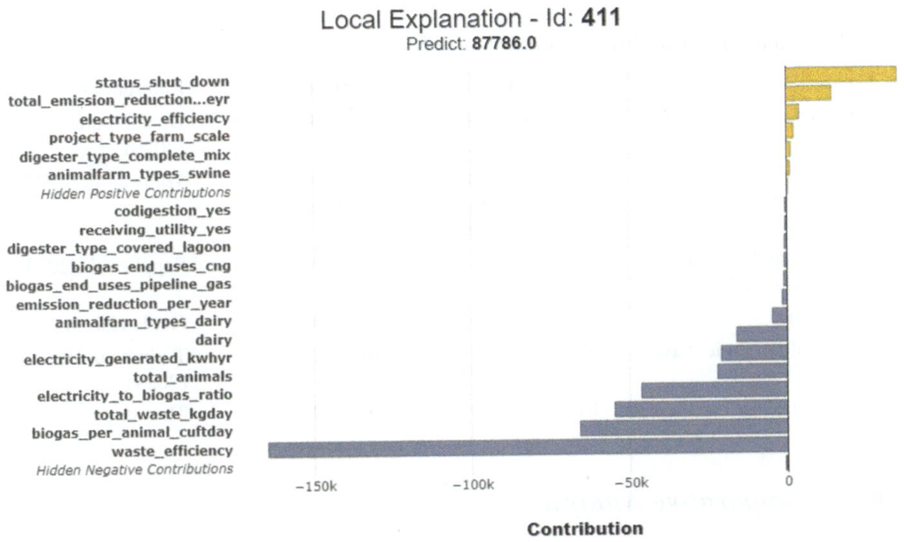

Fig. 14 Local explanation of a random Id: 411

That means waste_efficiency impacts most predicted biogas production. Though position 9 animalfarm_types_dairy and emission_reduction_per_year have the same average value, we have picked animalfarm_types_dairy for its twice appearance. The same procedure was done for status_shut_down and year_operational.

Table 6 Result of the average features ranking based on different XAI analysis

Features	Eli5	SHAP	Shapash	Average	Rank
waste_efficiency	1	1	1	1	1
total_waste_kgday	2	2	2	2	2
electricity_to_biogas_ratio	3	3	4	3.33	3
biogas_per_animal_cuftday	4	4	3	3.66	4
total_animals	5	5	5	5	5
electricity_generated_kwhyr	6	6	6	6	6
dairy	7	7	7	7	7
total_emission_reductions_mtco2eyr	8	8	8	8	8
animalfarm_types_dairy	–	9	9	9	9
emission_reduction_per_year	9	–	–	9	10
status_shut_down	–	10	10	10	11
year_operational	10	–	–	10	12

Table 7 Comparative analysis of different models

Dataset	Best model	Performance	References
Hainan dataset Shenzhen data	kNN XGBoost	$R^2 = 0.86$ $R^2 = 0.66$	De Clercq et al. (2020)
Primary data is taken from the East Bay Municipal Utility District	Tree-Based Pipeline Optimization Tool	$R^2 = 0.72$, RMSE = 247	Wang et al. (2021)
Operational data from the AD process of Tyrol	kNN	$R^2 = 0.72$	Sappl et al. (2023)
Unknown	RF	$R^2 = 0.62$	Gaida (2023)
Household organic waste	RF	$R^2 = 0.88$	Tryhuba et al. (2024)
U.S. biogas	LightGBM	$R^2 = 0.89$	This study

4.4 *Comparative Analysis*

As previously noted, the prediction of biogas using interpretable ML methods has recently gained recognition in academic research, though the number of such studies remains limited. Many earlier studies relied on single models to predict biogas production, but these models often underperformed due to inherent limitations and specific characteristics. In contrast, the model proposed in this research demonstrates significantly better performance. Table 7 below provides a comparative analysis of the most effective models from prior studies on biogas prediction, highlighting the superior performance of the model proposed in this study. However, direct comparison in this table is challenging due to differences in feature selection, sampling methods, data preprocessing, and other factors across the models.

5 Conclusion and Future Work

This chapter focuses on ML models for predicting daily biogas production using the U.S. biogas dataset from Kaggle. The study begins with the application of preprocessing techniques to eliminate unnecessary variables, followed by the implementation of ten ML models: RR, LR, KNNs, ER, CART, RF, XGBoost, LightGBM, GBM, and CatBoost. XGBoost and LightGBM are the most accurate and efficient biogas predictors among these models. XGBoost performs best at 80:20 and 50:50 ratios (RMSE: 0.091 and R^2: 0.847) and LightGBM at 70:30 (RMSE: 0.075 and R^2: 0.895). Improving the performance of the daily biogas prediction model can have significant practical implications, such as enhancing the efficiency and effectiveness of biogas production processes and addressing factors associated with gas production fluctuations. Furthermore, the study examines feature significance and dimension reduction by analyzing the contributions of various features to the predictions generated by interpretable methods. Interpretable methods were employed to identify the top eight features exerting the most influence on the prediction. These features were determined by analyzing the frequency of their appearance among the top ten features based on interpretability. Waste efficiency and total waste (kg/day) emerged as the most significant factors impacting biogas prediction. The above study recommends prioritizing these variables when developing biogas prediction systems or formulating organizational management policies to increase biogas production. By focusing on these critical factors, it is possible to enhance the predictive accuracy and overall productivity of biogas generation systems.

References

Alejo, L., Atkinson, J., Guzmán-Fierro, V., & Roeckel, M. (2018). Effluent composition prediction of a two-stage anaerobic digestion process: machine learning and stoichiometry techniques. *Environmental Science and Pollution Research, 25*, 21149–21163.

Cheon, A., et al. (2022). Application of various machine learning models for process stability of bio-electrochemical anaerobic digestion. *Processes, 10*, 158.

Daly, A., Dekker, T., & Hess, S. (2016). Dummy coding vs effects coding for categorical variables: Clarifications and extensions. *Journal of Choice Modelling, 21*, 36–41.

De Clercq, D. et al. (2017a). Application of DEA and statistical inference to model the determinants of biomethane production efficiency: A case study in south China. *Applied Energy, 205*, 1231–1243.

De Clercq, D., Wen, Z., & Fan, F. (2017b). Performance evaluation of restaurant food waste and biowaste to biogas pilot projects in China and implications for national policy. *Journal of Environmental Management, 189*, 115–124.

De Clercq, D., et al. (2019). Machine learning powered software for accurate prediction of biogas production: A case study on industrial-scale Chinese production data. *Journal of Cleaner Production, 218*, 390–399.

De Clercq, D. et al. (2020). Interpretable machine learning for predicting biomethane production in industrial-scale anaerobic co-digestion. *Science of the Total Environment, 712*, 134574.

Disha, R. A., & Waheed, S. (2022). Performance analysis of machine learning models for intrusion detection system using Gini impurity-based weighted random forest (GIWRF) feature selection technique. *Cybersecurity, 5*, 1.

Faruk, M. O., Reza, R. B., Hasan, M., & Rabbi, M. F. (2023). *Leveraging explainable AI to identify the hidden cause of it personnel stress* (pp. 1–6). IEEE.

Gaida, R. (2023). *Application of machine learning in biogas process*. Master's thesis, University of South-Eastern Norway.

Gaida, D., et al. (2012). State estimation for anaerobic digesters using the ADM1. *Water Science and Technology, 66*, 1088–1095.

González-Fernández, C., Méndez, L., Tomas-Pejó, E., & Ballesteros, M. (2019). Biogas and volatile fatty acids production: Temperature as a determining factor in the anaerobic digestion of spirulina platensis. *Waste and Biomass Valorization, 10*, 2507–2515.

Hasan, M., et al. (2023). *Ensemble based machine learning model for early detection of mother's delivery mode* (pp. 1–6). IEEE.

Hasan, M., Ahmed, T., Islam, M. R., & Uddin, M. P. (2024). Leveraging textual information for social media news categorization and sentiment analysis. *PLoS One, 19*, e0307027.

Hassan, M. M., Abrar, M. F., & Hasan, M. (2023). in *An explainable AI-driven machine learning framework for cybersecurity anomaly detection* (pp. 197–219). Routledge.

Hornik, K., Stinchcombe, M., & White, H. (1989). Multilayer feedforward networks are universal approximators. *Neural Networks, 2*, 359–366.

James, G., Witten, D., Hastie, T., Tibshirani, R., et al. (2013). *An introduction to statistical learning* (Vol. 112). Springer.

Mahedi Hassan, M., Fazle Rabbi, M., Hasan, M., & Roy, B. (2023). *An ensemble machine learning approach to classify Parkinson's disease from voice signal* (pp. 575–590). Springer.

Olatunji, K. O., Madyira, D. M., Ahmed, N. A., Adeleke, O., & Ogunkunle, O. (2023). Modeling the biogas and methane yield from anaerobic digestion of Arachis hypogea shells with combined pretreatment techniques using machine learning approaches. *Waste and Biomass Valorization, 14*, 1123–1141.

Prokhorenkova, L., Gusev, G., Vorobev, A., Dorogush, A. V., & Gulin, A. (2018). CatBoost: Unbiased boosting with categorical features. In *Advances in neural information processing systems* (Vol. 31).

Roy, P., Srijon, F. M. S., Hasan, M., Bhowmik, P., & Nitu, A. M. (2023). *An ensemble machine learning approach with hybrid feature selection technique to detect thyroid disease* (pp. 379–394). Springer.

Sajid, S. W., Anjum, K. R., Al-Shaharia, M., & Hasan, M. (2023). 8 investigating machine learning algorithms with model explainability for network intrusion detection. In *Cyber security and business intelligence: Innovations and machine learning for cyber risk management* (Vol. 121).

Sappl, J., Harders, M., & Rauch, W. (2023). Machine learning for quantile regression of biogas production rates in anaerobic digesters. *Science of the Total Environment, 872*, 161923.

Seelam, S. R., Kumar, K. H., Supritha, M. S., Gnaneswar, G., & Reddy, V. V. M. (2022). *Comparative study of predictive models to estimate employee attrition* (pp. 1602–1607). IEEE.

Sonwai, A., Pholchan, P., & Tippayawong, N. (2023). Machine learning approach for determining and optimizing influential factors of biogas production from lignocellulosic biomass. *Bioresource Technology, 383*, 129235.

Terradas-Ill, G., Pham, C. H., Triolo, J. M., Martí-Herrero, J., & Sommer, S. G. (2014). Thermic model to predict biogas production in unheated fixed-dome digesters buried in the ground. *Environmental Science & Technology, 48(6)*, 3253–3262.

Tryhuba, I., et al. (2024). Prediction of biogas production volumes from household organic waste based on machine learning. *Energies, 17*, 1786.

Tufaner, F., & Demirci, Y. (2020). Prediction of biogas production rate from anaerobic hybrid reactor by artificial neural network and nonlinear regressions models. *Clean Technologies and Environmental Policy, 22*, 713–724.

Wang, L., Long, F., Liao, W., & Liu, H. (2020). Prediction of anaerobic digestion performance and identification of critical operational parameters using machine learning algorithms. *Bioresource Technology, 298*, 122495.

Wang, Y., Huntington, T., & Scown, C. D. (2021). Tree-based automated machine learning to predict biogas production for anaerobic co-digestion of organic waste. *ACS Sustainable Chemistry & Engineering, 9*, 12990–13000.

Yildirim, O., & Ozkaya, B. (2023). Prediction of biogas production of industrial scale anaerobic digestion plant by machine learning algorithms. *Chemosphere, 335*, 138976.

Zareei, S., & Khodaei, J. (2017). Modeling and optimization of biogas production from cow manure and maize straw using an adaptive neuro-fuzzy inference system. *Renewable Energy, 114*, 423–427.

Md. Mahedi Hassan is a lecturer at the CSE Department of the World University of Bangladesh (WUB). He completed his MSc in Computer Science and Engineering from Hajee Mohammad Danesh Science and Technology University, Dinajpur, in 2023. Before that, he also completed his BSc (Engineering) in Computer Science and Engineering from the same university in 2021. Besides being a teacher, he has also devoted himself to research activities. His focal interest is in Machine Learning (ML), Deep Learning (DL), Artificial Intelligence (AI), Cybersecurity, and Data Science.

Mahira Shamim has completed BBA, and now she is pursuing MBA from the Department of Finance, University of Chittagong, Bangladesh. Her research interest is in Optimization, Machine Learning, Islamic Finance, and Financial Engineering and Derivatives. She has conducted some research projects on lease financing, insurance, financial inclusion, investment management, and strategic management. She aspires to collaborate with experienced researchers to solve complex business and economic problems and believes that an international academic environment will enhance her knowledge, skills, and cultural awareness.

Mahmudul Hasan is currently pursuing a PhD in Information Technology (IT) at Deakin University, Melbourne, Australia. He earned his BSc (Eng.) and MSc (Eng.) degrees in Computer Science and Engineering (CSE) from Hajee Mohammad Danesh Science and Technology University, Dinajpur, Bangladesh, in 2021 and 2023, respectively. He previously served as a Lecturer in the Department of CSE at the University of Creative Technology, Chittagong (UCTC), Bangladesh. He is the Founder and Director of the Center for Multidisciplinary Research and Development (CeMRD) and a moderator of "Be Researcher BD," the largest online research forum in Bangladesh. Additionally, he has taught online as a Data Science instructor to students in the USA, Italy, Denmark, South Korea, and Australia. He is also the founder of the online educational platform "MHM Academy." His research interests include federated learning, machine learning, deep learning, cybersecurity, health informatics, renewable energy, computational sociology, and business intelligence.

Md. Amir Hamja is currently pursuing his MSc in Statistics from the Department of Statistics at Hajee Mohammad Danesh Science and Technology University, Dinajpur, Bangladesh, and he completed BSc (Hons.) in Statistics from the same university in 2023. Currently, he is a Research Assistant in the Center for Multidisciplinary Research and Development (CeMRD). His research interests include Federated Learning, Machine Learning, Deep Learning, Cyber Security, Health Informatics, Business Intelligence, Time Series, Public Health, and Biostatistics.

Kanij Fatema is currently pursuing her BSc (Eng.) degree in Computer Science and Engineering (CSE) from Hajee Mohammad Danesh Science and Technology University, Dinajpur, Bangladesh. She is also working as a Research Assistant (RA) in Center for Multidisciplinary Research and Development (CeMRD). Additionally, she is working as a freelance teacher for secondary school students in Dinajpur, Bangladesh. She is a moderator on the online educational platform "MHM Academy." Her research interests include machine learning, deep learning, cybersecurity, health informatics, and educational data mining.

Sudipto Roy Pritom is currently working as a Research Assistant in the Center for Multidisciplinary Research and Development (CeMRD). He has completed BSc (Eng.) in Computer Science and Engineering (CSE) From American International University of Bangladesh in 2024. As a Data Science enthusiast, he worked on several projects in different domains. His research interests include Machine Learning, Deep Learning, and Cyber Security.

Application of Machine Learning Techniques in the Analysis of Sustainable Energy Finance

Riadul Islam Rabbi, Ekramul Haque Tusher, Mahmudul Hasan, and Md Rashedul Islam

1 Introduction

Today's business models must be adjusted to the changing characteristics of contemporary digital surroundings. Based on UN projects $5 trillion would need to be invested by 2020 in order to accomplish the sustainable development goals (SDGs) (Musleh Al-Sartawi et al., 2022). There must be a global shift toward renewable energy to battle climate change and achieve sustainability goals. This shift is led by sustainable energy financing (SEF), which involves funding and investments in renewable energy projects including wind, solar, power, and hydropower. Figure 1 displays the overall SDG scores for different countries in the year 2023, with Finland leading the rankings. On the other Fig. 2 shows the SDG index scores for various countries from 2000 to 2022, with Sweden achieving the highest score. Nordic and Western European countries generally dominate the top. Nevertheless, the industry has several constraints that prevent it from developing further. Therefore, innovative solutions are required to overcome major obstacles such as variable regulatory

R. I. Rabbi
Faculty of Engineering and Technology, Multimedia University, Ayer Keroh, Melaka, Malaysia

E. H. Tusher
Faculty of Computing, University Malaysia Pahang Al-Sultan Abdullah, Pekan, Pahang, Malaysia

M. Hasan (✉)
School of Information Technology, Deakin University, Geelong, VIC, Australia

Department of Computer Science and Engineering, Hajee Mohammad Danesh Science and Technology University, Dinajpur, Bangladesh

M. R. Islam
Department of Computer Science and Engineering, Hajee Mohammad Danesh Science and Technology University, Dinajpur, Bangladesh

M. Z. Abedin, W. Yong (eds.), *Machine Learning Technologies on Energy Economics and Finance*, International Series in Operations Research & Management Science 367, https://doi.org/10.1007/978-3-031-94862-6_10

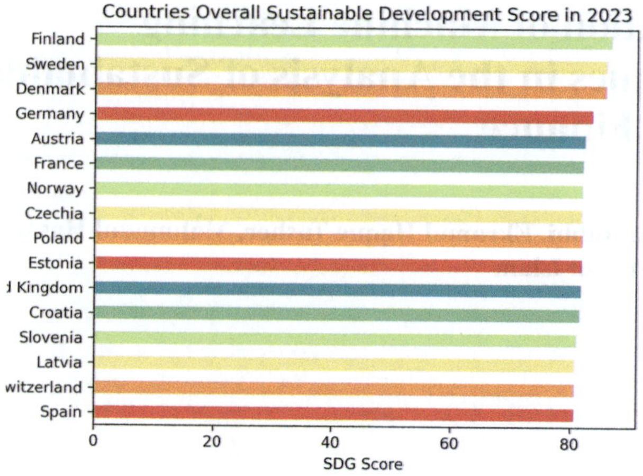

Fig. 1 Overall SDG score in 2023 by countries

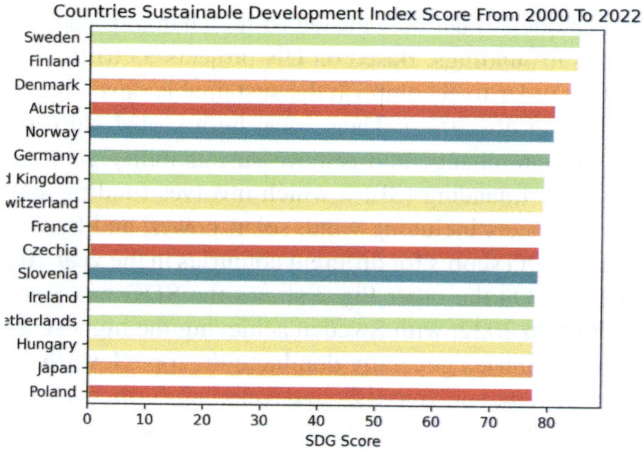

Fig. 2 SDG index score from 2000 to 2022 by countries

environments, high initial expenses, and unidentified financial returns (Maria et al., 2023).

On the SDG graph, Fig. 3 shows a density distribution of SDG index scores, ranging from approximately 30 to 90. The distribution is multimodal, with a primary peak around 65–70 and secondary peaks around 50 and 75. There are notable dips in the distribution around scores of 40 and 60, creating a complex, nonsymmetric shape that suggests multiple subgroups or factors influencing the overall SDG index scores. On the other graph overall scores range from approximately 25 to 100. The distribution is roughly bell-shaped but slightly asymmetric, with a peak density of about 70 and a longer tail extending toward lower scores. There is a noticeable small dip in the curve around the 55–60 SDG score range before it rises to its maximum.

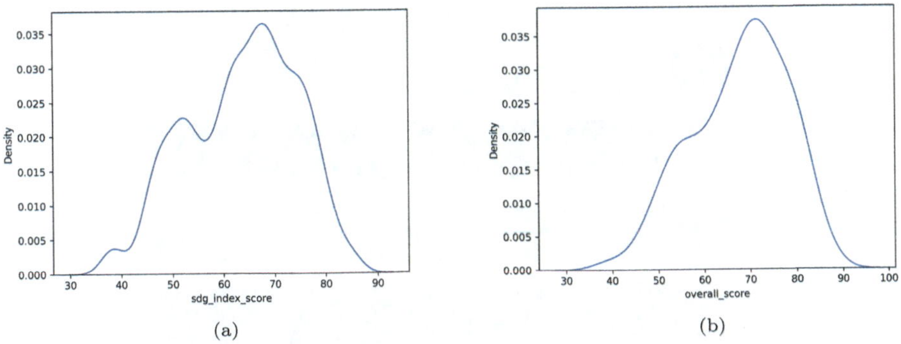

Fig. 3 Density of SDG index score and overall score in years 2023 and 2000–2022

The artificial intelligence (AI) field of machine learning (ML) has the potential to be revolutionary in tackling these difficulties. ML enables computers to learn from experience and make predictions or decisions without the need for explicit programming through the analysis and interpretation of complicated data using statistical models and algorithms (Mhlanga, 2021). Its integration into sustainable energy finance has the potential to transform risk assessment, investment decision-making, and market trend prediction in the sector.

ML algorithms can examine past financial data, patterns of energy usage, and market trends in order to forecast the financial success of renewable energy projects. This predictive capability assists investors in making better-informed decisions through the identification and mitigation of possible risks (Liu et al., 2021). Thus directing resources toward the most viable projects, ML can enhance investment portfolio optimization by analyzing diverse factors including profitability, ecological effects, and adherence to regulations. So, in Fig. 4 displays a list of seven "Green Growth Indicators" presented as colored bars. These indicators include air and water population, forest, biodiversity, water, climate change, energy, and urbanization, each represented by several colors and accompanied by a small circular icon.

Notwithstanding the potential benefits, the use of ML in sustainable energy financing is not without difficulties. Therefore, data availability and quality are crucial challenges, since reliable machine learning models require big and precise datasets. Furthermore, ensuring responsible ML technology use requires addressing key concerns, including data protection and algorithmic transparency, within regulatory and ethical frameworks (Mavlutova et al., 2022).

Moreover, policymakers and investors may make more strategic and well-informed decisions by using ML in sustainable energy finance (Gonzales Martínez, 2020). This integration advances the more general objective of building a resilient and sustainable energy future despite increasing the impact and efficiency of investment in renewable energy. With this inquiry, we intend to provide a comprehensive understanding of how ML might be used to get over obstacles in sustainable energy financing and expedite the move to a low-carbon economy.

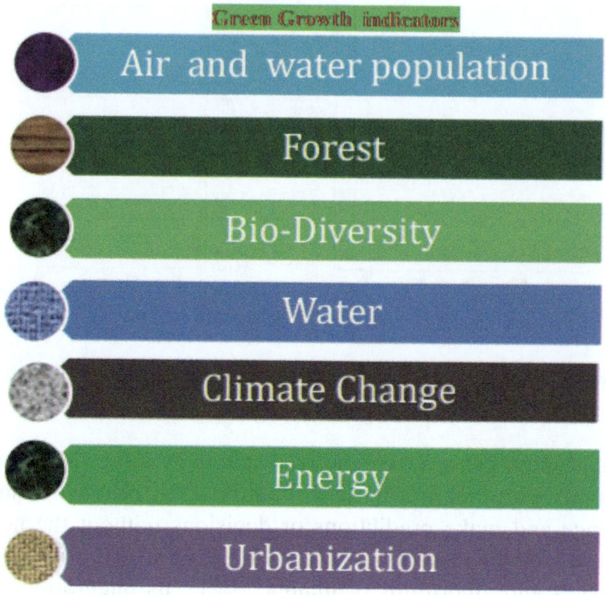

Fig. 4 Green growth indicators

Figure 5 shows the trend of renewable energy consumption as a percentage of total final energy consumption from 2011 to 2023. The line generally rose from 2011 to 2018, reaching a peak of around 12.8% before dropping sharply in 2019 and continuing a gradual decrease through 2023. So this research provides notable contributions to the topic of sustainable energy finances:

- This study provides a comprehensive comparison of multiple machine learning models, such as Decision Tree (DT), Random Forest (RF), Support Vector Machine (SVM), K-Nearest Neighbors (KNNs), Neural Network (NN), and XGBoost. This comparison research brings useful insights into the relative advantages and disadvantages of various architectures concerning the sustainability of energy financing.
- We establish the edge of our framework of investigation by utilizing metrics such as Mean Squared Error (MSE), Root Mean Squared Error (RMSE), Mean Absolute Error (MAE), Mean Absolute Percentage Error (MAEP), and R-Squared values when comparing all the models that were evaluated. This comprehensive assessment offers compelling evidence of the efficacy of our technique in enhancing the precision of forecasting.

This study seeks to provide insight into the possible benefits, methodologies, and future possibilities of this emerging topic, in exploring the interface between machine learning and sustainable energy. The rest of the chapter's structure is as follows: In Sect. 2 we will review the current applications of ML in sustainable energy finance. Section 3 discusses a description of the machine learning techniques,

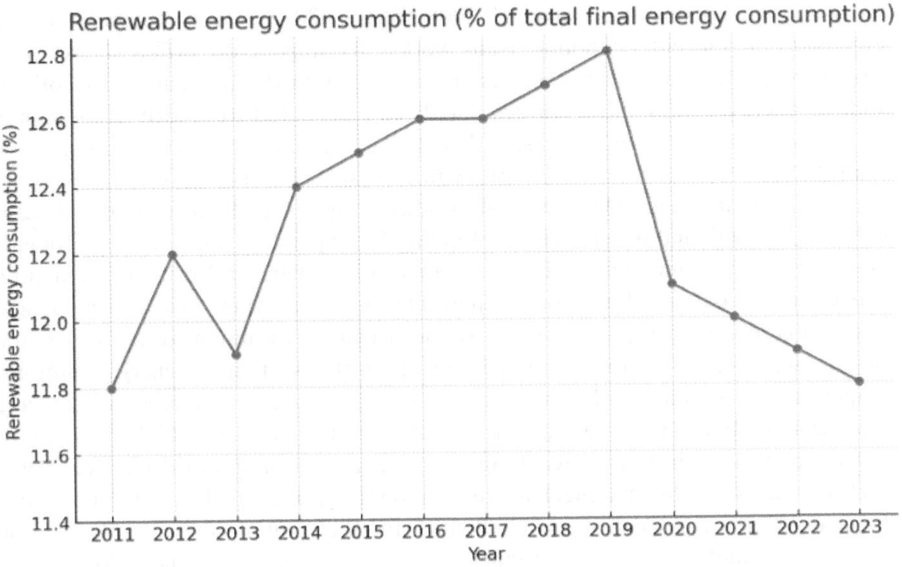

Fig. 5 Renewable energy consumption (% of total final energy consumption). Source: databank.worldbank.org

and Sect. 4 presents results and data analysis. Finally, we will outline the discussion, conclusion, future research directions, and policy implications.

2 Literature Review

The intent of the following section consists of providing an in-depth review of the predominant studies on the use of machine learning techniques to the task of predicting sustainable energy finances.

2.1 Sustainable Energy Finance

The term "Sustainable Energy Finance (SEF)" defines the financial and investment methods utilized to aid in the advancement and energy-efficient technologies, as well as deployment. In order to shift to a low-carbon economy, and accomplish sustainable development aims, is very necessary for this field. SEF is essential for mitigating climate change, maintaining energy security, and getting economic benefits and last but not least for social and environmental impact. Additionally, there has been a growing correlation between the global energy markets and the financial sectors, and energy prices have shown more characteristics since the worldwide financial collapse of 2008 (Zhang, 2018). Now, energy is a vital ingredient in current economic frameworks. Its impact on different facets of economic performance has been extensively examined. Conventionally, oil, as well as energy commodities' prices, has been considered to be driven by supply and demand in global marketplaces. What is energy finance?—energy finance is multidisciplinary by nature and starts with assessing the links between the energy and financial markets. Mahesh et al. studied sustainable finance that facilitates enhanced growth and provides better funding for expanding the economy (Kadaba et al., 2022). Sustainable development aims to safeguard and restock the natural ecosystem. It is crucial to foster renewable energy sharing, take on green and sustainable energy norms, and make sound decisions to maximize the utilization of natural resources. Moreover, to facilitate the shift to sustainable energy solutions, emphasize the vitality of designing global corporate green financing policies and plans that are both short and long terms in nature (Trivedi et al., 2023). Ultimately, SDGs have the objective of reaching the ideal and desired world (Bei & Wang, 2023). It has some finance mechanisms like public, private, hybrid, and innovative financing that have faced many challenges for high initial costs, regulatory and policy uncertainty, technical and market risks, and access to finance. So we have to say that from previous studies they are a very important component of the across-the-glove transition to a low-carbon economy.

2.2 *Machine Learning in Finance*

According to an analysis of recent academic literature, nonlinear econometric models and machine learning models have replaced linear econometric models based on the study of forecasting oil prices. As an area of AI, ML is a way of creating algorithms that analyze, interpret, and forecast data to make decisions. In the banking sector, ML is transferring with its revolutionary effects on productivity, accuracy, and client experiences. In Fig. 6 we can see that there are a lot of machine learning methods for applying native energy communities such as supervised learning, unsupervised learning, and reinforcement learning. These algorithms have worked on several applications of machine learning in finance which are for algorithmic trading, fraud detection and prevention, credit scoring and risk assessment, and market sentiment analysis. S.B. Jabeur et al. investigated in their study on oil price predicting crashes during the 2019 novel coronavirus (COVID-19) pandemic. This study employed several advanced ML algorithms to reduce the influence of the COVID-19 pandemic on oil prices using a precise forecasting methodology that takes into account the pattern of changes in oil prices (Jabeur et al., 2021). In another paper, M. Mohsin et al. suggested a novel approach for predicting crude oil prices depending on several kinds of sociopolitical and economic variables by applying the Least Absolute Shrinkage and Selection Operator (Lasso) model within the framework of green finance (Mohsin & Jamaani, 2023). Based on the systematic review A. Hernandez et al. presented the data produced by the power system and its

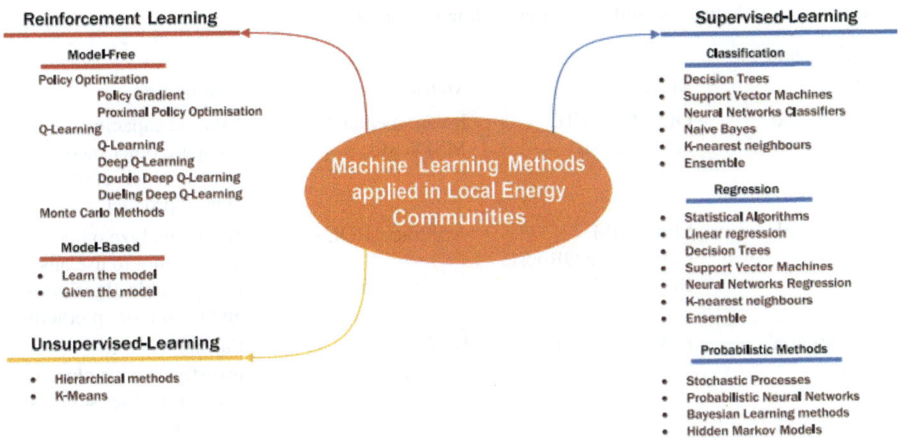

Fig. 6 Machine learning algorithms' use for energy communities

customers utilizing statistical learning theory where machine learning algorithms are data-driven models (Hernandez-Matheus et al., 2022).

In the distribution grid, energy communities are evolving into fresh types of organizations for prosumers and consumers. In addition, R. Rastogi et al. conducted to describe how renewable energy policies affect the financial performance of renewable energy firms and highlight trends in their economic performance (Rastogi et al., 2020). P. Sadorsky et al. studied to forecast the direction of clean energy stock prices using machine learning techniques. In this study, they found support vector machines achieved higher prediction accuracy than Lasso or Naïve Bayes (Sadorsky, 2022, 2021). Furthermore, I.M. Black discussed systematically, which is very imperative to ascertain the exact state of an asset to achieve the anticipated operating duration and efficiency. This requires a determination of equipment faults (Black et al., 2021). Likewise, A. Chang et al. aimed to determine how the SDGs influence the ICT industry's ability to anticipate corporate financial performance (CFP) (Chang et al., 2024). They supposed that several factors can improve the earnings per share (EAR) forecast, including return on total assets, adoption of the SDGs, and whether the company has created KPIs for SDG accomplishments. Table 1 is the literature review based on the last 5 years which are the most representative machine learning algorithms for predicting on finance sector.

2.3 *Intersection of Machine Learning and Sustainable Energy Finance*

The topic of ML and SEF is both dynamic and promising, combining sophisticated data analytics with investments in renewable energy. Financial organizations and

Table 1 Model, metrics, and limitations in finance sectors

Authors and reference	Model	Metrics	Limitations
Mohsin and Jamaani (2023)	OLS, GARCH, ANN, Lasso	Mean, median, MSPE, std.	Limited capacity, not to compare other models, an unknown subset of energy product
Jabeur et al. (2021)	LightGBM, CatBoost, XGBoost, RF, and NN	Accuracy, ROC curve	Not to analyze other metrics, not to use any interpretable algorithms, and need more predictive
Rastogi et al. (2020)	K-Means Cluster	ROE	Further, need to investigate each cluster, and study other sources of energy
Sadorsky (2022)	Extra Trees, SVM, RF, GBM, NB, Lasso	Accuracy, kappa, F1-value	Need to expand the predictor space, and extend the number of methods
Chang et al. (2024)	DT, RF, SVM	MAE, MSE, RMSE, MAPE	Only correlate between reports and financial performance in the ICT industry and not be confined to particular assumptions in ML
Sadorsky (2021)	Logit Model, Random Forest	Accuracy, Gini	Only three models are used and the analysis of additional technical indicators
Zhang et al. (2023)	Regression, Medication Effect, Threshold	T-values, coef.	Not to compare other models, need to extend digital transformation
Nguyen et al. (2021)	OLS, ElasticNet, NN, KNN, RF, XGBoost	MAE	Need to be improved prediction accuracy by incorporating additional variables
Xin et al. (2024)	XGBoost	R-Square, MAE, MSE, RMSE, MAPE	Only use China's prefecture-level cities, no consent on the concept of inclusive growth
May et al. (2022)	ANN, GMDH, ANFIS	RMSE, MAPE, R-Square	Need to compare other models and metrics

energy providers may increase efficiency, accelerate the expansion of sustainable energy projects, and optimize decision-making processes by utilizing machine learning algorithms. A significant benefit of ML application for SEF is improved data-driven decision-making capacity (Bashir et al., 2022). Therefore, ML algorithms may analyze large datasets, such as past energy output, weather patterns, market pricing, and financial performance. This link increases decision-making and risk assessment. Further, ML can maximize renewable energy distribution and production. The following example is that predictive algorithms can foresee equipment breakdowns before they happen by analyzing data from sensors on solar panels or wind turbines. SEF and ML have seven main themes of research likely socially responsible investing, climate financing, green financing, impact investing, carbon financing, energy financing, and governance of sustainable financing and investing (Kumar et al., 2025). So it has a strong correlation between SEF and ML, as well as very crucial for energy sources (Li & Umair, 2023). In the field of finance, machine learning algorithms can develop precise financial projections and valuations for renewable energy projects (Pincet et al., 2019). Stakeholders may improve financial results, streamline operations, and make better decisions, propelling the world's shift to sustainable energy through leveraging machine learning.

3 Methodology

In this study, we aim to predict the long-term viability of energy finances by utilizing comprehensive fuel data and employing various machine learning methods. We especially prioritize certain machine learning algorithms that have been enhanced to increase the sustainability of energy budgets. To assess the efficacy of our machine learning model, we conduct a comparative analysis with many well-established models in the finance domain.

3.1 Explanation of the Approach

A thorough machine learning workflow is depicted in Fig. 7, commencing with data collection, which is the process of gathering unprocessed data to serve as the project's basis. After that, Data Preprocessing is performed, which includes Data Cleaning to address missing or inconsistent data, the use of a MinMax Scalar to normalize features within a specified range, and Data Visualization to comprehend data patterns and distributions. The preprocessed data is subsequently partitioned into training data and test data. The training data is utilized for modeling a range of machine learning models, such as Decision Tree, Random Forest, Support Vector Machine, K-Nearest Neighbors, Neural Network, and XGBoost. Throughout the training process, the models' performances are assessed using various Performance Metrics, including MSE, RMSE, MAE, MAPE, and R-Squared. These criteria assist in the selection of the optimal model for making predictions. At the end the most effective model is implemented for the purpose of Result Prediction, wherein it is utilized to provide precise forecasts, and the resulting outcomes are thoroughly examined. This methodical methodology guarantees a methodical and effective procedure for creating, assessing, and implementing machine learning models.

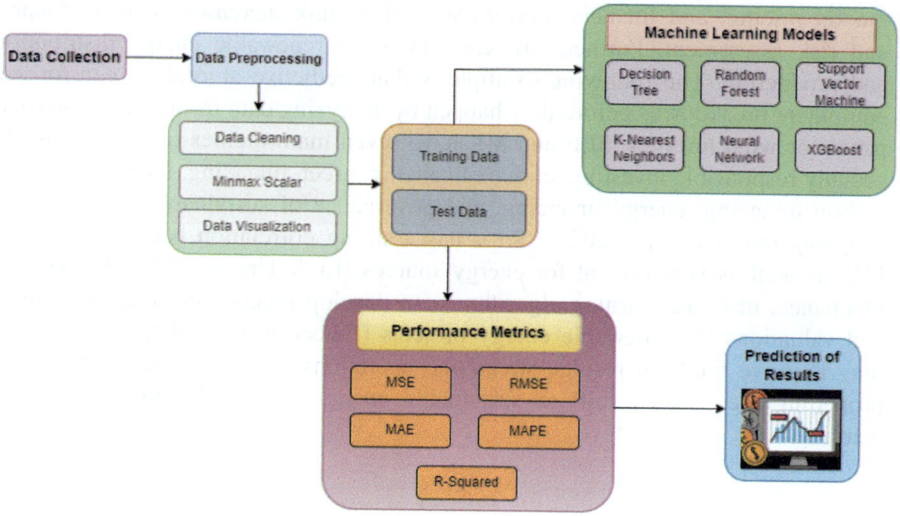

Fig. 7 Overview of ML-based energy financing sustainability

3.2 *Machine Learning Models*

There are a lot of machine learning techniques, and that we have chosen six algorithms for this study such as decision tree, random forest, support vector machine, k-nearest neighbors, neural networks, and XGBoost algorithm. We have described them below:

3.2.1 Decision Tree

DT is one of the best machine learning algorithms of supervised learning. It is nonparametric, and there are two different types of trees that are available: regression trees and classification trees, for the intent of classifying continuous and categorical variables. Both trees utilize a recursive partitioning approach, working from the top down. The splitting process continues until the desired level of uniformity is achieved (Abedin et al., 2025). Throughout the procedure, our data might be overfitting which will classify it too broadly. The approximated function can be described as follows:

$$G(Q_m, \theta) = \frac{n_m^{\text{left}}}{n_m} H(Q_m^{\text{left}}(\theta)) + \frac{n_m^{\text{right}}}{n_m} H(Q_m^{\text{right}}(\theta)) \quad (1)$$

Here, in this equation, Q_m represents each node of m with n_m samples and partitions the data into subsets $Q_m^{\text{left}}(\theta)$ and $Q_m^{\text{right}}(\theta)$. It varies based on whether the task is classification or regression, and the effectiveness of a quality of split node m is calculated using an impurity function or a loss function

3.2.2 Random Forest

The RF algorithm is commonly used as a machine learning approach for handling categorization problems. This is the most popular algorithm that has been used increasingly day by day in various fields like environmental protection, marketing, and finance. It is a set based on trees and is supplemented with a measure of the projection's average value derived at each tree's conclusion, reducing the absence of robustness in one tree (Hasan et al., 2023b). The predicted model can be expressed as follows:

$$\hat{Y} = \frac{1}{q} \sum_{i=1}^{q} f_x(X) \tag{2}$$

In this predicted function, *f*(x) represents a set of kth trainee random trees, where x is the input feature vector. RF is a meta-estimator that balances and overfitting and uses averaging to increase prediction accuracy by fitting numerous decision tree classifiers on different subsets of the dataset. According to a previous study, the Random forest algorithm is better than other machine learning algorithms.

3.2.3 Support Vector Machine

SVM has become the most effective and trustworthy algorithm for classification and regression in different kinds of application fields. The main objective is to classify the optimal hyperplane for separating the data points into various categories (Cervantes et al., 2020). It is very effective in high-dimensional spaces, memory efficient, and versatile. It has many kernel functions such as linear kernel, polynomial kernel, and radial basis function kernel (Hasan et al., 2024a). The SVM's predicted function is as follows:

$$\hat{Y} = \begin{cases} 0 & \text{if } W^T \cdot X + b < 0 \\ 1 & \text{if } W^T \cdot X + b \geq 0 \end{cases} \tag{3}$$

Using such an equation, $\hat{Y}$ is the predicted class for the input feature vector of X. W is the weight vector that is learned from the training data, and b is the bias term which is also trained in the data and aids in shifting the hyperplane. So it is suitable for both classification and regression problems.

3.2.4 K-Nearest Neighbors

KNN algorithm is a straightforward, nonparametric, and incremental learning technique for regression and classification applications. This algorithm mainly has two steps in classification of a learning step and an evaluation of the categorization. Identifying the classes to which its neighbors belong is categorized using the closet neighbor approach for new unlabeled data (Hasan et al., 2024b). The KNN's predicted output is as follows:

$$\hat{Y} = \frac{1}{k}\sum_{i=1}^{k} y_{NN(i)} \tag{4}$$

Here, $\hat{Y}$ is the predicted value in which *yNNi* denotes the target values of the k adjacent values. So KNN is the simplicity, comprehensibility, and scalability of each domain.

3.2.5 Neural Networks

Neural network is a kind of computing model that resembles the structure and functions of the human brain to identify patterns and resolve difficult problems. It is a widely used technique for classification and regression problems such as logistic regression or discriminant analysis (Rabbi et al., 2023). It is constructed of layers of networked nodes, or neurons, that investigate the input data and give an output. The estimated model has the following expression:

$$\hat{Y} = f\left(f\left(\sum_{i=1}^{N} \mu_{ij}x_i + b_i\right)\cdot\left(\sum_{j=1}^{k}\mu_j\right) + b\right) \tag{5}$$

Here, μ is the matrix of network weights, the neuronal activation function is represented by f, the number of features is n, and the deep layers of the number of neurons are denoted by k. So it is a more powerful model able to learn from data, identify patterns, and make predictions.

3.2.6 XGBoost Algorithm

XGBoost (Extreme Gradient Boosting) is a powerful machine learning technique that can help you better understand your data and decision-making. It is a scalable and highly effective gradient boosting method for supervised learning tasks. This algorithm is mainly designed for optimizing both computational and model performance, and it is an appealing choice for many machine learning challenges and practical applications (Hasan et al., 2023a). The final score is calculated by using this formula:

$$\hat{Y} = \sum_{h=1}^{H} g_h(X_i) \tag{6}$$

Here, the score of leaf trees is denoted by K in this equation, while H represents the number of trees. So it is the most popular gradient boosting framework that is effective, efficient, and versatile for issues.

4 Results and Data Analysis

4.1 Data and Variables

In this study, we used a historical dataset on fuels and energy like oil and gas from Yahoo Finance to extract the data which is collected from Kaggle (https://www.kaggle.com/datasets/guillemservera/fuels-futures-data/data). This dataset has provided comprehensive and up-to-date information on futures related to oil, gas, and fuels. Futures are financial agreements that commit the seller to sell a particular amount of a certain fuel at a defined price at a later date and the buyer to acquire it. It has eight features and five categories for predicting variables. Table 2 is a description of features.

4.2 Data Analysis

Table 3 represents the descriptive statistics for fuel market data, encompassing opening high, low, and closing prices, as well as trading volume. We discuss the

Table 2 Feature description on historical Yahoo finance dataset

Column	Descriptions
Date	The date when the data was documented. Format: YYYY-MM-DD
Open	Market's opening price for the day
High	Peak price during the trading window
Low	Lowest traded price during the day
Close	Price at which the market closed
Volume	Number of contracts exchanged during the trading period
Ticker	The unique market quotation symbol for the future
Commodity	Specifies the type of fuel the future contract pertains to (e.g., crude oil, natural gas)

Table 3 Descriptive statistics for finance dataset

Features	Mean	Std.	Min.	25%	50%	75%	Max.
Open	27.28	36.08	-14.00	2.03	3.37	54.88	1.46e+02
High	27.67	36.53	0.50	2.06	3.45	55.74	1.47e+02
Low	26.87	35.59	-40.3	1.99	3.30	53.90	1.44e+02
Close	27.28	36.08	-37.6	2.03	3.37	54.88	1.46e+02
Volume	105981.7	148442.4	0.000	26410.0	49032.0	114720.0	2.28e+06

following statistics: mean, standard deviation, the first, second, and third quartile, minimum, and maximum (refer to Table 3). The price-related features (Open, High, Low, and Close) show similar patterns, with mean values of around 27 and standard deviations of about 36, indicating high volatility. Interestingly, minimum values for open, low, and close are negative, which is unusual for price data and may suggest unique market conditions or data anomalies. The price ranges are wide, spanning from negative values to highs around 146–147. Trading volume statistics reveal a highly skewed distribution, with a mean of 105,981.7 and a standard deviation of 148,442.4. The volume ranges from 0 to a maximum of 2.28 million, with the median (49,032) being significantly lower than the mean, further highlighting the right-skewed nature of the volume data. Overall, these statistics paint a picture of a volatile fuel market with wide price fluctuations and highly variable trading volumes. On the contrary, the pairwise correlation coefficients between the original values in our investigation are shown in Fig. 8. A graphical representation of the correlation between the variables of finance variables is presented in Fig. 7. In the correlation matrix, coefficients revealed that several commodities on the finance dataset volume are highly connected with all other variables, and also there is some negative correlation on fuel energy datasets.

This section has discussed data patterns. Figure 9 depicts crude oil (a), heating oil (b), natural gas (c), gasoline (d), and Brent crude oil (e) prices from 2000 to 2024, showing significant fluctuations over time. Notable features include a sharp price spike around 2008, in the crude oil plot followed by a sharp decline, a period of relative stability from 2011 to 2014, and an unprecedented price crash to negative values in 2020 only for Covid, before recovering and fluctuating in subsequent years. Heating oil prices started low in 2000. Prices remained volatile but generally high from 2011 to 2014 and sharply dropped in 2020. The highest peak appears in 2022. Other natural gas main price spikes occurred in 2001, 2005, and 2008. So the highest peak was in 2005. There was a period of low prices from 2016 to 2020. In gasoline prices, a major drop in 2020 (likely due to COVID-19), and a dramatic rise to peak prices in 2022, followed by a decline toward 2024. The last fuel Brent crude oil prices started low in 2000 and remained high from 2011 until 2014. There was a dramatic crush in 2020. Prices moderated but stayed volatile toward 2024. The overall trend shows increasing prices over the 24 years.

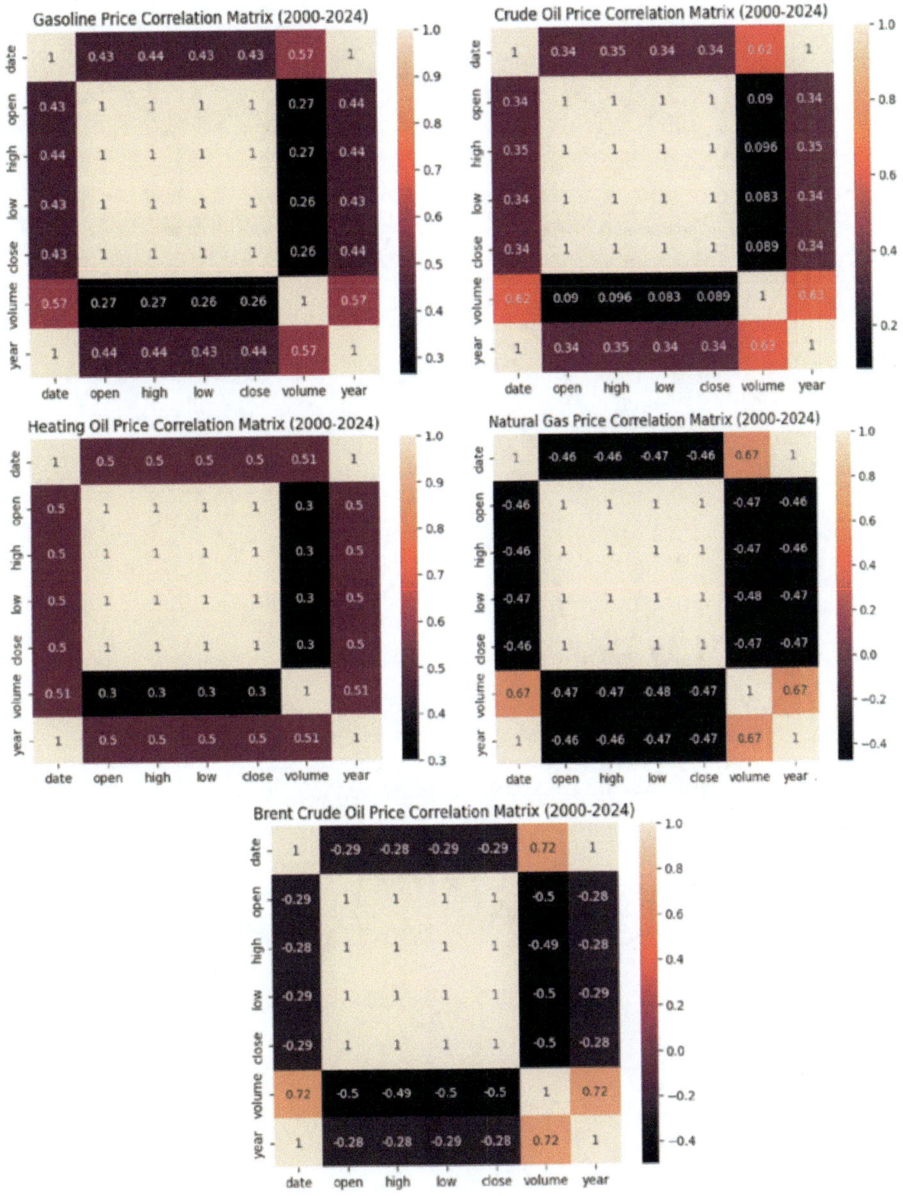

Fig. 8 Correlation matrix for five fuel energy datasets

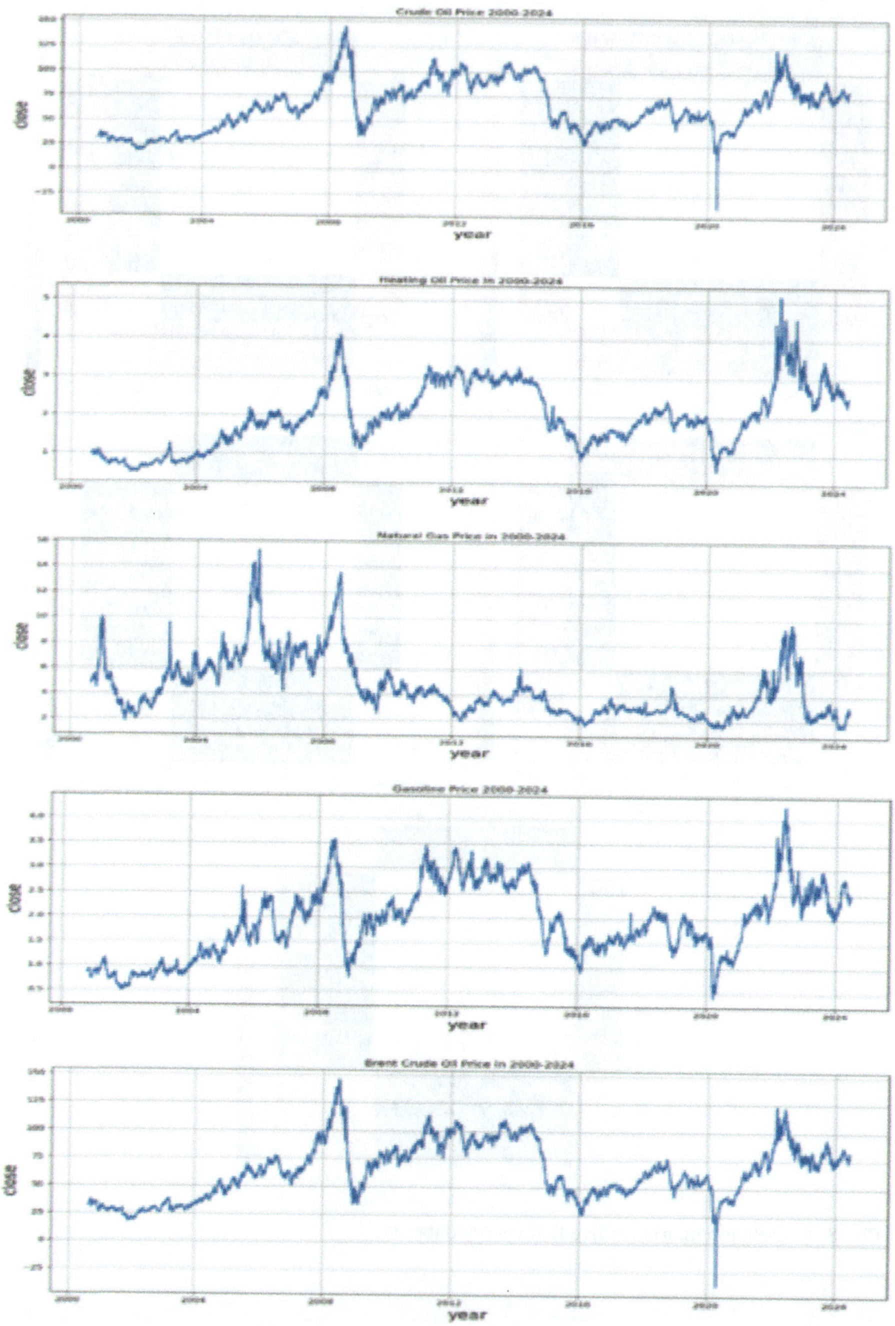

Fig. 9 (**a**) Crude oil, (**b**) heating oil, (**c**) natural gas, (**d**) gasoline, and (**e**) Brent crude oil close price from 2000 to 2024

4.3 Results

A comparison of the predicting abilities of the various machine learning models is given in this section. During this investigation to ascertain the effectiveness of each unique model for the validation process, the relationship between the initial characteristics and the predictor variables is constructed by five commodities. We used five several metrics for calculating model performance. Table 4 shows the regression performance measured by mean squared error, root mean squared error, mean absolute error, mean absolute percentage error, and R-Squared error. These can be calculated as follows:

$$\mathrm{MSE} = \frac{1}{n}\sum_{i=1}^{n}(Y_i - \hat{Y}_i)^2 \tag{7}$$

$$\mathrm{RMSE} = \sqrt{\frac{\sum_{i=1}^{n}(\hat{Y}_i - Y_i)^2}{n}} \tag{8}$$

$$\mathrm{MAE} = \frac{\sum_{i=1}^{n}|\hat{Y}_i - Y_i|}{n} \tag{9}$$

$$\mathrm{MAPE} = \frac{1}{n}\sum_{t=1}^{n}\left|\frac{Y_t - \hat{Y}_t}{Y_t}\right| \tag{10}$$

$$R^2 = 1 - \frac{\sum_i (Y_i - \hat{Y}_i)^2}{\sum_i (Y_i - \bar{Y})^2} \tag{11}$$

In this equation, n is the number of data points, Y_i is the observed values, and $\hat{Y}_i$ is the predicted values, respectively.

The data in Table 4 allows for the inference of several conclusions from historical finance datasets. Across all commodities, the RF model consistently demonstrates the highest accuracy, with the lowest error rates (MSE, RMSE, MAE, MAPE) and the highest R-Squared values, typically 0.998 or 0.999. This indicates that RF is extremely effective at predicting prices for these energy commodities, explaining nearly all of the variance in the data. The DT and XGBoost models also perform exceptionally well, often matching or coming very close to the RF model's performance. These three models (RF, DT, and XGBoost) consistently outperform the others across all commodities. In contrast, SVM and KNN models generally show the poorest performance, with significantly higher error rates and lower R-Squared values. This suggests that these models may not be as well suited for forecasting energy commodity prices compared to the tree-based models. The Neural Network model's performance varies considerably across different commodities. It performs reasonably well for crude oil and Brent crude oil but shows poor performance for natural gas, where it actually has a negative square value, indicating that it performs worse than a horizontal line for predicting natural gas prices. Overall, the tree-

Table 4 Performance of several metrics of different prediction models by testing datasets

Commodity	Algorithm	MSE	RMSE	MAE	MAPE	R-Square
Crude oil	Decision Tree	0.979	0.989	0.630	0.010	0.998
	Random Forest	0.495	0.704	0.478	0.008	0.998
	Support Vector Machine	359.4	18.95	14.65	0.276	0.440
	K-Nearest Neighbors	376.0	19.39	15.20	0.287	0.415
	Neural Network	7.119	2.668	1.824	0.033	0.988
	XGBoost	1.958	1.399	0.554	0.010	0.966
Heating oil	Decision Tree	0.001	0.037	0.019	0.010	0.998
	Random Forest	0.000	0.024	0.015	0.008	0.998
	Support Vector Machine	0.580	0.761	0.606	0.372	0.185
	K-Nearest Neighbors	0.696	0.834	0.669	0.440	0.022
	Neural Network	0.702	0.837	0.640	0.323	0.014
	XGBoost	0.000	0.028	0.016	0.008	0.998
Natural gas	Decision Tree	0.014	0.118	0.067	0.013	0.997
	Random Forest	0.007	0.088	0.053	0.011	0.998
	Support Vector Machine	3.813	1.952	1.353	0.314	0.243
	K-Nearest Neighbors	4.069	2.017	1.470	0.369	0.192
	Neural Network	174.3	13.20	12.68	3.376	-33.58
	XGBoost	0.010	0.101	0.058	0.012	0.997
Gasoline	Decision Tree	0.000	0.027	0.018	0.010	0.998
	Random Forest	0.000	0.021	0.014	0.008	0.999
	Support Vector Machine	0.425	0.651	0.524	0.327	0.229
	K-Nearest Neighbors	0.480	0.693	0.558	0.369	0.128
	Neural Network	0.053	0.232	0.180	0.100	0.902
	XGBoost	0.000	0.022	0.015	0.009	0.909
Brent crude	Decision Tree	0.894	0.945	0.629	0.011	0.998
	Random Forest	0.493	0.702	0.474	0.008	0.999
	Support Vector Machine	359.4	18.95	14.65	0.276	0.440
	K-Nearest Neighbors	376.0	19.39	15.20	0.287	0.415
	Neural Network	6.354	2.520	1.974	0.033	0.990
	XGBoost	1.958	1.399	0.554	0.010	0.996

based models (RF, DT, and XGBoost) appear to be the most reliable and accurate for forecasting energy commodity prices across all the commodities analyzed. The performance of other models, particularly SVM, KNN, and Neural Networks, is generally inferior and inconsistent across different commodities.

5 Conclusion and Future Directions

5.1 Conclusion

The research study has shown that although the availability of a literature review for many features pertaining to the financial, energy, and economy markets. But there are no friendly sustainable energy sectors like crude oil, heating oil, gasoline, etc. Therefore, this study assisted in determining the model performance of finance sectors that could contribute to this domain. In this research investigation, machine learning algorithms were used on the historical finance datasets. According to the findings, DT, RF, and XGBoost algorithms are given the best performance and most reliable, as well as trustworthy on financial data. So, the global market will grow with emerging technologies so that nations can reduce energy consumption over the past decade. As a result, the study of research also contributed to the development of a more accurate model that could be used for obtaining other energy sources.

5.2 Implications

Implementing policies effectively to gain immediate attention is considered crucial. Therefore, policymakers should aim to mitigate the observed volatilities among other features through more effective policy design. So for this outline some key implications of using machine learning for sustainable energy finance analysis are:

i. Improved risk assessment: The risks associated with sustainable energy projects can be analyzed in large datasets to better assess using machine learning models.
ii. Optimization of energy systems: The integration of renewable energy sources into existing grids can assist in improving efficiency and reducing costs using machine learning. As a result, these sustainable energy projects may become more profitable.
iii. Identification of investment opportunities: Machine learning techniques are able to analyze market trends, technological developments, and policy changes to recognize promising investment possibilities in the sustainable energy sector.
iv. Identification of investment opportunities: Machine learning techniques are able to analyze market trends, technological developments, and policy changes to recognize promising investment possibilities in the sustainable energy sector.
v. Automation of due diligence: Parts of the due diligence process for sustainable energy finance could get automated with ML algorithms, which could expedite investment decisions and reduce costs.

vi. Improved fraud detection: In order to protect investors and maintain the integrity of the green finance markets, machine learning models could help detect fraudulent behavior in the financing of sustainable energy.
vii. Personalized financial products: Machine learning makes it possible to create more individualized financial products which could lead to an increase in investment in this field.
viii. Policy impact assessment: For the purposes of facilitating evidence-based policymaking, ML models can assist assess the potential impact of several policy scenarios on financing for sustainable energy.

5.3 Limitations, Challenges, and Future Directions

Very little earlier research was carried out in the past on the sustainable energy sector with different features. Hence, few literature reviews were attained for this research study area, and a lot of characteristics have to be reconsidered. This research study solely explored the financing of sustainable energy to reduce costs. However, in this sector, there are many limitations and challenges. Firstly, data quality and availability are crucial for improving this field, but there is a lack of standardized privacy concerns limiting inconsistent or incomplete data across different regions and after that regulatory compliance issues when using complex ML models for financial decisions. On the other hand, historical datasets may contain biases that could be perpetuated by machine learning models and also the complexity of energy systems when integrating multiple variables into ML models. Finally, it needs high computational requirements for processing large datasets and running complex models. Nevertheless, for future research, develop more transparent machine learning models to improve trust and meet regulatory requirements in finance. Implement transfer learning knowledge gained from data-rich markets in data-poor regions, improving the global applicability of ML models. So for future studies, create specialized ML models that integrate climate science, energy technology, and financial data for more accurate long-term projections.

Acknowledgments We extend our deepest gratitude to the Center for Multidisciplinary Research and Development (CeMRD) for their resources, guidance, and support. The expertise and encouragement from CeMRD members have been invaluable. This research was made possible by the collaborative environment and cutting-edge facilities at CeMRD.

References

Abedin, M. Z., Moon, M. H., Hassan, M. K., & Hajek, P. (2025). Deep learning-based exchange rate prediction during the covid-19 pandemic. *Annals of Operations Research, 345*(2), 1335–1386.

Bashir, M. F., Benjiang, M., Hussain, H. I., Shahbaz, M., Koca, K., & Shahzadi, I. (2022). Evaluating environmental commitments to cop21 and the role of economic complexity, renewable energy, financial development, urbanization, and energy innovation: Empirical evidence from the RCEP countries. *Renewable Energy, 184*, 541–550.

Bei, J., & Wang, C. (2023). Renewable energy resources and sustainable development goals: Evidence based on green finance, clean energy and environmentally friendly investment. *Resources Policy, 80*, 103194.

Black, I. M., Richmond, M., & Kolios, A. (2021). Condition monitoring systems: A systematic literature review on machine-learning methods improving offshore-wind turbine operational management. *International Journal of Sustainable Energy, 40*(10), 923–946.

Cervantes, J., Garcia-Lamont, F., Rodríguez-Mazahua, L., & Lopez, A. (2020). A comprehensive survey on support vector machine classification: Applications, challenges and trends. *Neurocomputing, 408*, 189–215.

Chang, A., Lee, T.-S., & Lee, H.-M. (2024). Applying sustainable development goals in financial forecasting using machine learning techniques. *Corporate Social Responsibility and Environmental Management, 31*(3), 2277–2289.

Gonzales Martínez, R. (2020). Which social program supports sustainable grass-root finance? Machine-learning evidence. *International Journal of Sustainable Development & World Ecology, 27*(5), 389–395.

Hasan, M., Haque, A., Islam, M. M., & Al Amin, M. (2023a). How much do the features affect the classifiers on unsw-nb15? An XAI equipped model interpretability. In *Cyber security and business intelligence* (pp. 137–153). Routledge.

Hasan, M., Marjan, M. A., Uddin, M. P., Afjal, M. I., Kardy, S., Ma, S., & Nam, Y. (2023b). Ensemble machine learning-based recommendation system for effective prediction of suitable agricultural crop cultivation. *Frontiers in Plant Science, 14*, 1234555.

Hasan, M., Ahmed, T., Islam, M. R., & Uddin, M. P. (2024a). Leveraging textual information for social media news categorization and sentiment analysis. *PloS One, 19*(7), 0307027.

Hasan, M., Abedin, M. Z., Hajek, P., Coussement, K., Sultan, M. N., & Lucey, B. (2024b). A blending ensemble learning model for crude oil price forecasting. *Annals of Operations Research* 1–31.

Hernandez-Matheus, A., Löschenbrand, M., Berg, K., Fuchs, I., Aragüés-Peñalba, M., Bullich-Massagué, E., & Sumper, A. (2022). A systematic review of machine learning techniques related to local energy communities. *Renewable and Sustainable Energy Reviews, 170*, 112651.

Jabeur, S. B., Khalfaoui, R., & Arfi, W. B. (2021). The effect of green energy, global environmental indexes, and stock markets in predicting oil price crashes: Evidence from explainable machine learning. *Journal of Environmental Management, 298*, 113511.

Kadaba, D. M. K., Aithal, P., & KRS, S. (2022). Impact of sustainable finance on MSMEs and other companies to promote green growth and sustainable development. *International Journal of Applied Engineering and Management Letters, 6*(1), 60–76.

Kumar, S., Sharma, D., Rao, S., Lim, W. M., & Mangla, S. K. (2025). Past, present, and future of sustainable finance: Insights from big data analytics through machine learning of scholarly research. *Annals of Operations Research, 345*(2), 1061–1104.

Li, C., & Umair, M. (2023). Does green finance development goals affects renewable energy in china. *Renewable Energy, 203*, 898–905.

Liu, Y., Zhang, S., Chen, M., Wu, Y., & Chen, Z. (2021). The sustainable development of financial topic detection and trend prediction by data mining. *Sustainability, 13*(14), 7585.

Maria, M. R., Ballini, R., & Souza, R. F. (2023). Evolution of green finance: A bibliometric analysis through complex networks and machine learning. *Sustainability, 15*(2), 967.

Mavlutova, I., Spilbergs, A., Verdenhofs, A., Natrins, A., Arefjevs, I., & Volkova, T. (2022). Digital transformation as a driver of the financial sector sustainable development: An impact on financial inclusion and operational efficiency. *Sustainability, 15*(1), 207.

May, E.C., Bassam, A., Ricalde, L. J., Soberanis, M. E., Oubram, O., Tzuc, O.M., Alanis, A.Y., & Livas-García, A. (2022). Global sensitivity analysis for a real-time electricity market forecast

by a machine learning approach: A case study of Mexico. *International Journal of Electrical Power & Energy Systems, 135*, 107505.
Mhlanga, D. (2021). Artificial intelligence in the industry 4.0, and its impact on poverty, innovation, infrastructure development, and the sustainable development goals: Lessons from emerging economies? *Sustainability 13*(11), 5788.
Mohsin, M., & Jamaani, F. (2023). Green finance and the socio-politico-economic factors' impact on the future oil prices: Evidence from machine learning. *Resources Policy, 85*, 103780.
Musleh Al-Sartawi, A. M., Hussainey, K., & Razzaque, A. (2022). The role of artificial intelligence in sustainable finance. *Journal of Sustainable Finance & Investment* 1–6.
Nguyen, Q., Diaz-Rainey, I., & Kuruppuarachchi, D. (2021). Predicting corporate carbon footprints for climate finance risk analyses: A machine learning approach. *Energy Economics, 95*, 105129.
Pincet, A., Okabe, S., & Pawelczyk, M. (2019). Linking aid to the sustainable development goals—a machine learning approach.
Rabbi, M. F., Sultan, M. N., Hasan, M., & Islam, M. Z. (2023). Tribal dress identification using convolutional neural network. *Journal of Information Hiding and Multimedia Signal Processing, 14*(3), 72–80.
Rastogi, R., Jaiswal, R., & Jaiswal, R. K. (2020). Renewable energy firm's performance analysis using machine learning approach. *Procedia Computer Science, 175*, 500–507.
Sadorsky, P. (2021). A random forests approach to predicting clean energy stock prices. *Journal of Risk and Financial Management, 14*(2), 48.
Sadorsky, P. (2022). Using machine learning to predict clean energy stock prices: How important are market volatility and economic policy uncertainty? *Journal of Climate Finance, 1*, 100002.
Trivedi, J., Chakraborty, D., & Nobanee, H. (2023). Modelling the growth dynamics of sustainable renewable energy—flourishing green financing. *Energy Policy, 183*, 113846.
Xin, C., Fan, S., & Guo, Z. (2024). Can digital finance promote inclusive growth to meet sustainable development in China? A machine learning approach. *Environment, Development and Sustainability, 26*(10), 26647–26677.
Zhang, D. (2018). *Energy finance: Background, concept, and recent developments*. Taylor & Francis.
Zhang, X., Ai, X., Wang, X., Zong, G., & Zhang, J. (2023). A study on the effects of digital finance on green low-carbon circular development based on machine learning models. *Mathematics, 11*(18), 3903.

Riadul Islam Rabbi is presently pursuing his Master of Engineering Science (Research Mode) in the Faculty of Engineering and Technology (FET) at Multimedia University (MMU) in Ayer Keroh, Melaka, Malaysia. He is also working as a Research Assistant (RA) in the Centre for Advanced Analytics (CAA) at MMU. He achieved his BSc degree in Computer Science and Engineering (CSE) from the International Islamic University Chittagong (IIUC), Chittagong, Bangladesh. His research interests are mainly Machine Learning, Deep Learning, Medical Image Analysis, and Explainable AI.

Ekramul Haque Tusher received the BSc degree in computer science from International Islamic University Chittagong (IIUC). He is currently pursuing his master's by research in Soft Computing and Intelligent Systems at University Malaysia Pahang AlSultan Abdullah (UMPSA) in Pekan, Pahang, Malaysia. Mr. Ekram has been working as a research assistant in the Machine Intelligence Research Group (MIRG) at UMPSA since 2023. His current research interests are in the areas of Machine Learning methods, Deep Learning, Fuzzy System, and Explainable AI.

Mahmudul Hasan is currently pursuing a PhD in Information Technology (IT) at Deakin University, Melbourne, Australia. He earned his BSc (Eng.) and MSc (Eng.) degrees in Computer Science and Engineering (CSE) from Hajee Mohammad Danesh Science and Technology University, Dinajpur, Bangladesh, in 2021 and 2023, respectively. He previously served as a Lecturer in the Department of CSE at the University of Creative Technology, Chittagong (UCTC), Bangladesh. He is the Founder and Director of the Center for Multidisciplinary Research and Development (CeMRD) and a moderator of "Be Researcher BD," the largest online research forum in Bangladesh. Additionally, he has taught online as a Data Science instructor to students in the USA, Italy, Denmark, South Korea, and Australia. He is also the founder of the online educational platform "MHM Academy." His research interests include federated learning, machine learning, deep learning, cybersecurity, health informatics, renewable energy, computational sociology, and business intelligence.

Md Rashedul Islam currently serves as a Computer Programmer in the Information Technology Cell at Hajee Mohammad Danesh Science and Technology University (HSTU) in Dinajpur, Bangladesh. He earned his MSc (Eng.) in Computer Science and Engineering from HSTU in 2013, following his BSc (Eng.) from the same institution. His research interests encompass machine learning, deep learning, data science, and cybersecurity, among other areas. Additionally, he possesses a strong passion for software development.

Machine Learning and Deep Learning Strategies for Sustainable Renewable Energy: A Comprehensive Review

Md Raihanul Islam Tomal, Alamgir Kabir, Mahmudul Hasan, Sayed Mahmudul Haque, and Md Mehedi Hasan Jony

1 Introduction

Electricity is integral to modern life, akin to two sides of the same coin. Accordingly human daily activities are deeply intertwined with electrical devices such as mobile phones, computers, televisions, and internet connections, among others. The list is virtually endless. Numerous companies thrive on the use of electricity. A contemporary and prominent topic is the electric car, which many researchers predict will dominate the future (Bhatti et al., 2021). Hence, a reliable supply of electricity is essential for sustained progress. However, generating electricity is a complex task, unlike mining coal directly from the ground. It is produced from secondary energy sources derived from primary sources, including fossil fuels. Fossil fuels such as oil, coal, and natural gas are usually used to generate power and are known as non-renewable or conventional energy sources (Hassan et al., 2021). These

M. R. I. Tomal
Faculty of Computing, University Malaysia Pahang Al-Sultan Abdullah, Pekan, Pahang, Malaysia

A. Kabir
Faculty of Economics, Prince of Songkla University (PSU), Songkhla, Thailand

S. M. Haque
Department of Computer Science and Engineering, Hajee Mohammad Danesh Science and Technology University, Dinajpur, Bangladesh

M. Hasan (✉)
Department of Computer Science and Engineering, Hajee Mohammad Danesh Science and Technology University, Dinajpur, Bangladesh

School of Information Technology, Deakin University, Geelong, VIC, Australia

M. M. H. Jony
School of Information Technology, University of Technology Sydney, Ultimo, NSW, Australia

M. Z. Abedin, W. Yong (eds.), *Machine Learning Technologies on Energy Economics and Finance*, International Series in Operations Research & Management Science 367, https://doi.org/10.1007/978-3-031-94862-6_11

non-renewable energy sources are high-density, allowing for quicker electricity generation compared to renewable sources. Unfortunately, harmful greenhouse gases are released by the combustion of fossil fuels into the environment, which is the cause of global warming and greenhouse effect (Zhang et al., 2024).

Therefore, it is imperative to seek alternatives to non-renewable energy to protect our planet and its inhabitants. Renewable energy presents a viable solution as it is environmentally friendly and emission-free, thereby not negatively impacting the earth. Renewable energy sources are able to meet demands of the current generations barring the future generation's needs. Consequently, global energy usage is on the rise, with researchers estimating a 56% increase by 2040 (Maamoun et al., 2020). To mitigate global warming, it is crucial to reduce carbon dioxide emissions and decrease reliance on non-renewable energy, while maximizing the renewable energy uses. Wind similarly to sunlight is a source of natural renewable energies and is increasingly being adopted worldwide.

The integration of artificial intelligence learning methods for conversion of energy, forecasting, and power prediction plays an important role in the advancement of sustainable energy sources. Accurate power prediction in electrical networks is crucial for the cost-effective combination of renewable power resources. The demand for sustainable forecasting energy sources is increasing daily, aiding a variety of applications from small-scale to large-scale power grids. It is anticipated that the solar and wind turbines installation, particularly in offshore locations, will reach unprecedented levels in the coming decades. Power output, such as solar and wind variability, is influenced by environmental factors and significantly impacts applications related to these energy sources (Alkhayat & Mehmood, 2021; Hasan et al., 2024a).

Hydropower energy prediction is another critical area, with hydropower being recognized for its efficiency, achieving around 90%. Many countries support hydropower as a simple and major source of renewable energy due to its ability to generate electrical energy by harnessing potential energy from higher to lower elevations. Hydropower is also known for its cost-effectiveness compared to other renewable energies, making it more economical (May et al., 2020; Hasan et al., 2024b). Geothermal energy, another renewable resource, relies on heat from within the earth, harnessed by injecting wells with water and drilling or antifreeze materials. Despite the potential, technological limitations and insufficient capital investment have left many geothermal plants underdeveloped. However, as opposed to the power plants that based on fossil fuel the geothermal plants release significantly fewer greenhouse gases. Advancements in machine learning and deep learning technologies have also enhanced prediction of power in ocean and tidal energy. Biomass, generated from organic materials such as plants and animal waste, produces 3×10^6 kcal mg^{-1} of heating value when used as an alternative source of energy. Despite the challenges, these various sources of renewable energy—biomass, tidal, geothermal, wind, and solar—hold immense potential for forecasting, prediction of power, and energy conversion. Properly trained deep learning and machine learning models can address these challenges effectively.

In this study, we evaluate comprehensive data and real-time metrics for the previously mentioned renewable energy sources using several applications of machine learning and deep learning techniques. Additionally, we discuss the challenges faced by these approaches in an effective manner. Our objectives are outlined below for better understanding.

- This review comprehensively examines well-known sources of sustainable energy such as solar, tidal, hydropower, and wind, focusing on the usage techniques, including forecasting, energy conversion, and power prediction, utilizing recent deep learning and machine learning techniques.
- This study gives a comprehensive exploration on forecasting, energy conversion, and power prediction approach according to the source of energy tidal, solar, hydropower, and wind, highlighting the advantages as well as disadvantages of using deep learning and machine learning techniques.
- This research aims to assist future researchers by identifying the challenges in existing studies, thereby facilitating the development of advanced technologies based on robust models for improved efficiency and application in renewable energy systems.

2 Methodology

The aim of this study is to classify energy resources based on their availability for long-term, categorizing them as non-renewable and renewable energy. The primary focus is on sustainable renewable energy sources, for instance hydropower, tidal, wind, and solar energy. This study presents a comprehensive survey of various approaches to forecasting, power prediction, and energy conversion to promote a sustainable environment. Figure 1 provides a diagrammatic taxonomy representation, illustrating the different types of sustainable sources of energy. This research centers on renewable energy sources, emphasizing forecasting, power prediction, and energy conversion of well-known renewable sources like hydro, wind, and solar energy.

2.1 Search String (Keywords)

In order to assure the relevance and comprehensiveness of the literature review, this study employed a list of relevant keywords to gather multiple research publications from different fields. Initially, keywords such as "Renewable Energy," "Sustainable Energy," "Solar Energy," "Wind Energy," "Tidal Energy," and "Hydro Energy" were used to identify papers based on their titles. Subsequently, the collected papers were further refined using additional keywords like "Power Prediction," "Energy Prediction," and "Forecasting." As the focus of this study is on machine learning

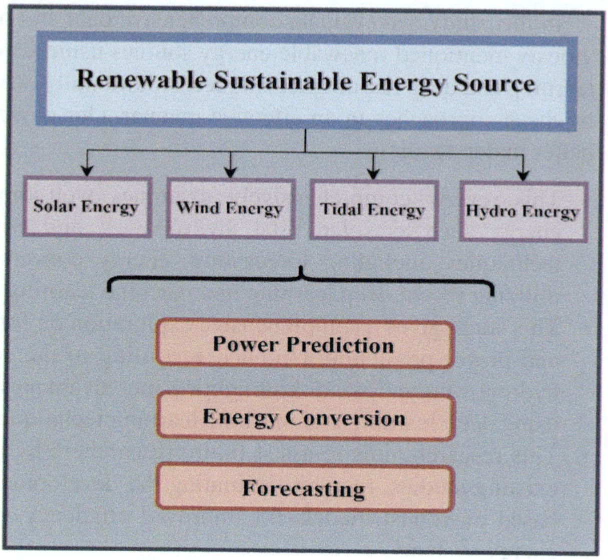

Fig. 1 Classification of machine learning approach in sustainable energy sources

and deep learning applications, the primary keywords for the literature search were "Machine Learning" and "Deep Learning." This systematic approach ensured the inclusion of pertinent research papers, providing a robust foundation for the study's objectives.

2.2 Databases and Paper Selection

This study primarily compiled documents from numerous sources that were relevant to the research objectives. The intention of document collection was to acquire high-impact journals, especially those included in Web of Science (WoS) and Scopus. The study utilized databases and platforms like IEEE, Springer, ResearchGate, and ScienceDirect to carry out this search.

2.3 Data Extraction and Synthesis

Using a standard extraction form, data extraction was carried out methodically. After selecting the appropriate keywords and setting up a database, the search was conducted, and the relevant articles and data were entered into an MS Excel spreadsheet to improve analysis and synthesis. According to the research subject matter, the data were quantitatively synthesized and summarized. Figure 2 displays the paper selection process, during which a total of 215 documents were chosen for primary screening. After selecting the relevant documents, a thorough screening

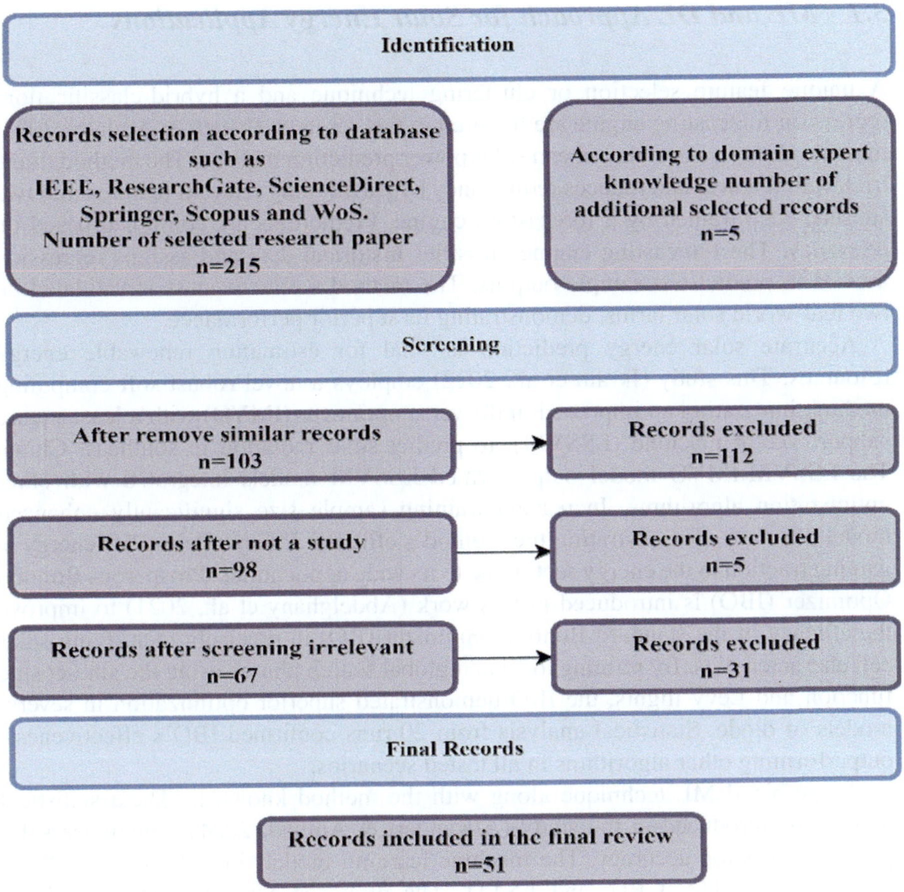

Fig. 2 Paper selection method

was performed utilizing the relevance of the study aims and purpose, and certain documents were excluded. Finally, a total of 51 articles were retrieved for review.

3 Descriptive Results

This section discusses the effectiveness in several sustainable energy sources such as tidal, hydro, wind, and solar which is based on recent research.

3.1 ML and DL Approach for Solar Energy Applications

A unique feature selection or clustering technique and a hybrid-classification-regression forecasting engine are featured in this research (Nejati & Amjady, 2022) that introduces a novel day-ahead solar power prediction method. The method filters irrelevant features and reduces redundancy by partitioning relevant features into two subsets, each trained by a forecasting engine. Predictions are combined based on relevancy. The forecasting engine classifies historical data and assigns regression models to predict test sample outputs. The method's effectiveness is validated on two real-world solar farms, demonstrating its superior performance.

Accurate solar energy prediction is vital for estimating renewable energy resources. This study (Ikram et al., 2022) employs a novel robust soft computing method, integrating an improved multi-verse optimizer (IMVO) with a least square support vector machine (LSSVM), to predict solar radiation in southeast China. The LSSVM-IMVO model outperformed LSSVM models integrated with other optimization algorithms. Increasing training sample size significantly enhanced model accuracy, demonstrating the method's efficacy. Photovoltaic (PV) energy is gaining traction in the energy sector due to its wide applications. Prosperous Bonobo Optimizer (IBO) is introduced in this work (Abdelghany et al., 2021) to improve the efficacy of the standard Bonobo Optimizer (BO) in precisely identifying solar cell characteristics. By refining local and global search phases using the sine-cosine function and Levy flights, the IBO demonstrated superior optimization in several models of diode. Statistical analysis from 20 runs confirmed IBO's effectiveness, outperforming other algorithms in all tested scenarios.

A combined ML technique along with the method known as Theta statistical method is introduced in the study (AlKandari & Ahmad, 2024) to enhance solar power forecasting accuracy. The machine learning models include the new Auto-GRU, Auto-LSTM, GRU, and LSTM. The proposed Statistical Hybrid Model (MLSHM) and Machine Learning utilize structural and data diversity and integrate predictions using four combining methods. Validated on datasets from Shagaya, Kuwait, and Cocoa, USA, the MLSHM demonstrated superior accuracy over traditional models, proving the effectiveness of integrating statistical methods with machine learning. However, this study (Munawar & Wang, 2020) develops a framework to evaluate and identify the perfect combinations of ML models then feature selection methods to forecast short-term solar power, essential for renewable energy integration. It examines few models such as XGBoost, artificial neural network, and random forest, alongside feature importance and principal component analysis (PCA) techniques. The research finds that XGBoost with PCA-selected features provides the best forecasting performance for solar power in Hawaii, US. The framework offers a robust method for selecting optimal ML techniques to forecast solar. In the study conducted in (Almeshaiei et al., 2020), researchers introduced innovative strategies where they assess micro-scale PV panel's performance for specific applications, which are combined with neural networks, short-term real data, and empirical lab testing. The method evaluates power output under various conditions, including seasonal, hourly, temperature, dust accumulation, and tilt angle. The approach was tested in Kuwait and demonstrated a maximum error of 23% compared to actual data, with correlation values between 87.3% and 91.9%. These findings suggest the method can provide rapid, accurate assessments, aiding manufacturers in decision-making and reducing investment risks. In order to address the lack of observation stations and the complex spatial patterns, this work (Koo et al., 2019) presents a novel machine learning approach in China that calculates the monthly average daily solar radiation using an advanced model known as k-means clustering and case-based reasoning (A-CBR). Data from 97 cities over 10 years (2006–2015) were utilized, achieving a prediction accuracy of 93.23%. The approach can be generalized using interpolation methods like kriging in GIS, aiding decision-makers in effectively implementing solar energy systems by determining optimal locations, sizes, and forms. This study (Mehrpooya et al., 2021) explores an integrated energy conversion system combining modeled in AspenTech v9.1, a coal-fueled molten carbonate fuel cell (MCFC) coupled with a gas turbine and solar thermochemical water-splitting hydrogen production. The zinc/zinc-oxide cycle enhances efficiency by directly using solar reactors, while the MCFC utilizes syngas from coal gasification. The system achieves an overall 85% approximately efficiency, electricity producing 13.63 MW, with the HHV efficiency, LHV 61% and MCFC showing 63%. Sensitivity analysis identifies current density, voltage, and fuel cell pressure as key performance factors. The challenges and benefits of implementing big data analytics in sustainable energy power stations within smart grids are addressed in the study of (Mostafa et al., 2022). Using a dataset of 60,000 instances and 12 variables, a five-step solution is described that uses several machine learning techniques to forecast the stability of the smart grid. The penalized linear regression model yielded an accuracy of 96%, while the random forest model yielded 84%, the decision tree model produced 78%, and the gradient boosted decision tree and CNN models produced 87%. The main limitation is the relatively small dataset, suggesting future research should involve larger, more diverse datasets across multiple countries. This study (Tercan et al., 2022) explores the techno-economic advantages of using partition energy reserve to enhance photovoltaic self-consumption in varying penetration rates of prosumer community. The desirable energy reserve was achieved through the application of the Best New Algorithm also technical performance simulations conducted along with genetic algorithm using PSS Sincal. Economic feasibility was assessed by considering residual energy and various incentives, utilizing point of reference internal rate of return, net present value, and payback period. The implementation of shared energy storage resulted in an increase in self-consumption by up to 11%, providing substantial economic advantages and enhancing power quality. In order to achieve practically zero-energy communities, this study (Liu et al., 2022) proposes an innovative distributed energy system (DES) that integrates cutting-edge solar energy technology and hybrid energy reserve (containing heat, ice, and electricity storage). The DES is optimized for environmental and economic factors, employing a new operational strategy to enhance system performance. Evaluation metrics include carbon emissions reduction and net interaction improvements compared to traditional systems, demonstrating potential benefits in achieving zero-energy targets, particularly for office buildings. Equipment costs, electricity price, and carbon tax for sensitivity analysis further support the system's viability and sustainability.

3.2 ML and DL Approach for Wind Energy Applications

In order to integrate volatile renewable wind power into sustainable energy systems, this research (Zhao & You, 2022) introduces an assurance framework—a robust and creative unit. This frame of work employs vagueness sets of data-driven partitive, leveraging machine learning techniques to manage uncertain intermittent power outputs effectively. It utilizes K-means and DBSCAN clustering methods to organize uncertainty data, constructing disjunctive sets from multiple basic uncertainty types. The approach is applied to a two-stage adaptive robust unit commitment model with a tailored optimization algorithm, demonstrating significant reductions in robustness costs and computational time compared to traditional methods. Report on 118-bus systems and IEEE 39-bus validate their effectiveness in enhancing economic performance while ensuring reliable power system operations. Using signal processing techniques, this research (Zhang & Chen, 2022) introduces a novel way to increase the speed of wind prediction accuracy. It accomplishes singular value decomposition (SVD) and complete ensemble empirical mode decomposition with adaptive noise (CEEMDAN) to preprocess data, followed by prediction using autoregressive integrated moving average model (ARIMA) and Elman neural network optimized by particle swarm optimization (PSO). The model enhances prediction effectiveness, reduces errors, and supports stable operation of wind farms and grid-connected power plants. Results demonstrate its potential to contribute significantly to sustainable wind energy utilization and environmental conservation efforts. For short-term wind turbine power forecasting in a utility-scale wind farm, this study (Meka et al., 2021) presents a robust deep learning approach using temporal convolutional networks (TCNs). The TCN model is optimized using an orthogonal array tuning method based on Taguchi design, demonstrating superior performance across various wind speeds compared to existing methods. Validation with 12 months of data from an 86-turbine wind farm confirms the efficacy of the proposed TCN model in capturing temporal dynamics and meteorological relationships for accurate power predictions.

This research (Wang et al., 2021) introduces SIRAE (Stacked Independently Recurrent Autoencoder), an innovative DL framework tailored for ultra-short-term wind power prediction. Utilizing variational mode decomposition for data preprocessing, SIRAE employs independent recurrent autoencoders (IRAE) to capture structural features and temporal dependencies in power of wind data. The exploratory outcome indicates that SIRAE significantly outperforms existing models, achieving notable improvements into the root mean square error across different months comparing with the persistence model. The approach is highlighted for its effective and stable forecasting performance, showcasing its potential in enhancing grid operation reliability. In order to enhance wind energy efficiency, this study (Aksoy & Selbaş, 2021) employs machine learning algorithms to predict energy production using wind turbine data from 2015. Achieving 90% accuracy, a mathematical model estimates energy output using temperature, wind speed, and direction inputs. A user-friendly computer program was developed to disseminate these results, emphasizing practical application and potential efficiency gains in wind energy production. In contrast, this paper (Fathy et al., 2022) addresses the challenge of optimizing wind energy generation under varying weather conditions by proposing an Archimedes Optimization Algorithm (AOA) for Maximum Power Point Tracking (MPPT). The system integrates a wind turbine with a constant magnet synchronous generator and employs a boost converter controlled by AOA to maximize electrical output power. Evaluations across the speed of real wind, variable and fixed in Saudi Arabia demonstrate superior performance of AOA-MPPT compared to other algorithms like electric charged particle optimization, grasshopper optimization, and cuckoo search, validating its robustness in wind energy systems. This study (Rushdi et al., 2020) focuses on harnessing wind energy using kites, specifically a kite system introduced through Kyushu University which traction power is 7 kW. Experimental data from the system were taken advantage of to train ML regression models for predicting tether forces. Key input parameters were identified through sensitivity analysis, and various regression models, including neural networks, were evaluated for accuracy in predicting tether forces. The results demonstrate promising capabilities in accurately forecasting tether forces for new input combinations, potentially facilitating optimal design and power generation improvements.

Deep reinforcement learning (DRL) is the focus of this research (Yang et al., 2020), which attempts at enhancing revenue generation for wind power producers (WPPs) in deregulated contexts. The method employs a data-driven controller that utilizes electricity prices and wind generation forecasting to determine optimal steps such as reserve purchase schedules and energy storage system (ESS) operations. Implemented with the Rainbow algorithm, the approach improves upon traditional DRL methods by accommodating continuous input states, thereby optimizing control strategies effectively amidst uncertainties. Simulation results demonstrate significant revenue benefits for WPPs under varying conditions of electricity price and wind power uncertainties. This study (Emrani et al., 2022) introduced a novel methodology for optimizing the design and arrangement of a hybrid PV-Wind plant within a gravity energy storage (GES) system to enhance technical and economic competitiveness. Using a genetic optimization algorithm, the study aims to minimize construction costs while ensuring structural integrity against mechanical loads. A case study validates the approach, revealing optimal dimensions of 48 m height, 24 m diameter, and 3 m wall thickness, with a total construction cost of 6.7 M€. Integration with the hybrid plant facilitates efficient renewable energy dispatching, mitigating issues of overcharging/discharging. Compared to battery storage, GES demonstrates superior performance with regard to discharge depth, lifetime, also efficiency, making it a promising solution for renewable energy integration. This study (Wang et al., 2022) introduces a novel self-adjusted triboelectric nanogenerator (SA-TENG) designed for efficient harvesting of random wind energy. The SA-TENG dynamically adjusts its driving-torque to match varying wind speeds (5.0–13.2 m s^{-1}), achieving a peak power output of 7.69 mW. Compared to conventional TENGs and electromagnetic generators, SA-TENG exhibits significantly improved power growth rates and energy conversion efficiencies, demonstrating its potential as a distributed energy source for environmental monitoring sensors. Similarly, this study (Angadi et al., 2022) introduces an innovative maximum power tracking algorithm for hill climbing, designed for a stand-alone self-excited induction generator primarily powered by wind energy, which can drive an induction motor pump. By utilizing a single voltage source converter (VSC) with the VSC operating frequency as the control variable and incorporating a feed-forward hill-climbing algorithm, the system significantly enhances stability and efficiency without relying on speed sensors. Both simulation and experimental results validate the algorithm's effectiveness under varying wind and load conditions, presenting a robust and economical solution for remote stand-alone applications.

3.3 ML and DL Approach for Hydro and Tidal Energy Applications

The energy sector faces challenges like rising demand, efficiency issues, and changing supply patterns. This research paper (Chen et al., 2021) proposes an Artificial Intelligence-based Evaluation Model (AIEM) for forecasting renewable energy's impact on the economy and enhancing energy efficiency. The study uses AI to address challenges such as consumer selection, competitive pricing, scheduling, facility management, and incentivizing demand response. The AIEM model aims to boost energy efficiency to 97.32% and optimize renewable energy utilization, providing significant economic insights and improvements. This study (Mostafa et al., 2020) introduces an evaluation model which can evaluate by considering various approach and their technical characteristics of short-term, medium-term, and long-term energy storage. The model integrates economic factors, for example disposal costs, replacement, capital, maintenance, and operation. Main magnitudes, including the reckoned annually the levelized cost of energy (LCOE) and life cycle cost of storage (LCCOS), are for guide energy storage functional decisions. A sensitivity analysis further aids in assessing the economic viability of energy storage systems, providing a robust decision-making tool. The paper (Sahu et al., 2022) presents a powerful comptroller for regulating the frequency in an microgrid with various uncertainties known as Tilt Fuzzy Cascade. The microgrid, incorporating renewable energy sources with low inertia, suffers from frequency stability issues. To address this, fuzzy cascade controller based on a tilt is employed, optimized using a novel deep Q-network (DQN) algorithm. Comparative assessments validate the controller's effectiveness, demonstrating that significantly enhances frequency of the DQN-optimized tilt fuzzy cascade controller regulation in microgrids. This study (Ma et al., 2021) proposes two hybrid thermal energy storage systems (HTESS), outlet temperature control is indicated through HTESS-OTC and thermocline storage is known as HTESS-TS. Comparative analysis shows that HTESS-TS and HTESS-OTC improve utility factors by 12.5% and 22.1%, respectively, over single-tank thermal energy storage systems, with HTESS-OTC reducing unit costs by 8.6%. Additionally, annual electricity generation increases by 9.8% and 14.1%, respectively, demonstrating enhanced performance and economic benefits.

This work (Fonseca et al., 2021) addresses the growing need for climate change mitigation and energy security by proposing a strategy for decentralized power plant deployment, which is mentioned as multi-criteria. The approach includes various energy vectors and considers the time-varying operations and seasonal storage system behaviors. Economic, environmental, and social aspects are evaluated, focusing on annual cost, dependence on grid and CO2. Accuracy shows significant benefits of decentralized generation over centralized systems, with potential emission reductions up to 89% of CO2 and self-sufficiency improvements up to 81%, power plant structure and policy highlighting the influence of assessed criteria. This study (Naik et al., 2022) addresses the need for an effective strategy of power management in a DC microgrid (MG) accomplishing micro hydro power plant (MHPP) sources, battery and photovoltaic. Due to technical constraints, such as MHPP's battery C-rate limitations and mechanical response time, load dynamics cannot be instantly compensated. SPMS, known as Supervised Power Management Scheme, optimizes MHPP's contribution during load transients while considering battery limitations. The SPMS's effectiveness is validated through hardware-in-loop experiments, demonstrating stable power flow control and voltage stability in the DC MG during load transients. The challenges of predicting renewable energy levels in light of their fluctuation are addressed in this study (Abd El-Aziz, 2022) by integrating the Cat Boost algorithms with Support Vector Regression and Multilayer Perceptron. This hybrid approach aims to enhance the predictability and performance of sustainable energy consumption. Evaluations of the system described at both train and test levels show that it outperforms other current methods, offering high prediction accuracy, lower costs, and improved overall system performance. This paper (Zhao & Kok Foong, 2022) explores a hybrid approach for predicting the electric power (PE) output of combined cycle power plants (CCPP) using an artificial neural network (ANN) with an electrostatic discharge algorithm (ESDA). Considering factors like relative humidity, atmospheric pressure, exhaust vacuum, and ambient temperature, a $4 \times 9 \times 1$ network structure is employed. The ESDA-ANN hybrid demonstrated superior performance compared to conventionally trained ANNs, including the Levenberg-Marquardt algorithm. The study concludes that the ESDA-ANN is a robust and reliable tool for PE modeling, offering improved prediction accuracy and computational efficiency.

The above discussion outlines the analysis and findings using ML and DL approach related to predicting power, forecasting, also conversion of energy. Despite the promising outcomes from these approaches, they faced certain exceptions, particularly with regard to prediction accuracy, improper load balancing, and power management during energy conversion. Addressing these limitations, future work aims to build productive and robust methodologies to conquer the issues present in current models.

4 Methodology-Based Results

4.1 Renewable Solar Energy

Solar energy is one of the most prominent and widely utilized renewable energy sources, renowned for its purity and lack of carbon emissions. The Earth receives approximately 140 PW (petawatts) of power from sunlight, though only about 36 PW is feasibly harnessed for practical use. There are two primary methods for harvesting energy from solar: Concentrated Solar Power (CSP) and Photovoltaic (PV) systems. Anyhow, solar energy does have certain limitations, including reduced efficiency during cloudy weather and nighttime. PV panels capture radiation from sunlight, which consists of direct radiation, diffuse radiation, and ground-reflected radiation, with direct and diffuse radiation contributing the most to the total solar radiation. Eqs. (1)–(4) delineate the methods used to accurately evaluate solar radiation, with Eq. 1 specifically representing the evaluation method for direct radiation.

$$I_{br} = G_{sc}P^{M} \tag{1}$$

Here, G_{sc} represents the solar constant, P represents the transparency factor as well as M represents the mass of the air. The following Eq. (2) represents the methods of evaluating M.

$$M = \frac{1}{\mathrm{Sin}\alpha} \tag{2}$$

In this equation, α represents solar altitude angle. The following Eqs. (3) and (4) represent direct radiation of the tilted plane and the horizontal plane.

$$I_{bH} = I_{br}\sin\alpha \tag{3}$$

$$I_{b\beta} = I_{br}\,(\sin\alpha\ \ \cos\beta + \cos\alpha\ \ \cos\gamma\ \ \sin\beta) \tag{4}$$

According to the equation, β represents tilt angle, γ represents azimuth angle. Now the horizontal plan, titled plan, and ground-reflected radiation are represented by I_{dH}, $I_b\beta$, and I_p, respectively; moreover, tilted angle represented by β and p represent diffused ground reflectance, the diffused radiation is evaluated through Eqs. (5) to (7) as follows:

$$I_{dH} = 0.5\ G_{sc}\frac{1 - p^M}{1 - 1.4\ln P}\sin\alpha \tag{5}$$

$$I_{d\beta} = \cos^2\frac{\beta}{2} I_{dH} \tag{6}$$

$$I_p = H_p \left(\frac{1 - \cos\beta}{2}\right) \tag{7}$$

Here H is evaluated by Eq. (8) as follows:

$$H = I_{bH} + I_{dH} \tag{8}$$

The sum of the $I_{b\beta}$, $I_d\beta$, I_p is the total solar radiation from the sun that evaluated through Eq. (9):

$$I_R = I_{b\beta} + I_{d\beta} + I_p \tag{9}$$

There are some stages that are undergone to convert solar energy into electrical energy that are mentioned below:

The early step in harnessing solar energy involves the absorption of solar radiation by the cells of a solar panel. Each solar cell comprises two thin layers of silicon semiconductor material, which can function as both insulators and conductors. These layers are classified as P-type and N-type materials, representing positively and negatively charged layers, respectively. When sunlight strikes the surface of the solar panel, it interacts with small energy packets known as photons. The interaction between photons and the PV material generates electricity. Table 1 presents solar energy related recent research which utilizes ML and DL techniques.

The main consumers of solar energy are mostly from household usage rather than the application of industry. However, Table 1 describes very recent research of solar energy-based ML and DL techniques that are used in the process of predicting power, forecasting and conversion energy.

4.2 *Renewable Wind Energy*

To produce electrical power, wind energy utilizes the force of the wind. Here's a brief overview of how it works:

- **Wind Turbine Energy:** This energy which consists of large blades mounted on a tower captures wind energy. The kinetic energy of the wind turns the turbine blades.

Table 1 Solar energy research based on ML and DL for forecasting, conversion of energy, and power prediction

Reference	Proposed method	Advantage	Limitation
(Rajasundrapandiyanleebanon et al., 2023)	The paper proposes utilizing ML and DL approaches to forecast global solar radiation (GSR). It highlights a shift from conventional mathematical methods to intelligent AI techniques for predicting solar energy availability	AI techniques significantly enhance the reliability of modeling complex and unpredictable relationships between input and output data. This approach improves data-driven decision-making in the solar energy sector, making planning more judicious and functional	While AI techniques offer improved accuracy and reliability, they can be intensive computationally also large datasets are required for training. Additionally, the implementation of these methods may require specialized knowledge in AI and machine learning, which can be a barrier for widespread adoption
(Lin et al., 2020)	In this work, an enhanced algorithm introduced known as moth-flame optimization. This algorithm is introduced for SVM prediction of photovoltaic power generation. The improvements include an inertia weighting strategy and the Cauchy mutation operator. The inertia weighting strategy balances the search and mining capabilities at the population location search equation, while the Cauchy mutation operator enhances population diversity and prevents entrapment in local optima. Additionally, gray relational analysis for the model prediction is used optimistically input data	The proposed research demonstrates Australia's power station, showing superior optimization performance, as verified through several test functions and real-world data from a PV power station. By accurately predicting PV output, the method helps maintain power system stability, minimizing the influence of PV penetration of the power into the grid which ensures reliability of the system	While the proposed method shows improved prediction accuracy and system reliability, its effectiveness may still be influenced by the weather conditions unpredictability and the inherent variability, which can affect PV power generation. Furthermore, the complexity of the algorithm may require significant computational resources and expertise for implementation

(Stoean et al., 2023)	This study proposes using Bidirectional LSTM (BiLSTM) models and Long Short-Term Memory (LSTM), enhanced with hyper parameter tuning via an improved Reptile Search Algorithm (RSA), to forecast solar energy generation. The data used includes both information of corresponding weather and energy generation from solar through time series values	This research method approach outperforms several state-of-the-art metaheuristic optimizations on the similar task. It achieves a high R2R^2R2 value of 0.604 and a normalized MSE value of 0.014, marking a progress of near 13% over consecutive ML models	The proposed method's effectiveness may depend on the specific dataset and test case. Additionally, the complexity of hyperparameter tuning and need for extensive computational resources can pose challenges for practical implementation
(Kaushaley et al., 2023)	The paper proposes a model named Optimized Artificial Neural Network (OANN) for solar power forecasting, incorporating optimization techniques such as the Seagull Optimization Algorithm (SOA-ANN) and the Crow Search Algorithm (CSA-ANN). These models are compared with a standard Artificial Neural Network (ANN) for long, mid-, and short-term solar power predictions using temperature, irradiation, and time as input parameters	The SOA-ANN model significantly improves prediction accuracy for solar power generation over traditional ANN and CSA-ANN models, especially for long-term and mid-term forecasts. This is demonstrated through statistical parameters such as R^2, MAPE, MSE, and MAE, showing improvements of 3.61%, 1.67%, 16.05%, and 6.54%, respectively	While the SOA-ANN model shows improved performance, the CSA-ANN model does not perform as well overall, particularly for higher frequency values. Additionally, the ANN model struggles with mid-term and long-term predictions, indicating that further refinement and testing may be necessary for comprehensive forecasting solutions

(continued)

Table 1 (continued)

Reference	Proposed method	Advantage	Limitation
(Khan et al., 2022)	The paper introduces a generally applicable algorithm for forecasting accurate solar energy, called DSE-XGB, a stacked ensemble algorithm. This model combines two DL algorithms, long short-term memory and artificial neural network, as base models. The predictions improve forecast accuracy. The model's performance is assessed on several data sets that are based on solar generation	The DSE-XGB method offers improved accuracy and consistency in solar energy forecasting compared to individual models like ANN, LSTM, and Bagging. It demonstrates a 10–12% improvement in R^2 value over these models, making it highly effective in dealing with weather variations. The use of the shapely additive explanation (SHAP) framework provides deeper insights into the learning mechanism of the algorithm	Details about the specific limitations do not mention in this research, still potential limitations could include the complexity of implementing ensemble models, the computational resources required for training, and the need for extensive data to achieve optimal performance. Additionally, the robustness of the model in entirely different climatic or geographical conditions may need further validation
(Lee, 2022)	The study proposes using two tree-based ensemble ML models, XGBoost and Random Forest, in order to predict the open-circuit voltage (Voc) of non-fullerene acceptor-based organic solar cells with reasonable accuracy. The models leverage intrinsic electronic parameters to make predictions. Moreover, the Shapley Additive Explanations analysis is employed to interpret the XGBoost model and visualize the correlation between the frontier orbital energies and Voc	The proposed ML techniques provide an accurate experimental method to predict the Voc of NFAs-OSCs, facilitating the efficient design of new donor-NFA combinations. This method surpasses the conventional statistical models by accounting for the complex, nonlinear relationships between electronic parameters and Voc, thereby enhancing prediction accuracy and interpretability	The study acknowledges that predicting Voc using a numerical method remains challenging due to the vast number of possible donor-NFA combinations and the uncoordinated relationship between HOMO(D)-LUMO(A) and Voc offsets. Additionally, the application of these machine learning models might require extensive computational resources and a substantial amount of data to achieve optimal performance

(Bouzgou & Gueymard, 2019)	The study introduces the Wrapper Mutual Information Methodology (WMIM), which combines mutual information with an Extreme Learning Machine (ELM) for forecasting solar irradiance time series. This approach is applied to short forecast horizons ranging from 5 min to 3 h ahead and is evaluated against three scenarios such as principal component analysis, partial space, and full space	The WMIM demonstrates superior forecasting performance compared to other strategies, such as PCA, by effectively reducing the historical input space, which enhances prediction accuracy. Specifically, WMIM achieves better coefficients of determination (R^2) and Normalized lower Mean Absolute Percentage Error (MAPE) and Mean Squared Error (NMSE) across various forecast horizons. Additionally, the ELM technique used in WMIM is significantly more computationally efficient than conventional methods like the Multi-Layer Perceptron (MLP)	The method's performance is evaluated using data from only two arid sites, which may limit its generalizability to other climatic regions. Furthermore, while WMIM shows improved accuracy and efficiency, the complexity of the mutual information-based variable selection and the dependency on the quality of input data could pose challenges in diverse practical applications

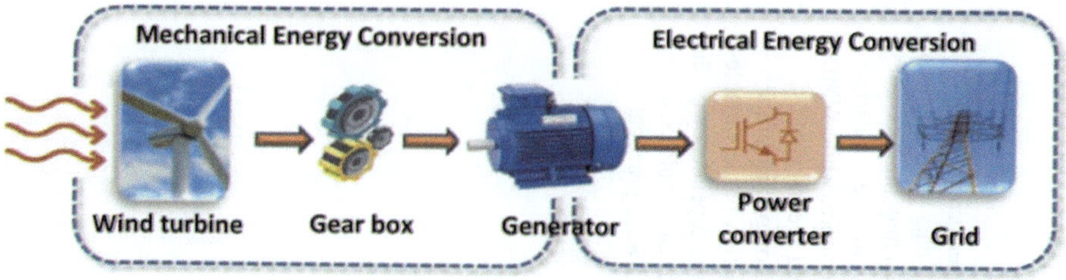

Fig. 3 Wind energy conversion system schematic diagram (Abdelateef Mostafa et al., 2023)

- **Rotor and Shaft:** The movement of the blades spins a rotor connected to a main shaft inside the turbine.
- **Gearbox:** The main shaft turns into a gearbox that increases the rotational speed for electricity generation.
- **Generator**: A generator known as the gearbox powers, which transforms the mechanical energy into electrical energy.
- **Transformer and Grid Connection**: The generated electricity is then passed through a transformer to match the voltage level of the grid. Finally, it is transmitted and distributed through the electricity grid to homes and businesses.
- **Control Systems:** Modern wind turbines have control systems that adjust the blade angle and direction to ensure safe operation and optimize energy capture during differing wind conditions.

This process of converting kinetic energy of wind into electrical energy is a clean, renewable way to generate power. Figure 3 emerges the entire process. Table 2 presents recent research studies on wind energy that utilize ML and DL approach.

Moreover, Table 2 presents recent research on wind energy utilizing ML and DL techniques for predicting power, forecasting, and conversion of energy. Wind energy undergoes several critical stages before it can be converted into electricity. During the generation process, power losses occur due to heat production, which affects the reliability of wind turbines. By appropriately utilizing ML and DL approach, it is possible to solve these issues and enhance power generation capabilities. These advanced methods can improve durability and forecast accuracy for electricity generation and pumping applications.

4.3 Tidal and Hydro Energy

Tidal energy is a type of hydropower that transforms the energy derived from tides into practical forms of power, mainly electricity. Use the kinetic energy of moving water to turn turbines, like how wind turbines work with air. This energy production is environmentally friendly and low emission, but the initial infrastructure cost is high. Hydro energy, or hydropower, is a form of sustainable energy that harnesses

Table 2 Wind energy research based on ML and DL for forecasting, conversion of energy, and power prediction

Reference	Proposed method	Advantage	Limitation
(Rajesh et al., 2022)	The proposed method combines Recalling Enhanced Recurrent Neural Network (RERNN) and Dynamic Differential Annealed Optimization (DDAO) to create a hybrid strategy called D2AORERN2. DDAO optimizes input parameters such as time to minimize error in rectifier power, DC current and rectifier DC voltage, generating a dataset for training. RERNN then utilizes this dataset to find the desirable solution, adjusting the DC rectifier to control inverter switch pulses	The proposed D2AORERN2 approach achieves higher efficiency (99%) compared to existing methods like Hill Climb Search (HCS) (89%), 81% efficiency in Genetic Algorithm and 85% from Particle Swarm Optimization (PSO), demonstrating superior performance in tracking the system of conversion wind energy for maximum power	The implementation and complexity of the hybrid method might be higher compared to traditional techniques, potentially requiring more computational resources and sophisticated tuning of parameters to ensure optimal performance in various operating conditions
(Li et al., 2023)	The proposed approach merges deep reinforcement learning (DRL) with federated learning to create a scheme for forecast ultra-short-term wind power scheme called federated deep reinforcement learning (FedDRL). This method enhances prediction accuracy using the deep deterministic policy gradient (DDPG) algorithm as the core forecasting model. FedDRL embeds this model within the federated learning framework, enabling the sharing of model parameters rather than private data, thereby preserving data privacy and addressing the issue of data silos	To forecast accuracy, FedDRL outperforms traditional prediction methods. It effectively protects data privacy by sharing only model parameters rather than private data, and it reduces communication pressure compared to centralized forecasting methods. This decentralized approach ensures robust and accurate predictions while safeguarding sensitive information	Implementing FedDRL requires sophisticated integration of federated learning and DRL, which may involve complex setup and higher computational resources. Additionally, the performance and robustness of the method can depend on the selection and tuning of federated learning parameters, which may require significant expertise and experimentation

(continued)

Table 2 (continued)

Reference	Proposed method	Advantage	Limitation
(He et al., 2022)	The proposed method combines numerical weather prediction (NWP) analysis with DL models for forecasting short-term wind power. The approach uses the lowest abundance of the highest topical methods to select main component from the data of NWP. These factors are then analyzed to categorize weather patterns. CNN and LSTM, two DL models are applied for different weather types. The final forecast results are integrated from both models	This method enhances forecasting accuracy and extends the forecast period, addressing the demands of economic dispatching and day-ahead power markets. It effectively handles different weather circumstances and outperforms traditional methods such as Radial Basis Function (RBF), Extreme Learning Machine (ELM) and Support Vector Machine (SVM), for accurate forecasting	The implementation of this combined model requires significant computational resources and expertise in both deep learning and numerical weather prediction. Additionally, the effectiveness of the approach depends on the accurate selection and analysis of weather patterns, which may introduce complexity and potential for error. The integration of multiple models may also increase the overall system's complexity and maintenance requirements
(Bin Abu et al., 2024)	The research assesses the global situation of wind and solar energy reception by examining their historical growth, current trends, and emerging technologies like vertical-axis wind turbines and floating solar. It also explores the role of energy storage alternatives and smart grid technology in improving renewable energy productivity. Additionally, the integration of Electric Vehicles (EVs) into modern smart grids and the application of ML for optimizing wind and solar energy generation are assessed	The review highlights the possibility of wind and solar energy to significantly climb overall energy demands sustainably. It underscores the benefits of advanced technologies and smart grids in improving energy efficiency and reliability. Furthermore, it points out the economic benefits, recent technological advancements, and the environmental and social implications of renewable energy adoption, providing valuable insights for researchers, industry leaders, and policymakers	The assessment identifies challenges in the sustainable energy industry, for example the variability of wind and solar energy and the significance for improved storage solutions of energy. In addition, it points out cost and the complexity of implementing smart grid technologies and integrating machine learning applications. Additionally, the evolving nature of renewable energy technologies may pose uncertainties and require continuous adaptation and innovation

(Zhang et al., 2021)	The study introduced method for the online self-tuning of Power System Stabilizer (PSS) parameters called sparsity-promoting adaptive control method. This method uniquely combines the sensitivity analysis theory and Deep Deterministic Policy Gradient (DDPG) algorithm to train an agent to learn a sparse correlated policy control for multiple PSSs. The trained agent can apply control signals only when necessary and the primary PSS parameters which significantly affect system stability	The proposed method enhances the damping of low-frequency oscillations and improves robustness against the variability of wind energy. By applying control signals selectively and sparingly, it optimizes the use of resources and maintains system stability more effectively than traditional PSS tuning methods or other adaptive control approaches	The complexity of training the DDPG agent and the need for accurate sensitivity analysis may pose implementation challenges. Additionally, the method's performance ponderously depends on the training data quality and the accuracy of the system model, which might limit its applicability in highly dynamic or poorly modeled systems
(Zhang & Li, 2021)	The research introduces a new method of downsizing based on the bidirectional gated recurrent unit (BiGRU) to study future offshore wind energy resources in China. This method is applied to downscale the result of simulation. The new downsizing technique is compared and validated with traditional methods, focusing on improving the accuracy of wind speed predictions in shallow water and coastal areas	The BiGRU-based downscaling techniques significantly reduce biases and improve the spatial consistency of downscaled wind speed data compared to traditional methods. This results in more accurate projections of offshore wind speeds and wind power density, providing valuable insights for the planning and siting of wind farms. The method's application over a long-term period (2021–2100) and its high resolution (0.25°) offer detailed and reliable data for future wind energy resource assessment	While the new method shows improved accuracy, it may still face challenges related to the inherent uncertainties of climate models and future scenarios. The reliance on CMIP6 projections means that any inaccuracies or biases in these models could affect the downscaled results. Additionally, the method's effectiveness in regions with highly variable or complex meteorological conditions needs further validation

(continued)

Table 2 (continued)

Reference	Proposed method	Advantage	Limitation
(Shirzadi et al., 2022)	The research introduced an approach that optimizes the daily operational cost of a power system is known as Mixed-Integer Linear Programming (MILP) approach contains wind turbines, batteries, and a conventional grid. This method also focuses on increasing system resilience. The study employs deep learning and statistical models, along with a novel hybrid model, to forecast load demand and wind power output 3 days in advance. For load dispatch of the urban microgrid usually MLP model is applied	The hybrid model improves prediction accuracy for load demand and wind speed, reducing 22% to 44% error of root mean squared according to forecast load and 10.5% to 16.6% according to the prediction of wind speed. The inclusion of battery storage reduces the grid-connected daily operational cost significantly, turning a cost of \$8.4/day into an income of \$109.8/day. The approach enhances system resilience by providing an off-grid mode and minimizes battery degradation cost, thus extending battery life	The off-grid mode compared to the power of wind curtailment cost; the degradation cost of batteries becomes a more significant portion of the operational costs. The study's findings are specific to the tested urban microgrid and may need adaptation for different configurations or larger-scale implementations. The MILP model, while effective, might require substantial computational resources and time for larger, more complex systems

the power of water in motion, such as flowing rivers or waterfalls, to generate electricity. The process of generating electricity is quite interesting. First, water stored in reservoirs is released through turbines, generating electricity as it flows down. Second, it utilizes the natural flow of rivers without large reservoirs. Third, turbines are placed in the flow of the river to generate power. Lastly, it uses two different elevations of water reservoirs. Water is pumped to the higher reservoir during low demand and released to generate electricity during peak demand. The main advantage of this research is we can get clean energy sources and high efficiency, with 90% conversion rates. It can be used for irrigation and flood control. Despite that, the main limitation is the high initial capital costs for dam construction. Table 3 represents recent research studies on tidal and hydro energy that utilize ML and DL approach.

5 Conclusion

In recent times, artificial intelligence-based learning applications have demonstrated significant potential in addressing real-world challenges, particularly those related to sustainable environments. Electricity generation from sustainable sources of energy, especially wind and solar encounters limitations, including minimal electricity production and substantial financial investment requirements. This survey provides a comprehensive analysis of various DL and ML approaches applied to sustainable energy. It highlights models that can forecast energy, predict power output, and aid in energy conversion processes. Furthermore, the tabulated research findings serve as a valuable resource for future studies in this domain. The review concludes that the development and implementation of advanced AI-based techniques and hybridized approaches can effectively address existing limitations, offering promising solutions for enhancing the efficiency and viability of sustainable energy systems.

Table 3 Tidal and hydro energy research based on ML and DL for forecasting, conversion of energy, and power prediction

Reference	Proposed method	Advantage	Limitation
(Ravinuthala et al., 2022)	The proposed method for harnessing tidal energy involves leveraging its inherent uniformity and predictability to generate power. This includes the development and deployment of advanced tidal turbine technology and infrastructure that can efficiently capture the kinetic energy from tidal waves. The chapter aims to address these challenges and propose solutions to make tidal energy a viable and clean renewable energy source	Sustainability and Renewability: Tidal energy is a renewable source with a steady and predictable output, making it a reliable addition to the energy mix Environmental Benefits: Utilizing tidal energy helps in reducing carbon dioxide emissions, contributing to green innovation and lessening the global carbon footprint Localized Benefits: Tidal energy projects can be especially advantageous for regions near coastlines, potentially making this energy source more cost-effective and accessible for these areas	Environmental Impact: The development of tidal energy projects can have significant ecological effects, particularly on marine life and local ecosystems, which may hinder project implementation High Initial Costs: The financial investment required for establishing tidal energy infrastructure is substantial, making it a cost-intensive option compared to more established renewable energy sources like wind Efficiency Challenges: Ensuring the efficiency and reliability of tidal energy systems remains a technical challenge, requiring ongoing innovation and improvement

(Monahan et al., 2023)	The proposed method for tidal currents prediction using short-term online is a hybrid model called Harmonic Residual Analysis (HRA). This model enhances existing numerical schemes by forecasting residual errors. It utilizes Singular Spectrum Analysis (SSA) techniques from Fractal Theory and Information to implement a novel component selection criterion. This approach removes true noise from the residual time series and decomposes the signal into components suitable for high-order fuzzy time series (HOFTS) and linear-recurrent forecasting (LRF)	Enhanced Reliability and Accuracy: The HRA method has demonstrated superior accuracy and reliability in predicting tidal currents for 6 min and 1 h forecast horizons across various sites. Noise Reduction: The use of SSA with a novel component selection criterion effectively removes noise, improving forecast precision. Versatility: The model is viable across sites with varying degrees of nonlinearity, indicating its robustness and adaptability to different tidal conditions	Complexity and Computational Demand: The hybrid model's complexity and the integration of multiple advanced techniques may require significant computational resources and expertise to implement and maintain Dependency on Accurate Initial Numerical Schemes: The HRA model augments existing numerical schemes, meaning its effectiveness is partially dependent on the accuracy of these initial methods Limited Empirical Data: While the model has been tested with simulated and real data, further empirical validation across diverse geographic locations and tidal conditions may be necessary to fully establish its generalizability

(continued)

Table 3 (continued)

Reference	Proposed method	Advantage	Limitation
(Yang et al., 2024)	The proposed method for forecasting daily ocean tidal energy is a multi-stage system called ICEEMDAN-RCMDE-WSOTVFEMD-PDESSAGRU–CNN–EC. This system uses a series of advanced techniques to decompose, analyze, and predict tidal energy data. The process involves decomposing tidal energy into intrinsic mode functions (IMFs) using ICEEMDAN, separating these into high- and low-complexity components with RCMDE, further decomposing high-complexity components using WSOTVFEMD, and predicting each component with PDESSA. Finally, error correction is performed using CNN, which involves decomposing prediction errors into error IMFs and predicting these with the CNN	High Accuracy Prediction: The multi-stage system demonstrates high prediction accuracy, outperforming 13 comparative models in case studies involving American cities like San Francisco, Sitka, and Wauna Handling Nonstationarity and Nonlinearity: The method effectively addresses the nonstationary and nonlinear characteristics of tidal energy, enhancing the reliability of forecasts Comprehensive Error Correction: The use of CNN for error correction significantly improves the overall prediction results by addressing residual errors	Complexity and Computational Demand: The multi-stage forecasting system involves multiple sophisticated techniques and stages, which may require considerable computational resources and expertise to implement and maintain Dependency on Component Decomposition: The effectiveness of the method heavily relies on the accuracy of the initial decomposition of tidal energy into IMFs, which might vary based on the input data quality and the specific characteristics of the tidal energy being analyzed

(Shen et al., 2023)	The study proposes a creative method that replaces the classical mechanical seal with a consistent seal known as integrated permanent magnet generator (IPMG) to improve power generation efficiency and operational stability. To improve IPMG arrangement based on finite element method improved techniques is used, followed by the manufacturing of an IPMG prototype. The prototype's performance was tested under the impacts of main parameters, water velocities and several rotation speeds for investigating numerical performance	Elimination of Friction Loss and Reduced Leakage Risk: The use of a permanent magnet generator in the IPMG design eliminates friction loss and reduces the risk of water leakage, leading to better performance Extended Operating Hours and Increased Power Generation: The IPMG effectively reduces the startup water flow velocity, extending total operating hours by 35% and increasing annual power generation by 35.8% compared to traditional generators	Implementation Complexity: The novel design and optimization process may involve higher complexity and require more precise manufacturing and testing compared to traditional systems Specific Application: The performance benefits are particularly quantified for urban water supply pipelines, which may limit the generalizability to other types of water pipeline systems without further validation

(continued)

Table 3 (continued)

Reference	Proposed method	Advantage	Limitation
(da Silva et al., 2024)	The research evaluates and compares three models— Temporal Convolutional Networks (TCN), LSTM and Temporal Fusion Transformer (TFT)—for predicting 14-day water inflows into the Tucuruí hydroelectric plant. The TFT model is further hypertuned using Optuna to create an optimized structure (h-TFT)	Superior Prediction Accuracy: The h-TFT model demonstrates a mean absolute percentage error of 13.1 and a Nash–Sutcliffe efficiency of 0.96, outperforming both its initial version and the bidirectional LSTM model Improved Decision-Making: The high accuracy of the h-TFT model offers better insights for the National Electric System Operator, aiding in more informed decision-making for the scheduling and operation of Brazil's power system	Model complexity: The optimization and tuning process of the TFT model using Optuna may add complexity and require significant computational resources Specific Application: The study focuses on a single hydroelectric plant, which might limit the generalizability of the findings to other plants or regions without further validation
(Stefenon et al., 2024)	The paper proposes a model that can predict dam level rise in hydroelectric power plants during floods by using an automatic hyperparameter tuning temporal fusion transformer (AutoTFT)	High Accuracy: The AutoTFT model achieves a root mean square error (RMSE) of 2.78 for short-term forecasting and 1.72 for median-term forecasting, surpassing other DL techniques such as adaptive neuro-fuzzy inference system, LSTM, bagged, boosted, and stacked generalization ensemble learning methods Operational Efficiency: The accurate predictions provided by the AutoTFT technique can improve working logic, ensure safety, and optimize generating energy during flood events, aiding in long-term energy planning and emergency decision-making	Model complexity: The automatic hyperparameter tuning and implementation of the AutoTFT model might add complexity and require significant computational resources Specific Context: Performance of the model is evaluated in the context of predicting dam level rise during floods, which may limit its generalizability to other time series prediction tasks without further validation

References

Abd El-Aziz, R. M. (2022). Renewable power source energy consumption by hybrid machine learning model. *Alexandria Engineering Journal, 61*(12), 9447–9455. https://doi.org/10.1016/j.aej.2022.03.019

Abdelateef Mostafa, M., El-Hay, E. A., & ELkholy, M. M. (2023). Recent trends in wind energy conversion system with grid integration based on soft computing methods: Comprehensive review, comparisons and insights. *Archives of Computational Methods in Engineering, 30*(3), 1439–1478. https://doi.org/10.1007/s11831-022-09842-4

Abdelghany, R. Y., Kamel, S., Sultan, H. M., Khorasy, A., Elsayed, S. K., & Ahmed, M. (2021). Development of an improved bonobo optimizer and its application for solar cell parameter estimation. *Sustainability, 13*(7), 3863. https://doi.org/10.3390/su13073863

Aksoy, B., & Selbaş, R. (2021). Estimation of wind turbine energy production value by using machine learning algorithms and development of implementation program. *Energy Sources, Part A Recovery, Utilization, and Environmental Effects, 43*(6), 692–704. https://doi.org/10.1080/15567036.2019.1631410

AlKandari, M., & Ahmad, I. (2024). Solar power generation forecasting using ensemble approach based on deep learning and statistical methods. *Applied Computing and Informatics, 20*(3/4), 231–250. https://doi.org/10.1016/j.aci.2019.11.002

Alkhayat, G., & Mehmood, R. (2021). A review and taxonomy of wind and solar energy forecasting methods based on deep learning. *Energy and AI, 4*, 100060. https://doi.org/10.1016/j.egyai.2021.100060

Almeshaiei, E., Al-Habaibeh, A., & Shakmak, B. (2020). Rapid evaluation of micro-scale photovoltaic solar energy systems using empirical methods combined with deep learning neural networks to support systems' manufacturers. *Journal of Cleaner Production, 244*, 118788. https://doi.org/10.1016/j.jclepro.2019.118788

Angadi, S., Yargatti, U. R., Suresh, Y., & Raju, A. B. (2022). Speed sensorless maximum power point tracking technique for SEIG-based wind energy conversion system feeding induction motor pump. *Electrical Engineering, 104*(5), 2935–2948. https://doi.org/10.1007/s00202-022-01519-2

Bhatti, G., Mohan, H., & Raja Singh, R. (2021). Towards the future of smart electric vehicles: Digital twin technology. *Renewable and Sustainable Energy Reviews, 141*, 110801. https://doi.org/10.1016/j.rser.2021.110801

Bin Abu Sofian, A. D. A., Lim, H. R., Siti Halimatul Munawaroh, H., Ma, Z., Chew, K. W., & Show, P. L. (2024). Machine learning and the renewable energy revolution: Exploring solar and wind energy solutions for a sustainable future including innovations in energy storage. *Sustainable Development*. https://doi.org/10.1002/sd.2885

Bouzgou, H., & Gueymard, C. A. (2019). Fast short-term global solar irradiance forecasting with wrapper mutual information. *Renewable Energy, 133*, 1055–1065. https://doi.org/10.1016/j.renene.2018.10.096

Chen, C., Hu, Y., Karuppiah, M., & Kumar, P. M. (2021). Artificial intelligence on economic evaluation of energy efficiency and renewable energy technologies. *Sustainable Energy Technologies and Assessments, 47*, 101358. https://doi.org/10.1016/j.seta.2021.101358

da Silva, E. C., Finardi, E. C., & Stefenon, S. F. (2024). Enhancing hydroelectric inflow prediction in the Brazilian power system: A comparative analysis of machine learning models and hyperparameter optimization for decision support. *Electric Power Systems Research, 230*, 110275.

Emrani, A., Berrada, A., & Bakhouya, M. (2022). Optimal sizing and deployment of gravity energy storage system in hybrid PV-wind power plant. *Renewable Energy, 183*, 12–27. https://doi.org/10.1016/j.renene.2021.10.072

Fathy, A., Alharbi, A. G., Alshammari, S., & Hasanien, H. M. (2022). Archimedes optimization algorithm based maximum power point tracker for wind energy generation system. *Ain Shams Engineering Journal, 13*(2), 101548. https://doi.org/10.1016/j.asej.2021.06.032

Fonseca, J. D., Commenge, J.-M., Camargo, M., Falk, L., & Gil, I. D. (2021). Multi-criteria optimization for the design and operation of distributed energy systems considering sustainability dimensions. *Energy, 214*, 118989. https://doi.org/10.1016/j.energy.2020.118989

Hasan, M., Abedin, M. Z., Hajek, P., Coussement, K., Sultan, M. N., & Lucey, B. (2024a). A blending ensemble learning model for crude oil price forecasting. *Annals of Operations Research*, 1–31.

Hasan, M., Sahid, M. A., Uddin, M. P., Marjan, M. A., Kadry, S., & Kim, J. (2024b). Performance discrepancy mitigation in heart disease prediction for multisensory inter-datasets. *PeerJ Computer Science, 10*, e1917.

Hassan, M. H., Kamel, S., Abualigah, L., & Eid, A. (2021). Development and application of slime mould algorithm for optimal economic emission dispatch. *Expert Systems with Applications, 182*, 115205. https://doi.org/10.1016/j.eswa.2021.115205

He, B., et al. (2022). A combined model for short-term wind power forecasting based on the analysis of numerical weather prediction data. *Energy Reports, 8*, 929–939. https://doi.org/10.1016/j.egyr.2021.10.102

Ikram, R. M. A., Dai, H.-L., Ewees, A. A., Shiri, J., Kisi, O., & Zounemat-Kermani, M. (2022). Application of improved version of multi verse optimizer algorithm for modeling solar radiation. *Energy Reports, 8*, 12063–12080. https://doi.org/10.1016/j.egyr.2022.09.015

Kaushaley, S., Shaw, B., & Nayak, J. R. (2023). Optimized machine learning-based forecasting model for solar power generation by using crow search algorithm and seagull optimization algorithm. *Arabian Journal for Science and Engineering, 48*(11), 14823–14836. https://doi.org/10.1007/s13369-023-07822-9

Khan, W., Walker, S., & Zeiler, W. (2022). Improved solar photovoltaic energy generation forecast using deep learning-based ensemble stacking approach. *Energy, 240*, 122812. https://doi.org/10.1016/j.energy.2021.122812

Koo, C., Li, W., Cha, S. H., & Zhang, S. (2019). A novel estimation approach for the solar radiation potential with its complex spatial pattern via machine-learning techniques. *Renewable Energy, 133*, 575–592. https://doi.org/10.1016/j.renene.2018.10.066

Lee, M.-H. (2022). Identifying correlation between the open-circuit voltage and the frontier orbital energies of non-fullerene organic solar cells based on interpretable machine-learning approaches. *Solar Energy, 234*, 360–367. https://doi.org/10.1016/j.solener.2022.02.010

Li, Y., Wang, R., Li, Y., Zhang, M., & Long, C. (2023). Wind power forecasting considering data privacy protection: A federated deep reinforcement learning approach. *Applied Energy, 329*, 120291. https://doi.org/10.1016/j.apenergy.2022.120291

Lin, G.-Q., Li, L.-L., Tseng, M.-L., Liu, H.-M., Yuan, D.-D., & Tan, R. R. (2020). An improved moth-flame optimization algorithm for support vector machine prediction of photovoltaic power generation. *Journal of Cleaner Production, 253*, 119966. https://doi.org/10.1016/j.jclepro.2020.119966

Liu, Z., et al. (2022). A novel distributed energy system combining hybrid energy storage and a multi-objective optimization method for nearly zero-energy communities and buildings. *Energy, 239*, 122577. https://doi.org/10.1016/j.energy.2021.122577

Ma, Z., Li, M.-J., Zhang, K. M., & Yuan, F. (2021). Novel designs of hybrid thermal energy storage system and operation strategies for concentrated solar power plant. *Energy, 216*, 119281. https://doi.org/10.1016/j.energy.2020.119281

Maamoun, N., Kennedy, R., Jin, X., & Urpelainen, J. (2020). Identifying coal-fired power plants for early retirement. *Renewable and Sustainable Energy Reviews, 126*, 109833. https://doi.org/10.1016/j.rser.2020.109833

May, R., Nygård, T., Falkdalen, U., Åström, J., Hamre, Ø., & Stokke, B. G. (2020). Paint it black: Efficacy of increased wind turbine rotor blade visibility to reduce avian fatalities. *Ecology and Evolution, 10*(16), 8927–8935. https://doi.org/10.1002/ece3.6592

Mehrpooya, M., Raeesi, M., Pourfayaz, F., & Delpisheh, M. (2021). Investigation of a hybrid solar thermochemical water-splitting hydrogen production cycle and coal-fueled molten carbonate fuel cell power plant. *Sustainable Energy Technologies and Assessments, 47*, 101458. https://doi.org/10.1016/j.seta.2021.101458

Meka, R., Alaeddini, A., & Bhaganagar, K. (2021). A robust deep learning framework for short-term wind power forecast of a full-scale wind farm using atmospheric variables. *Energy, 221*, 119759. https://doi.org/10.1016/j.energy.2021.119759

Monahan, T., Tang, T., & Adcock, T. A. (2023). A hybrid model for online short-term tidal energy forecasting. *Applied Ocean Research, 137*, 103596.

Mostafa, M. H., Abdel Aleem, S. H. E., Ali, S. G., Ali, Z. M., & Abdelaziz, A. Y. (2020). Techno-economic assessment of energy storage systems using annualized life cycle cost of storage (LCCOS) and levelized cost of energy (LCOE) metrics. *Journal of Energy Storage, 29*, 101345. https://doi.org/10.1016/j.est.2020.101345

Mostafa, N., Ramadan, H. S. M., & Elfarouk, O. (2022). Renewable energy management in smart grids by using big data analytics and machine learning. *Machine Learning with Applications, 9*, 100363. https://doi.org/10.1016/j.mlwa.2022.100363

Munawar, U., & Wang, Z. (2020). A framework of using machine learning approaches for short-term solar power forecasting. *Journal of Electrical Engineering and Technology, 15*(2), 561–569. https://doi.org/10.1007/s42835-020-00346-4

Naik, K. R., Rajpathak, B., Mitra, A., Sadanala, C., & Kolhe, M. L. (2022). Power management scheme of DC micro-grid integrated with photovoltaic - battery - micro hydro power plant. *Journal of Power Sources, 525*, 230988. https://doi.org/10.1016/j.jpowsour.2022.230988

Nejati, M., & Amjady, N. (2022). A new solar power prediction method based on feature clustering and hybrid-classification-regression forecasting. *IEEE Transactions on Sustainable Energy, 13*(2), 1188–1198. https://doi.org/10.1109/TSTE.2021.3138592

Rajasundrapandiyanleebanon, T., Kumaresan, K., Murugan, S., Subathra, M. S. P., & Sivakumar, M. (2023). Solar energy forecasting using machine learning and deep learning techniques. *Archives of Computational Methods in Engineering, 30*(5), 3059–3079. https://doi.org/10.1007/s11831-023-09893-1

Rajesh, P., Muthubalaji, S., Srinivasan, S., & Shajin, F. H. (2022). Leveraging a dynamic differential annealed optimization and recalling enhanced recurrent neural network for maximum power point tracking in wind energy conversion system. *Technology and Economics of Smart Grids and Sustainable Energy, 7*(1), 19. https://doi.org/10.1007/s40866-022-00144-z

Ravinuthala, S., Das, S. K., Nithya, R., & Das, S. P. (2022). Ocean, tidal and wave energy: Science and challenges. In S. J. Joshi, R. Sen, A. Sharma, & P. A. Salam (Eds.), *Status and future challenges for non-conventional energy sources* (Clean energy production technologies) (Vol. 1). Springer. https://doi.org/10.1007/978-981-16-4505-1_1

Rushdi, M. A., Rushdi, A. A., Dief, T. N., Halawa, A. M., Yoshida, S., & Schmehl, R. (2020). Power prediction of airborne wind energy systems using multivariate machine learning. *Energies, 13*(9), 2367. https://doi.org/10.3390/en13092367

Sahu, P. C., Baliarsingh, R., Prusty, R. C., & Panda, S. (2022). Novel DQN optimised tilt fuzzy cascade controller for frequency stability of a tidal energy-based AC microgrid. *International Journal of Ambient Energy, 43*(1), 3587–3599. https://doi.org/10.1080/01430750.2020.1839553

Shen, Z., Yao, Y., Wang, Q., Lu, L., & Yang, H. (2023). A novel micro power generation system to efficiently harvest hydroelectric energy for power supply to water intelligent networks of urban water pipelines. *Energy, 268*, 126694.

Shirzadi, N., Nasiri, F., El-Bayeh, C., & Eicker, U. (2022). Optimal dispatching of renewable energy-based urban microgrids using a deep learning approach for electrical load and wind power forecasting. *International Journal of Energy Research, 46*(3), 3173–3188. https://doi.org/10.1002/er.7374

Stefenon, S. F., Seman, L. O., da Silva, L. S. A., Mariani, V. C., & dos Santos Coelho, L. (2024). Hypertuned temporal fusion transformer for multi-horizon time series forecasting of dam level in hydroelectric power plants. *International Journal of Electrical Power & Energy Systems, 157*, 109876.

Stoean, C., et al. (2023). Metaheuristic-based hyperparameter tuning for recurrent deep learning: Application to the prediction of solar energy generation. *Axioms, 12*(3), 266. https://doi.org/10.3390/axioms12030266

Tercan, S. M., Demirci, A., Gokalp, E., & Cali, U. (2022). Maximizing self-consumption rates and power quality towards two-stage evaluation for solar energy and shared energy storage empowered microgrids. *Journal of Energy Storage, 51*, 104561. https://doi.org/10.1016/j.est.2022.104561

Wang, L., Tao, R., Hu, H., & Zeng, Y.-R. (2021). Effective wind power prediction using novel deep learning network: Stacked independently recurrent autoencoder. *Renewable Energy, 164*, 642–655. https://doi.org/10.1016/j.renene.2020.09.108

Wang, Y., et al. (2022). Driving-torque self-adjusted triboelectric nanogenerator for effective harvesting of random wind energy. *Nano Energy, 99*, 107389. https://doi.org/10.1016/j.nanoen.2022.107389

Yang, H., Wu, Q., & Li, G. (2024). A multi-stage forecasting system for daily ocean tidal energy based on secondary decomposition, optimized gate recurrent unit and error correction. *Journal of Cleaner Production, 449*, 141303.

Yang, J. J., Yang, M., Wang, M. X., Du, P. J., & Yu, Y. X. (2020). A deep reinforcement learning method for managing wind farm uncertainties through energy storage system control and external reserve purchasing. *International Journal of Electrical Power & Energy Systems, 119*, 105928. https://doi.org/10.1016/j.ijepes.2020.105928

Zhang, Y., & Chen, Y. (2022). Application of hybrid model based on CEEMDAN, SVD, PSO to wind energy prediction. *Environmental Science and Pollution Research, 29*(15), 22661–22674. https://doi.org/10.1007/s11356-021-16997-3

Zhang, G., Hu, W., Cao, D., Huang, Q., Chen, Z., & Blaabjerg, F. (2021). A novel deep reinforcement learning enabled sparsity promoting adaptive control method to improve the stability of power systems with wind energy penetration. *Renewable Energy, 178*, 363–376. https://doi.org/10.1016/j.renene.2021.06.081

Zhang, S., & Li, X. (2021). Future projections of offshore wind energy resources in China using CMIP6 simulations and a deep learning-based downscaling method. *Energy, 217*, 119321. https://doi.org/10.1016/j.energy.2020.119321

Zhang, Y., Li, L., Sadiq, M., & Chien, F. (2024). The impact of non-renewable energy production and energy usage on carbon emissions: Evidence from China. *Energy & Environment, 35*(4), 2248–2269. https://doi.org/10.1177/0958305X221150432

Zhao, Y., & Kok Foong, L. (2022). Predicting electrical power output of combined cycle power plants using a novel artificial neural network optimized by electrostatic discharge algorithm. *Measurement, 198*, 111405. https://doi.org/10.1016/j.measurement.2022.111405

Zhao, N., & You, F. (2022). Sustainable power systems operations under renewable energy induced disjunctive uncertainties via machine learning-based robust optimization. *Renewable and Sustainable Energy Reviews, 161*, 112428. https://doi.org/10.1016/j.rser.2022.112428

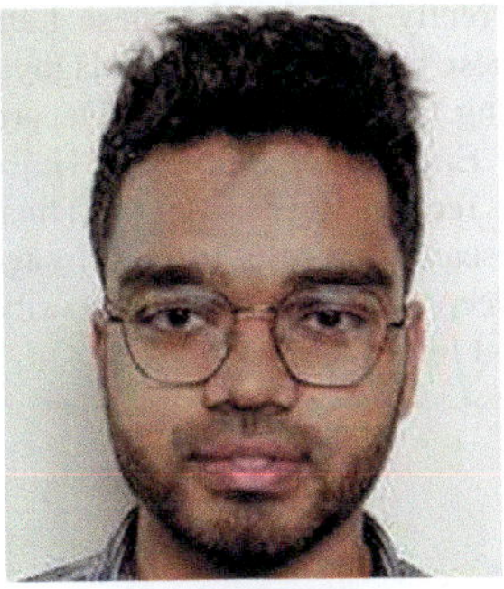

Md Raihanul Islam Tomal is currently pursuing his master's by research in the field of Machine Learning and NLP from the Universiti Malaysia Pahang Al-Sultan Abdullah (UMPSA), Pekan, Malaysia. He has completed his Bachelor's degree from the International Islamic University Chittagong, Bangladesh in the field of Computer Science and Engineering in 2022. Currently, he is a full-time research assistant at Data Science Simulation and Modeling lab at UMPSA. His research interest includes Natural Language Processing, Machine Learning, Deep Learning, and Image Processing.

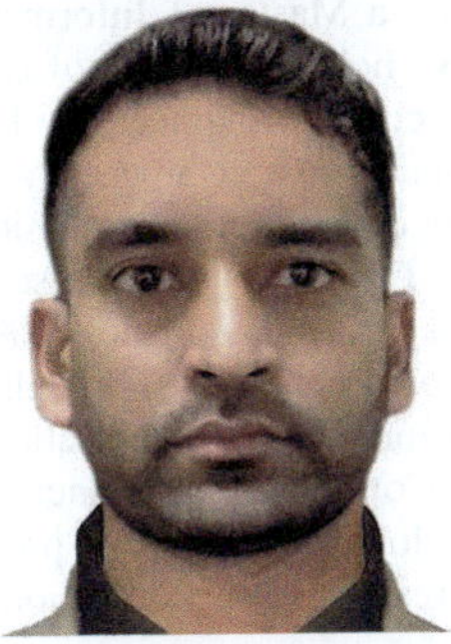

Alamgir Kabir is currently pursuing his Master of Business Administration in Agribusiness Management (MAB) at Faculty of Economics, Prince of Songkla University, Hat Yai Campus, Thailand. He completed his M.S.S. in Sociology from the Department of Sociology at Hajee Mohammad Danesh Science and Technology University, Dinajpur Bangladesh, in 2023 and B.S.S. in Sociology from the same university in 2021. His research interest includes E-commerce, Women Entrepreneurs, Poverty and Inequality, Indigenous Elderly, Mobile Financial Service (MFS), Computational Sociology, Sustainable Energy, and SDG.

Mahmudul Hasan is currently pursuing a Ph.D. in Information Technology (IT) at Deakin University, Melbourne, Australia. He earned his B.Sc. (Eng.) and M.Sc. (Eng.) degrees in Computer Science and Engineering (CSE) from Hajee Mohammad Danesh Science and Technology University, Dinajpur, Bangladesh, in 2021 and 2023, respectively. He previously served as a Lecturer in the Department of CSE at the University of Creative Technology, Chittagong (UCTC), Bangladesh. He is the Founder and Director of the Center for Multidisciplinary Research and Development (CeMRD) and a moderator of "Be Researcher BD," the largest online research forum in Bangladesh. Additionally, he has taught online as a Data Science instructor to students in the USA, Italy, Denmark, South Korea, and Australia. He is also the founder of the online educational platform "MHM Academy." His research interests include federated learning, machine learning, deep learning, cybersecurity, health informatics, renewable energy, computational sociology, and business intelligence.

Sayed Mahmudul Haque is currently learning Japanese Language at Remnant Academy Japanese Language School, Nagoya, Japan. He completed his BSc (Eng.) in CSE from the Department of Computer Science and Engineering (CSE) at Hajee Mohammad Danesh Science and Technology University, Dinajpur, Bangladesh, in 2022. His research interests include Image Processing, Machine Learning, Deep Learning, Cyber Security, Business Intelligence, and Artificial Intelligence.

Md Mehedi Hasan Jony is pursuing a Master of Information Technology degree in Australia. I also hold a Bachelor of Engineering degree in Electrical and Electronic Engineering. I am an associate member of the Australian Computer Society (ACS) (Member ID: 4411692), reflecting my dedication to professional growth and ethical practices in the IT industry. My research interests include machine learning, deep learning, business intelligence, AWS Cloud AI, and computational sociology. Through my work, I aim to explore innovative solutions at the intersection of technology, data science, and society. Committed to advancing IT expertise, I aspire to contribute to the global academic and professional community through impactful research and publications.

Efficient Gasoline Spot Price Prediction Using Hyperparameter Optimization and Ensemble Machine Learning Approach

Md. Amir Hamja, Md Rakinus Sakib, Mahmudul Hasan, and Md Sabir Hossain

1 Introduction

Energy is fundamental to economic growth and social progress, with prices at the heart of the energy market. Oscillations in energy prices significantly impact the distribution and movement of resources within the market, exerting considerable economic influence (Agbaji et al., 2023). Many countries face challenges related to excessive energy consumption across industries and economies. While energy conservation is widely seen as a key solution, determining the most effective strategies for conserving energy across various sectors remains difficult (Fathi et al., 2020). Energy price prediction involves using historical data to create models that forecast future prices by analyzing factors like market supply and demand, participant behavior, costs, the socioeconomic environment, and energy system structure (Khan et al., 2023). Energy price forecasting is crucial for three main reasons: (a) It enables dynamic cost control, (b) it helps in accurately understanding market trends and seizing opportunities, and (c) it provides a solid foundation for policymaking and market regulation (Lu et al., 2021). The prediction

M. A. Hamja · M. R. Sakib
Department of Statistics, Hajee Mohammad Danesh Science and Technology University, Dinajpur, Bangladesh

M. Hasan (✉)
School of Information Technology, Deakin University, Geelong, VIC, Australia

Department of Computer Science and Engineering, Hajee Mohammad Danesh Science and Technology University, Dinajpur, Bangladesh

M. S. Hossain
Department of Information and Computer Science, King Fahd University of Petroleum and Minerals (KFUPM), Dhahran, Kingdom of Saudi Arabia
e-mail: g202314790@kfupm.edu.sa

M. Z. Abedin, W. Yong (eds.), *Machine Learning Technologies on Energy Economics and Finance*, International Series in Operations Research & Management Science 367, https://doi.org/10.1007/978-3-031-94862-6_12

of energy prices has garnered significant interest from researchers, leading to a rapid increase in academic publications on the topic in recent years. But in the rapidly changing global energy markets, accurately forecasting gasoline prices is a persistent challenge. As economies expand and technology transforms energy consumption and production, the factors influencing gasoline prices grow more complex and volatile (Eliwa et al., 2024). Fuel that comes from crude oil and other petroleum-based liquids is called gasoline, and it is mostly utilized in car engines. Petroleum refineries and blending facilities generate it, while fueling stations sell it as finished motor gasoline (MultiMedia LLC, 2024). Since the pandemic began in December 2019 due to COVID situation, many countries imposed lockdowns and limited social interaction to curb the spread of the virus. This led to reduced consumption and travel, causing a drop in gasoline demand and prices. Recently, conflicts such as those between Russia and Ukraine, and Israel and Palestine, have driven energy prices up due to supply shortages. These events significantly impact crude oil demand and supply, leading to sharp fluctuations in the price of gasoline. It is vital to the economy, influencing the Consumer Price Index (CPI) and potentially triggering inflation and economic downturns. Gasoline prices are closely linked to macroeconomic activity, with oil price shocks often preceding economic recessions. Additionally, gasoline prices can affect foreclosure rates and house prices (Hamilton, 2009). Researchers have studied consumer responses to fluctuations in gasoline prices to gain insights into different economic behaviors, such as demand for automobiles (Allcott & Wozny, 2014), transportation choices (Knittel & Sandler, 2011), search patterns (Lewis & Marvel, 2011), and price stickiness (Borenstein & Shepard, 1996). Accurate gasoline price predictions are crucial for modeling the automobile market and analyzing environmental policies (Busse et al., 2013). As a key driver of the economy, gasoline prices influence overall market balance and the functioning of economic activities, directly affecting people's lives. Therefore, forecasting gasoline prices holds significant practical importance for the global economy. Traditional and AI-based energy forecasting models can be generally divided into two groups (Lu et al., 2021). With the rise of AI, many researchers have turned to AI algorithms for energy prediction. Various studies have reviewed these models from different perspectives. Early oil price forecasts often relied on statistical models like AutoRegressive Integrated Moving Average (ARIMA), Generalized AutoRegressive Conditional Heteroskedasticity (GARCH), Linear Regression (LR), random walk, and Vector Error Correction Model (VECM) (Hasan et al., 2024a; Jin & Xu, 2024; Yuan et al., 2023). While these models are effective for linear relationships and short-term predictions, they struggle with the nonlinear nature of gasoline prices (Abdollahi & Ebrahimi, 2020). Gasoline prices, being nonstationary time series, pose challenges for time series models, which also rely on assumptions of linearity and normal distribution, failing to capture the specific characteristics of gasoline prices. In contrast, AI models are better equipped to handle the nonlinearity and complexity of gasoline prices due to their flexible structures (Yuan et al., 2023). Popular AI models for gasoline price forecasting include support vector regression (SVR) (Hasan et al., 2024a) and artificial neural networks (ANNs). Among ANNs, models like

Extreme Learning Machine (ELM), Backpropagation Neural Network (BPNN), Random Vector Functional Link Network (RVFL), Recurrent Neural Network (RNN), LSTM, BiLSTM, GRU, and Bidirectional Gated Recurrent Unit (Bi-GRU) are frequently used (Zheng et al., 2024; Salamai, 2023; Dong et al., 2024; Li et al., 2020). Despite their effectiveness, artificial intelligence models can be sensitive to parameter settings and may face challenges such as local optimization and overfitting. In this study, we examine the two most commonly used gasoline spot price datasets for forecasting. In order to enable predictions over a range of temporal periods, we conduct necessary preprocessing and mode adjustments in addition to analyzing the statistical properties of the gasoline price time series data. The following is a summary of this chapter's main contributions:

- We create an analysis framework for gasoline spot prices that incorporates deep learning (DL), machine learning (ML), and ensemble learning models.
- To enhance the ML, DL, and ensemble ML models' prediction performance for more thorough analysis, we deepen our hyperparameter tuning.
- A set of multi-scale models using stacking ensemble learning is introduced to predict gasoline spot prices, addressing the limitations of conventional single-method time series decomposition analysis.
- It is demonstrated that the suggested stacking ensemble model performs better than the most advanced models currently in use, which are regarded as benchmarks for predicting gas spot prices.

The structure of the remaining sections of this chapter is outlined as follows. The related works are outlined in Sect. 2. Section 3 is dedicated to presenting our proposed methodology and the experimental setup. We detail the approach we have taken to address the research problem, including the methods, techniques, and tools employed in our study. In Sect. 4, we present the outcomes of our experiments. The chapter concludes in Sect. 5 with a summary of our findings and their significance. Additionally, we outline avenues for future research and development in this domain, emphasizing the potential directions for further exploration and enhancement.

2 Literature Review

Forecasting energy prices has advanced significantly in recent years. Numerous researchers have demonstrated promising outcomes through the analysis of various energy price time series using statistical, econometric, and ML techniques. This section highlights the latest developments in energy price forecasting, with a particular focus on ML approaches, including DL and hybrid and ensemble models.

Hasan et al. (2023b) use blended ensemble learning to create a forecasting model that combines support vector regression, ridge regression, linear regression, regression trees, and k-nearest neighbor regression (Hasan et al., 2024a). This model demonstrated greater accuracy in both short- and medium-term forecasts and was validated using multiple time series of crude oil prices, namely WTI and Brent. Variational mode decomposition (VMD), which divides data into low- and high-frequency components, is the basis of an interval-based framework that Zheng et al. (2024) suggested based on the "divide and conquer" theory (Zhang et al., 2021). An autoregressive conditional interval (ACI) model is used to predict the low-frequency component, and interval long short-term memory (iLSTM) networks are used to anticipate the high-frequency component. Combined predictions form the final interval-valued forecast, leading to improved forecasting and trading performance. Zhao et al. (2024) introduced a hybrid model that incorporates financial market factors and crude oil news, utilizing a two-layer multivariate decomposition to predict weekly WTI oil spot prices (Zhao et al., 2024). Benchmarks were greatly underperformed by this model. The Jaynes Weight Hybrid (JWH) model, which integrates Shannon information entropy with classical statistics, neural network, and deep learning models, is a unique combined forecasting approach with time-varying weights that Liu et al. (2024) introduced to predict crude oil prices (Liu et al., 2024). In order to anticipate crude oil prices, Qin et al. (2023) used Google Trends data using a stacking ensemble approach. They identified pertinent indicators and assessed their impact using Granger causality tests and co-integration tests (Qin et al., 2023). Multiple-model approaches were found to be more effective than single-model approaches in the study. Yuan et al. (2023) introduced a clustering-based weight assignment strategy to reduce outlier impact and balance the ensemble model's competitiveness and robustness, significantly improving forecasting accuracy for West Texas Intermediate oil prices (Yuan et al., 2023). Salamai (2023) developed a framework for predicting daily and weekly crude oil prices, using optimized variational mode decomposition (OVMD), a Tree-structured Parzen Estimator (TPE) algorithm, and enhanced AdaBoost with random forest (Salamai, 2023). The model captures spatial-temporal patterns with a Conv-Former module and stacked LSTM networks, showing superior performance in predicting Brent crude oil prices. Dong et al. (2024) proposed a model using VMD, PSR, CNN, and BiLSTM for crude oil price forecasting, achieving low MAPEs and MSEs (Dong et al., 2024). The model's superiority was confirmed by the Diebold-Mariano test. Li et al. (2024) introduced a hybrid forecasting method combining MEEMD and Mix-KELM, optimizing local and global kernel functions with a genetic algorithm, resulting in lower prediction errors for crude oil prices (Li et al., 2024). Zhang et al. (2024) proposed an attention-based PCA method to enhance oil price forecasting, integrating multiple attention mechanisms and diverse models (Zhang et al., 2024). The attention-PCA model significantly reduced MAPE, with the best combination model achieving a MAPE of 4.40%. Lastly, nonlinear autoregressive neural network models were employed by Jin and Xu (2024) to forecast monthly prices for New York Harbor No. 2 heating oil and Henry Hub natural gas, as well as daily prices for WTI and Brent crude oil (Jin & Xu, 2024). In exploring several model configurations, their work produced simplified models with strong accuracy on a variety of datasets. The relative root mean square error for WTI, Brent, New York Harbor No. 2 heating oil, Henry Hub natural gas, and other crude oils was 1.95%/1.80%, 9.51%, and 20.35% overall.

3 Methodology

3.1 Approach Overview

Outline of the suggested methodology is illustrated in Fig. 1. The collection and preprocessing of daily spot prices for gasoline in the United States and New York has been done to enhance computational efficiency and model performance. Consistent with prior research, the gasoline price datasets have been partitioned using a sequential validation approach. Specifically, an 80:20 ratio was used to divide the data into training and testing subsets. Various ML models, including Polynomial Regression, Linear Regression, Ridge, Lasso, SVR, Random Forest, Decision Tree, Gradient Boosting, XGBoost, LightGBM, KNN, and DL models such as MLP, GRU, LSTM, and BiLSTM, were then trained on both datasets. Using a variety of error criteria, the suggested and comparison models' performance was assessed, including R^2, MSE, RMSE, MAE, MAPE, sMAPE, and elapsed time in seconds. The next section details the methods applied in this study.

3.2 Description of Dataset and Variables

This study uses two common gasoline spot prices datasets such as U.S. Gulf Coast Conventional Gasoline Regular Spot Price and New York Harbor Conventional Gasoline Regular Spot Price which are obtained from the U.S. Energy Information Administration (MultiMedia LLC, 2024). In both datasets the prices are represented as dollars per gallon. The duration of the both gasoline price datasets is from June 02, 1986 to August 03, 2024, so the total number of observations for both is 9593.

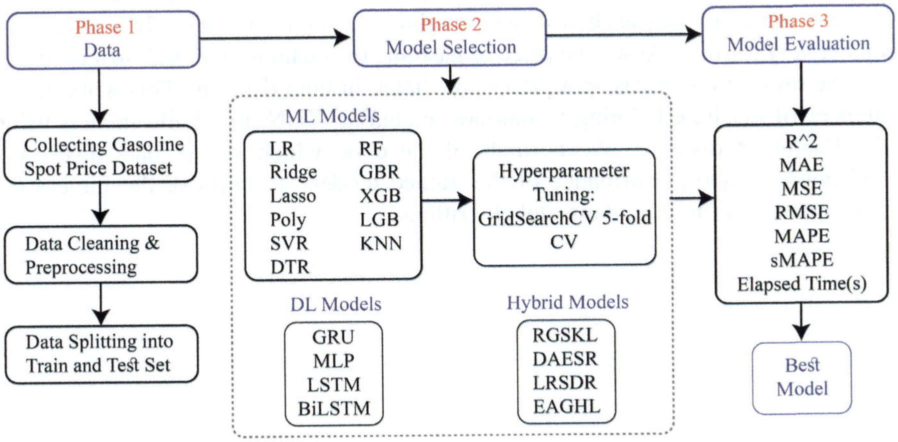

Fig. 1 Outline of the suggested methodology of energy price forecasting system

Table 1 Summary statistics of daily gasoline spot price time series

Dataset	No. of observations	Minimum	Maximum	Mean	Standard deviation	Variance	Skewness	Kurtosis
U.S. gasoline spot price	9593	0.270	4.873	1.364	0.855	0.731	0.615	− 0.766
New York gasoline spot price	9593	0.290	4.509	1.400	0.876	0.767	0.601	− 0.845

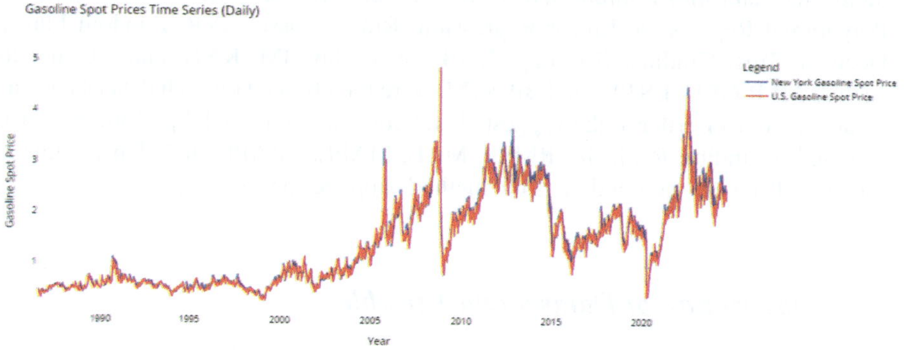

Fig. 2 The relationship between gas costs and time

Also there are no missing values for both datasets. The data was collected daily and are not adjusted for seasonal variations. The number of observations and additional descriptive statistics are detailed in Table 1.

For both datasets, the kurtosis was negative, indicating that outliers are not a major problem because the distributions appear to have lighter tails than a normal distribution. Both datasets had skewness values between 0.5 and 1, which suggests a moderately positive skew. The price of gasoline throughout time is shown in Fig. 2.

The time series of gas spot prices are shown in both datasets. This work uses a number of machine learning techniques, including KNN, that build models using the Euclidean distance. We normalized the data, which lowers data dispersion and improves the performance of the trained models, to improve the forecasting performance of these fundamental algorithms.

3.3 *Machine Learning Algorithms*

3.3.1 LR

LR is a fundamental statistical method often employed to forecast time series. The model forecasts future values by establishing a linear connection between the explained variable and one or more explanatory variables (Hyndman & Athanasopoulos, 2018; Hasan et al., 2024b). When applied to time series forecasting, linear regression can be expressed as follows:

$$Y_t = \beta_0 + \beta_1 X_{t-1} + \beta_2 X_{t-2} + \cdots + \beta_n X_{t-n} + \epsilon_t$$

where the values for the dependent variable at time t are represented by Y_t, the lagged values are represented by $X_{t-1}, X_{t-2}, \ldots, X_{t-n}$, and the intercept term is represented by β_0. The coefficients are represented by $\beta_1, \beta_2, \ldots, \beta_n$, and the error term is represented by ϵ_t. By minimizing the sum of squared differences between the observed and predicted values, the model estimates the coefficients under the assumption that the independent and dependent variables have a linear relationship.

3.3.2 Ridge

It is an extension of LR, which is particularly useful for time series forecasting when multicollinearity exists among the predictors. It enhances the linear regression model by incorporating a penalty term into the loss function, which aids in preventing overfitting and improves the model's generalization to new data (Hastie et al., 2005; Hasan et al., 2024a). The ridge regression model for time series forecasting can be formulated as

$$Y_t = \beta_0 + \beta_1 X_{t-1} + \beta_2 X_{t-2} + \cdots + \beta_n X_{t-n} + \lambda \sum_{i=1}^{n} \beta_i^2 + \epsilon_t$$

where the values for the dependent variable at time t are represented by Y_t, and the lagged values are represented by $X_{t-1}, X_{t-2}, \ldots, X_{t-n}$. The intercept term is denoted by β_0, and the coefficients are represented by $\beta_1, \beta_2, \ldots, \beta_n$. ϵ_t denotes the error term, and λ indicates the regularization parameter that regulates the severity of the penalty on the coefficients. The model is stabilized when the regularization parameter λ reduces the coefficients toward zero, lessening the influence of less significant predictors. Ridge regression is particularly beneficial in situations where the number of predictors is large, or when the predictors are highly correlated, as it can yield more accurate and reliable forecasts.

3.3.3 Lasso

Lasso is an LR technique enhanced by a regularization mechanism, making it particularly useful for time series forecasting when feature selection is crucial. It introduces a penalty term to the loss function that helps prevent overfitting and allows the model to perform automatic feature selection by reducing some coefficients to zero (Tibshirani, 1996). The lasso regression model for time series forecasting can be expressed as

$$Y_t = \beta_0 + \beta_1 X_{t-1} + \beta_2 X_{t-2} + \cdots + \beta_n X_{t-n} + \lambda \sum_{i=1}^{n} |\beta_i| + \epsilon_t$$

where the values for the dependent variable at time t are represented by Y_t, and the lagged values are represented by $X_{t-1}, X_{t-2}, \ldots, X_{t-n}$. The intercept term is represented by β_0. The coefficients are represented by $\beta_1, \beta_2, \ldots, \beta_n$. The error term is ϵ_t, and the regularization parameter λ determines how much of a penalty is applied to the total of the absolute values of the coefficients. The regularization parameter λ determines how much the coefficients are shrunk, with larger values of λ leading to more coefficients being reduced to zero, effectively selecting a simpler model. This characteristic makes lasso regression particularly effective in scenarios where there are many predictors, but only a subset is expected to have a significant impact on the forecast.

3.3.4 Poly

Poly is an enhancement of LR that models the relationship between the independent and dependent variables using an nth degree polynomial. This technique is particularly useful in time series forecasting when the data exhibits a nonlinear trend that cannot be captured by a simple linear model (Montgomery et al., 2021). It can be mathematically represented as

$$Y_t = \beta_0 + \beta_1 X_{t-1} + \beta_2 X_{t-1}^2 + \cdots + \beta_n X_{t-1}^n + \epsilon_t$$

where Y_t represents the values of the dependent variable at time t, X_{t-1} is the lagged value of the dependent or other predictors, β_0 indicates the intercept, $\beta_1, \beta_2, \ldots, \beta_n$ represents the coefficients for each polynomial degree, and ϵ_t is the error term. By including higher degree terms of the predictor variables, polynomial regression allows the model to capture the curvature in the data, making it suitable for forecasting complex, nonlinear time series patterns.

3.3.5 Decision Tree Regression (DTR)

The DTR is a nonparametric model utilized in time series forecasting, adept at capturing nonlinear relationships by recursively dividing the data into smaller subsets. It constructs a tree structure where each node corresponds to a decision based on a feature, and each leaf node represents a predicted value (Breiman, 2001; Hasan et al., 2024c). Mathematically this model can be represented as

$$Y_t = f(X_{t-1}, X_{t-2}, \ldots, X_{t-n})$$

where Y_t indicates the predicted value at time t, and $X_{t-1}, X_{t-2}, \ldots, X_{t-n}$ are the lagged predictors. Decision trees are highly interpretable and handle complex patterns well, though they may overfit if the tree is too deep. Pruning and ensemble methods like RF can mitigate this issue.

3.3.6 SVR

The SVR is a robust machine learning model used in time series forecasting, particularly adept at capturing complex relationships within the data which works by finding a hyperplane that best fits the data within a margin of tolerance, known as the epsilon-insensitive zone. The model aims to minimize prediction errors while ensuring the margin is as wide as possible (Smola & Schölkopf, 2004; Sajid et al., 2023). The SVR model can be expressed as

$$Y_t = \sum_{i=1}^{n} (\alpha_i - \alpha_i^*) K(X_{t-i}, X) + b$$

where the predicted value at time t is represented by Y_t, the Lagrange multipliers are denoted by α_i and α_i^*, the kernel function that converts input data into a higher dimensional space is denoted by $K(X_{t-i}, X)$, and the bias term is b. SVR may describe both linear and nonlinear patterns in time series data depending on the kernel function (linear, polynomial, or radial basis function, for example). SVR is recognized for its robustness and strong generalization capabilities, making it well suited for forecasting tasks, though it often requires careful tuning of hyperparameters.

3.3.7 RF

Because of its capacity to handle complicated, nonlinear data, the RF regressor is an ensemble learning technique that is frequently applied in time series forecasting. To improve accuracy and lessen overfitting, it builds several decision trees during training and averages their predictions. Every tree is constructed using a random subset of data and random feature selection to capture a variety of patterns (Breiman, 2001; Hasan et al., 2023b). This model can be represented as

$$Y_t = \frac{1}{M} \sum_{m=1}^{M} f_m(X_{t-1}, X_{t-2}, \ldots, X_{t-n})$$

where Y_t denotes the predicted value at time t, $X_{t-1}, X_{t-2}, \ldots, X_{t-n}$ are the lagged predictors, M is used as the number of trees, and f_m shows the prediction from the mth tree. By averaging the predictions of multiple trees, it enhances prediction accuracy and robustness while mitigating the risk of overfitting that might occur in individual decision trees.

3.3.8 KNN

The KNN regressor is a simple but effective model employed in time series forecasting, particularly known for its ability to model nonlinear relationships which predicts the target variable's value by averaging the values of the k closest neighbors in the training dataset. These neighbors are identified using a distance metric, most commonly the Euclidean distance (Altman, 1992). The KNN model can be mathematically described as

$$Y_t = \frac{1}{k} \sum_{i=1}^{k} Y_{N(i)}$$

where Y_t is the predicted value at time t, k is the number of nearest neighbors, and $Y_{N(i)}$ represents the values of the nearest neighbors. KNN is particularly useful for forecasting when the time series data is irregular or contains nonlinearity that traditional linear models cannot capture. Its simplicity and nonparametric nature make KNN a popular choice. However, it can be sensitive to the selection of k and the chosen distance metric, and it may face challenges when dealing with high-dimensional data.

3.3.9 AdaBoost

The AdaBoost is also an ensemble learning approach that enhances the accuracy of time series forecasting by integrating several weak learners into a powerful predictive model. It operates by iteratively training these weak learners on the dataset, with each new model paying more attention to the instances that earlier models struggled with. The final prediction is determined by a weighted sum of the predictions from all the weak learners (Freund & Schapire, 1997). This model can be represented as

$$Y_t = \sum_{m=1}^{M} \alpha_m f_m(X_{t-1}, X_{t-2}, \ldots, X_{t-n})$$

where Y_t is the predicted value at time t, α_m is the weight assigned to the mth weak learner, f_m, and M represents the total number of weak learners. The model adjusts the weights α_m to minimize the overall prediction error, with higher weights given to models that perform better. AdaBoost is particularly effective in enhancing the accuracy of weak learners and is robust against overfitting, making it suitable for complex time series data.

3.3.10 GBR

Gradient Boosting Regression (GBR) enhances prediction accuracy by sequentially incorporating weak learners to address the mistakes made by previous models. Each learner is fit to the residual errors of the combined predictions from earlier learners (Friedman, 2001). This model can be represented as

$$Y_t = \hat{Y}_t + \sum_{m=1}^{M} \alpha_m f_m(X_{t-1}, X_{t-2}, \ldots, X_{t-n})$$

where Y_t indicates the predicted value at time t, $\hat{Y}_t$ is the initial prediction (often the mean value), f_m represents the mth weak learner, α_m is the weight of the mth learner, and M is the total number of iterations. Each f_m is trained to minimize the residual errors of the model at the previous iteration. This iterative process allows it to refine its predictions progressively, making it effective for capturing complex patterns in data.

3.3.11 LGB

Light Gradient Boosting (LGB) regressor improves performance by using a histogram-based approach to bin continuous features and a leaf-wise growth strategy to build decision trees (Ke et al., 2017). This method enhances computational efficiency and accuracy. The model's prediction for a time series at stage m can be represented as

$$\hat{y}_i^{(m)} = \hat{y}_i^{(m-1)} + \eta \cdot h_m(x_i)$$

where $\hat{y}_i^{(m)}$ denotes the prediction at stage m, $\hat{y}_i^{(m-1)}$ is the prediction from the previous stage, $h_m(x_i)$ is the output of the mth decision tree, and η is the learning rate that controls the contribution of each tree. LightGBM optimizes the loss function $L(y_i, \hat{y}_i^{(m)})$ through gradient descent, with a focus on reducing the computation time. It can capture more intricate patterns with fewer trees since it employs leaf-wise growth for trees instead of level-wise growth. A very effective and scalable model is produced by adding together the predictions made by each tree to arrive at the final forecast.

3.3.12 XGB

XGB (eXtreme Gradient Boosting) is an efficient gradient boosting framework that builds an ensemble of decision trees to improve prediction accuracy (Hasan et al., 2023a). The prediction at stage m is given by

$$\hat{y}_i = \sum_{k=1}^{m} \eta \cdot h_k(x_i)$$

where $\hat{y}_i$ is the final prediction, $h_k(x_i)$ is the output of the kth tree, and η is the learning rate. XGB optimizes an objective function that includes a loss term and a regularization term to prevent overfitting:

$$\text{Objective} = L(y_i, \hat{y}_i) + \sum_{k=1}^{m} \Omega(h_k)$$

where $\Omega(h_k)$ accounts for tree complexity. Key features include regularization, parallel processing, and column subsampling, making XGB a powerful and efficient tool for predictive modeling.

3.4 Deep Learning Algorithms

3.4.1 MLP

The MLP is a kind of feedforward neural network that is used for time series forecasting. It consists of an input layer, one or more hidden layers, and an output layer, which are the different layers of neurons. Because every layer's neurons are fully coupled to every other layer's, the network can recognize complex patterns in the data. Training of the MLP model is carried out through backpropagation, which adjusts the network's weights to reduce prediction errors (Rumelhart et al., 1986; Hasan et al., 2023c). The MLP model can be represented as

$$Y_t = \sigma\left(W^L \cdot \sigma\left(W^{L-1} \cdots \sigma\left(W^1 \cdot X_{t-1} + b^1\right) \cdots + b^{L-1}\right) + b^L\right)$$

where Y_t is the predicted value at time t, $X_{t-1}, X_{t-2}, \ldots, X_{t-n}$ are the input features (lagged observations), W^l and b^l are the weights and biases for layer l, and σ denotes an activation function such as ReLU or sigmoid. MLP is flexible in modeling nonlinear relationships in time series data, but it requires careful tuning of hyperparameters, such as the number of layers and neurons, to achieve optimal performance.

3.4.2 LSTM

The LSTM network is a type of RNN designed for time series forecasting which effectively captures long-term dependencies through its gating mechanisms and manages the flow of data in memory cells (Cho et al., 2014; Rabbi et al., 2022). The LSTM equations are as follows: Forget:

$$f_t = \sigma(W_f \cdot [h_{t-1}, X_t] + b_f)$$

Input:

$$i_t = \sigma(W_i \cdot [h_{t-1}, X_t] + b_i)$$

$$\tilde{C}_t = \tanh(W_C \cdot [h_{t-1}, X_t] + b_C)$$

Cell state:

$$C_t = f_t \cdot C_{t-1} + i_t \cdot \tilde{C}_t$$

Output :

$$o_t = \sigma(W_o \cdot [h_{t-1}, X_t] + b_o)$$

$$h_t = o_t \cdot \tanh(C_t)$$

where C_t is called the cell state, h_t is called the hidden state, X_t is the input at time t, and σ is the sigmoid function. LSTMs excel in capturing long-term patterns in time series data but require careful tuning and are computationally intensive.

3.4.3 BiLSTM

The BiLSTM network extends the LSTM architecture by processing data in both directions (forward and backward), enabling it to grasp relationships from both historical and future contexts in forecasting. This approach increases the model's ability to learn from sequences where future context improves the prediction of past elements (Cho et al., 2014).The BiLSTM model can be expressed with the following components:

Forward LSTM:

$$\begin{aligned}
f_t^f &= \sigma(W_f^f \cdot [h_{t-1}^f, X_t] + b_f^f) \\
i_t^f &= \sigma(W_i^f \cdot [h_{t-1}^f, X_t] + b_i^f) \\
\tilde{C}_t^f &= \tanh(W_C^f \cdot [h_{t-1}^f, X_t] + b_C^f) \\
C_t^f &= f_t^f \cdot C_{t-1}^f + i_t^f \cdot \tilde{C}_t^f \\
o_t^f &= \sigma(W_o^f \cdot [h_{t-1}^f, X_t] + b_o^f) \\
h_t^f &= o_t^f \cdot \tanh(C_t^f)
\end{aligned}$$

Backward LSTM:

$$\begin{aligned}
f_t^b &= \sigma(W_f^b \cdot [h_{t+1}^b, X_t] + b_f^b) \\
i_t^b &= \sigma(W_i^b \cdot [h_{t+1}^b, X_t] + b_i^b) \\
\tilde{C}_t^b &= \tanh(W_C^b \cdot [h_{t+1}^b, X_t] + b_C^b) \\
C_t^b &= f_t^b \cdot C_{t+1}^b + i_t^b \cdot \tilde{C}_t^b \\
o_t^b &= \sigma(W_o^b \cdot [h_{t+1}^b, X_t] + b_o^b) \\
h_t^b &= o_t^b \cdot \tanh(C_t^b)
\end{aligned}$$

Output:

$$h_t = [h_t^f ; h_t^b]$$

where h_t^f and h_t^b are the hidden states from the forward and backward passes, respectively, and $[\cdot; \cdot]$ denotes concatenation. BiLSTM improves forecasting accuracy by utilizing both previous and upcoming information, making it particularly useful for sequences where the full context enhances prediction.

3.4.4 GRU

Reducing the number of parameters and computational complexity in comparison to LSTMs, the GRU network is a variation of the LSTM network that maintains effectiveness in capturing long-term dependencies while simplifying its architecture

(Cho et al., 2014). It does this by combining the input and forget gates into a single update gate and using a reset gate to control the information flow. The following equations can be used to represent the GRU model for time series forecasting:

Update Gate:

$$z_t = \sigma(W_z \cdot [h_{t-1}, X_t] + b_z)$$

Reset Gate:

$$r_t = \sigma(W_r \cdot [h_{t-1}, X_t] + b_r)$$

Candidate Activation:

$$\tilde{h}_t = \tanh(W_h \cdot [r_t \odot h_{t-1}, X_t] + b_h)$$

Final Hidden State:

$$h_t = (1 - z_t) \odot h_{t-1} + z_t \odot \tilde{h}_t$$

where h_t is the hidden state at time t, X_t indicates the input at time t, z_t is the update gate, r_t is the reset gate, $\tilde{h}_t$ is the candidate activation, and $\odot$ signifies element-wise multiplication. The update gate regulates the extent of past information to retain, whereas the reset gate dictates the amount of past information to discard. GRUs are efficient and effective for forecasting, with fewer parameters than LSTMs, making them suitable for tasks with limited computational resources.

3.5 *Stacking Ensemble Learning Model*

Stacking ensemble learning leverages the complementary strengths of various base models to enhance overall performance and generalization capabilities. This approach generally involves two stages: training the base models and training the meta-model (Wolpert, 1992). The original data is divided into training and testing sets in the first step. The k-fold cross-validation method is used to further divide the training set. Using this method, the training set is divided into k subsets, of which $k - 1$ subsets are utilized for training and each subset for testing. Every subset is used as a test set once during the k repetitions of the process. In the second stage, the predictions generated during k-fold cross-validation are collected and reassembled according to the original training dataset order, creating a new training set. The meta-model, which combines the outputs from the different basic models, is then trained using this new training set. The meta-model is trained on this combined dataset after the predictions from the underlying models on the test set are combined to create the test set. Let us have four base models, namely A, B, C, D, and a meta-

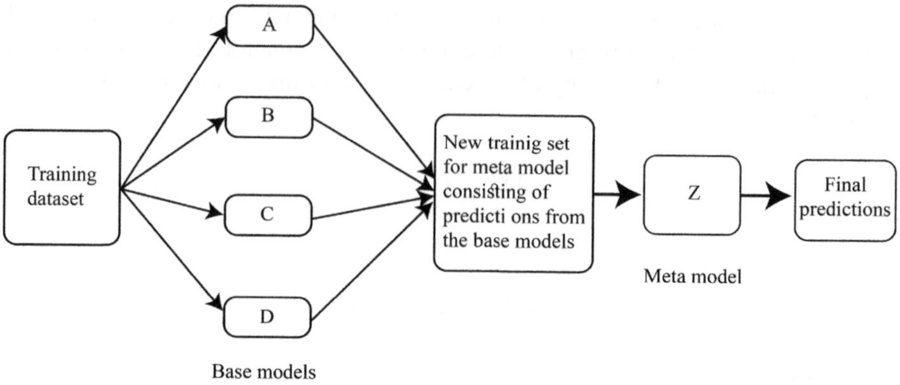

Fig. 3 Diagram of the working process of stacking ensemble learning models

model Z, then the diagram of stacking ensemble learning is presented as Fig. 3, and the algorithm can be expressed as the below algorithm.

Algorithm 1 Stacking ensemble learning model

```
Input: Training data D = {(x_i, y_i)}_{i=1}^n, Base models A, B, C, D, Meta-model Z
Output: Stacked model prediction
Step 1: Train Base Models
for each model M in {A, B, C, D} do
    Perform k-fold cross-validation on M to generate predictions
end for
Step 2: Train Meta-model
Create meta-training set D_meta from base model predictions
D_meta = {(ŷ_i^A, ŷ_i^B, ŷ_i^C, ŷ_i^D, y_i)}_{i=1}^n
Train meta-model Z on D_meta
Step 3: Make Final Predictions
Generate base models' predictions on test data
Create meta-test set D_meta,test
D_meta,test = {(ŷ_i^A, ŷ_i^B, ŷ_i^C, ŷ_i^D)}_{i=1}^m
Generate final predictions with Z
    return Final predictions from Z
```

This study creates four ensemble models which are listed below:

- Stacking Random-Gradient-SVR-KNN with Logistic Regression (RGSKL) model (base models: RF, GB, SVR, and KNN and meta-model: LR)
- Stacking Decision-AdaBoost-ElasticNet-SVR with Ridge (DAESR) model (base models: DT, AdaBoost, SVR, and ElasticNet and meta-model: Ridge)
- Stacking Logistic-Ridge-SVR-Decision Tree with Ridge (LRSDR) model (base models: LR, Ridge, SVR, and DT and meta-model: Ridge)
- Stacking ElasticNet-AdaBoost-GB-Huber with Logistic Regression (EAGHL) model (base models: EN, AdaBoost, GBR, and Huber and meta-model: LR)

3.6 Performance Measure Metrics

The observed value is represented by y_i, the predicted value by $\hat{y}_i$, the mean of the observed values by $\bar{y}$, and the total number of observations by n. Next, the MSE measures the average squared discrepancies between actual and anticipated values, indicating the size of the mistakes (Dong et al., 2024). It is expressed as

$$\text{MSE} = \frac{1}{n}\sum_{i=1}^{n}(y_i - \hat{y}_i)^2$$

The RMSE, which provides a measure in the same units as the dependent variable, is the square root of the MSE (Dong et al., 2024):

$$\text{RMSE} = \sqrt{\text{MSE}}$$

The MAE measures the average magnitude of errors in predictions (James, 2013), treating all errors equally:

$$\text{MAE} = \frac{1}{n}\sum_{i=1}^{n}|y_i - \hat{y}_i|$$

The MAPE measures the average magnitude of prediction errors as a percentage of the observed values (Dong et al., 2024):

$$\text{MAPE} = \sqrt{\frac{1}{n}\sum_{i=1}^{n}\left(\frac{y_i - \hat{y}_i}{y_i}\right)^2}$$

The sMAPE adjusts for the scale of errors and is symmetric (Dong et al., 2024):

$$\text{sMAPE} = \sqrt{\frac{1}{n}\sum_{i=1}^{n}\left(\frac{y_i - \hat{y}_i}{y_i + \hat{y}_i}\right)^2}$$

The percentage of the dependent variable's variance that the independent variables can account for is represented by the R^2 metric (James, 2013):

$$R^2 = 1 - \frac{\sum_{i=1}^{n}(y_i - \hat{y}_i)^2}{\sum_{i=1}^{n}(y_i - \bar{y})^2}$$

4 Result Analysis

4.1 Obtained Hyperparameters and Suitable Values for Model Training

The hyperparameters for the ML models, determined through a grid search across specified parameter values, are listed in Table 2.

We conducted this process using the scikit-learn library in Python. Using a PC with an Intel iRISxe graphics card, 8 GB of RAM, and a 1.30 GHz CPU, all tests were conducted on Google Colab with a reliable Internet connection. The "Grid Search" method was used to methodically adjust the hyperparameters. With this approach, a grid of potential hyperparameters is created, their corresponding values or ranges are specified, and cross-validation is applied to the model for every conceivable combination. The optimal set is the arrangement that yields the best results when evaluated using a particular evaluation metric. Grid Search is a rigorous and systematic method for optimizing hyperparameters, although it can be computationally intensive, particularly when working with huge parameter spaces.

Table 2 Hyperparameter values for the ML models

Models	Best hyperparameters	Range of value search
LR	N/A	N/A
Ridge	"alpha": 0.1	"alpha": [0.1, 1, 10, 100]
Lasso	alpha': 0.01	"alpha": [0.01, 0.1, 1, 10]
Poly	linear__fit_intercept': False, "poly__degree=2	poly__degree': [2, 3, 4], "linear__fit_intercept": [True, False]
SVR	"C": 0.1, "epsilon": 0.01, "kernel": "linear"	kernel': ["linear", "rbf"], "C": [0.1, 1, 10], "epsilon": [0.01, 0.1, 0.2]
DTR	"max_depth": 10, "min_samples_split": 10	max_depth': [None, 10, 20, 30], "min_samples_split": [2, 5, 10]
RF	max_depth': 10, "min_samples_split": 10, "n_estimator"=100	"n_estimators": [100, 200, 500], "max_depth": [None, 10, 20, 30], "min_samples_split": [2, 5, 10]
GBR	learning_rate': 0.01, "max_depth": 5, "n_estimator"=100	"n_estimators": [100, 200, 500], "learning_rate": [0.01, 0.1, 0.2], "max_depth": [3, 5, 7]
XGB	learning_rate': 0.1, "max_depth": 3, "n_estimator"=100	"n_estimators": [100, 200, 500], "learning_rate": [0.01, 0.1, 0.2], "max_depth": [3, 5, 7]
LGB	learning_rate': 0.2, "max_depth": -1, "n_estimator"=100	"n_estimators": [100, 200, 500], "learning_rate": [0.01, 0.1, 0.2], "max_depth": [-1, 10, 20]
KNN	"n_neighbors": 10, "weights": "uniform"	n_neighbors': [3, 5, 7, 10], "weights": ["uniform", "distance"]

4.2 Results of the U.S. Gasoline Spot Price Forecasting

4.2.1 Performance of the ML Models

The initial outcome demonstrates the accuracy of daily U.S. gasoline spot price predictions using various ML techniques without hyperparameter tuning which shows by Table 3. This indicates that the LR, Ridge, and Poly models outperformed the other ML models in this analysis prior to hyperparameter tuning. These models exhibited lower absolute and relative errors and achieved an R^2 of 97%, implying that they are better at capturing short-term fluctuations in the data and providing more accurate daily forecasts of the U.S. gasoline spot price. Additionally Lasso performed worst with higher errors and lower R^2 of only 7%. However, it is worth noting that the computation time for LR was slightly longer compared to the Ridge and Poly models. The performance metrics for the ML models after hyperparameter tuning, as shown in Table 4, reveal that LR, Ridge, and SVR outperformed the others, with relatively lower absolute and relative errors and an R^2 of 97%. This shows that these models are better at identifying short-term variations in the data and producing daily estimates of the current price of gasoline in the United States that are more accurate. However, the SVR model had a significantly higher computation time compared to the other two, making Ridge the optimal choice due to its lower elapsed time. Notably, after tuning, the Lasso model showed a remarkable improvement, achieving an R^2 of 96.9% and errors close to those of the optimal model. In contrast, GBR was the poorest performer among the tuned models. Figure 4 illustrates the actual versus predicted curves for the models, with the curves for the LR, Ridge, and SVR models closely aligning with the actual data. The GBR, XGB, LGB, and KNN models, on the other hand, were less successful and frequently greatly overestimated or underestimated the projected pricing.

Table 3 Performance measures of the ML models without hyperparameter tuning for forecasting daily gasoline spot prices

Models	Elapsed time (s)	MAE	MSE	RMSE	MAPE	sMAPE	R^2
LR	0.02	0.06	0.01	0.09	2.36	2.35	0.97
Ridge	0.01	0.06	0.01	0.09	2.36	2.35	0.97
Lasso	0.00	1.21	1.72	1.31	43.34	56.29	0.07
Poly	0.00	0.06	0.01	0.09	2.35	2.35	0.97
SVR	0.29	0.10	0.03	0.17	3.14	3.22	0.88
DTR	0.02	0.12	0.06	0.24	4.03	3.91	0.78
RF	1.78	0.09	0.02	0.14	3.25	3.23	0.92
GBR	0.94	0.11	0.04	0.21	3.48	3.39	0.83
XGB	0.29	0.10	0.03	0.18	3.24	3.32	0.87
LGB	0.17	0.10	0.03	0.17	3.20	3.27	0.88
KNN	0.01	0.09	0.02	0.14	3.10	3.14	0.92

Table 4 Performance measures of the ML models after hyperparameter tuning for forecasting daily gasoline spot prices

Models	Elapsed time (s)	MAE	MSE	RMSE	MAPE	sMAPE	R^2
LR	2.83	0.064	0.0075	0.087	2.358	2.353	0.970
Ridge	0.15	0.064	0.0075	0.087	2.358	2.353	0.970
Lasso	0.14	0.067	0.0079	0.089	2.446	2.458	0.969
Poly	0.20	0.068	0.0089	0.094	2.466	2.468	0.965
SVR	75.38	0.064	0.0075	0.087	2.354	2.347	0.970
DTR	0.51	0.086	0.0203	0.142	3.049	3.016	0.920
RF	277.69	0.079	0.0149	0.122	2.800	2.802	0.941
GBR	153.64	0.102	0.0351	0.187	3.361	3.303	0.861
XGB	14.18	0.099	0.0319	0.179	3.242	3.322	0.874
LGB	26.89	0.097	0.0293	0.171	3.199	3.266	0.884
KNN	0.48	0.096	0.0262	0.162	3.180	3.241	0.896

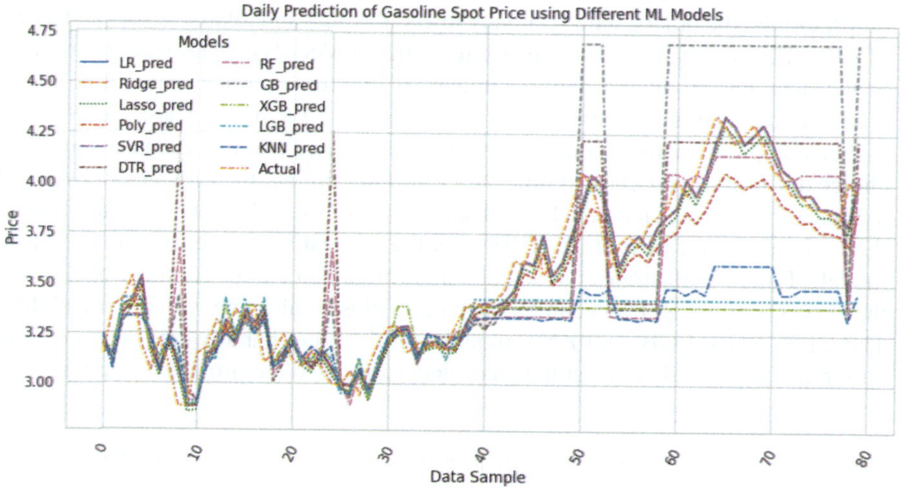

Fig. 4 Daily prediction of gasoline spot price using different ML models

4.2.2 Performance of the DL models

Similar to the ML models, Table 5 shows the performance of the DL models for predicting the daily U.S. gasoline spot price. LSTM and BiLSTM stood out with lower absolute and relative errors, achieving an R^2 of 96.4%, closely followed by the GRU model with an R^2 of 95.9%. However, LSTM had the advantage in terms of elapsed time, making it the more efficient choice. On the other hand, MLP performed poorly among the DL models, with significantly higher errors and a low R^2 of only 26.5%.

Similarly, Fig. 5 illustrates the daily forecasts of the U.S. gasoline spot price based on the predictions from the DL models. As with the ML models, LSTM

Table 5 Performance measures of the DL models for forecasting daily gasoline spot prices

Models	Elapsed time (s)	MAE	MSE	RMSE	MAPE	sMAPE	R^2
MLP	33.85	0.412	0.1861	0.431	15.192	16.486	0.265
LSTM	60.53	0.069	0.0090	0.095	2.486	2.481	0.964
BiLSTM	76.81	0.069	0.0091	0.095	2.482	2.487	0.964
GRU	60.20	0.073	0.0104	0.102	2.592	2.610	0.959

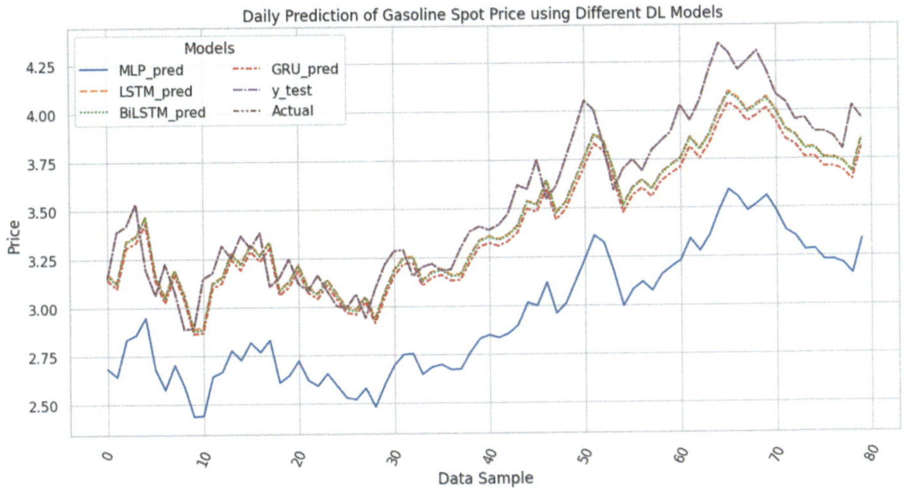

Fig. 5 Daily prediction of gasoline spot price using different DL models

and BiLSTM closely track the actual gasoline prices, providing almost accurate forecasts. However, MLP falls short in delivering accurate predictions.

4.2.3 Performance of the Ensemble Models

Lastly, the forecasting performance of four stacking ensemble models for predicting the daily U.S. gasoline spot price is shown in Table 6. Among these, the LRSDR model emerged as the top performer, surpassing all single ML, DL, and other stacking models with the lowest absolute and relative errors and achieving an impressive R^2 of 98%. Additionally, this model had the shortest elapsed time compared to the others. It was closely followed by two other stacking models, EAGHL and DAESR, which attained R^2 values of 96.7% and 95.1%, respectively.

Figure 6 displays the predicted versus actual curves generated by various stacking models. The stacking LRSDR model, in particular, closely mirrors the actual curve, demonstrating near-accurate predictions. The EAGHL and DAESR models are also close to the actual curve.

Based on all the performance metrics discussed in this study, the optimal model for forecasting the daily U.S. gasoline spot price is the stacking LRSDR model.

Table 6 Performance measures of the stacking models for forecasting daily gasoline spot prices

Models	Elapsed time (s)	MAE	MSE	RMSE	MAPE	sMAPE	R^2
RGSKL	11.21	0.110	0.0446	0.211	3.552	3.459	0.824
DAESR	7.00	0.076	0.0124	0.111	2.719	2.710	0.951
LRSDR	1.57	0.059	0.0067	0.081	2.266	2.261	0.980
EAGHL	6.32	0.067	0.0084	0.092	2.449	2.443	0.967

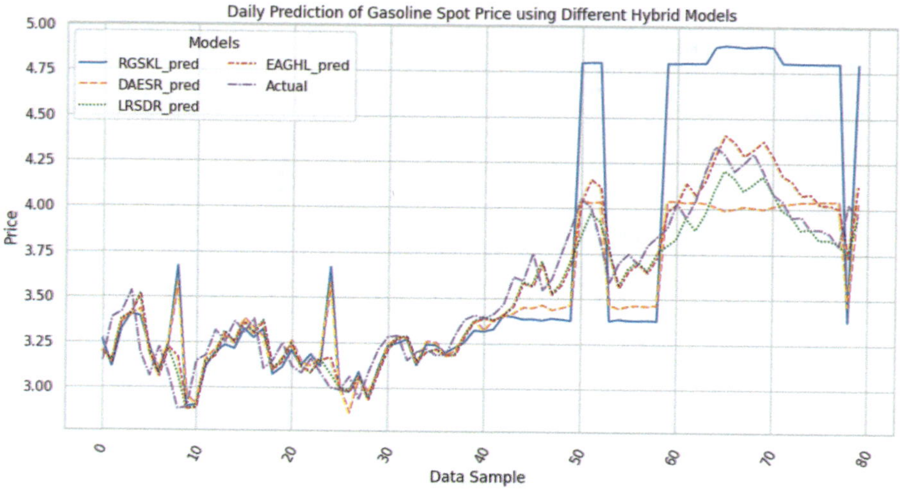

Fig. 6 Daily prediction of gasoline spot price using different stacking models

4.3 Results of the New York Gasoline Spot Price Forecasting

4.3.1 Performance of the ML Models

In line with the previous analysis, the initial results demonstrate the accuracy of daily New York gasoline spot price predictions using various ML techniques with hyperparameter tuning, as shown in Table 7. We did not include results without parameter tuning since the previous section established that model performance improves based on the hyperparameters listed in Table 2. The table indicates that the LR, Ridge, Lasso, Poly, and SVR models outperformed the others in this analysis, exhibiting lower absolute and relative errors and achieving an R^2 of 99.2%. This shows that these models are more successful in identifying brief variations in the data and producing daily predictions of the spot price of New York gasoline that are more accurate. Among them, the Lasso model can be considered optimal based on elapsed time, while SVR stands out for its performance in MAPE and sMAPE. The remaining models closely followed these leaders.

Figure 7 presents the actual versus predicted curves for these models, with LR, Ridge, Lasso, Poly, and SVR models closely tracking the actual data. In

Table 7 Performance measures of the ML models with hyperparameter tuning for forecasting daily New York gasoline spot prices

Models	Elapsed time (s)	MAE	MSE	RMSE	MAPE	sMAPE	R^2
LR	2.37	0.047	0.0045	0.067	2.256	2.244	0.992
Ridge	0.18	0.047	0.0045	0.067	2.256	2.244	0.992
Lasso	0.08	0.050	0.0049	0.070	2.354	2.349	0.992
Poly	0.14	0.047	0.0045	0.067	2.254	2.243	0.992
SVR	83.82	0.047	0.0045	0.067	2.253	2.240	0.992
DTR	0.51	0.069	0.0171	0.131	2.913	2.942	0.971
RF	299.45	0.068	0.0175	0.132	2.860	2.891	0.970
GBR	177.69	0.068	0.0198	0.141	2.802	2.845	0.966
XGB	14.72	0.070	0.0220	0.148	2.852	2.901	0.962
LGB	23.56	0.069	0.0210	0.145	2.818	2.863	0.964
KNN	0.26	0.070	0.0207	0.144	2.867	2.910	0.964

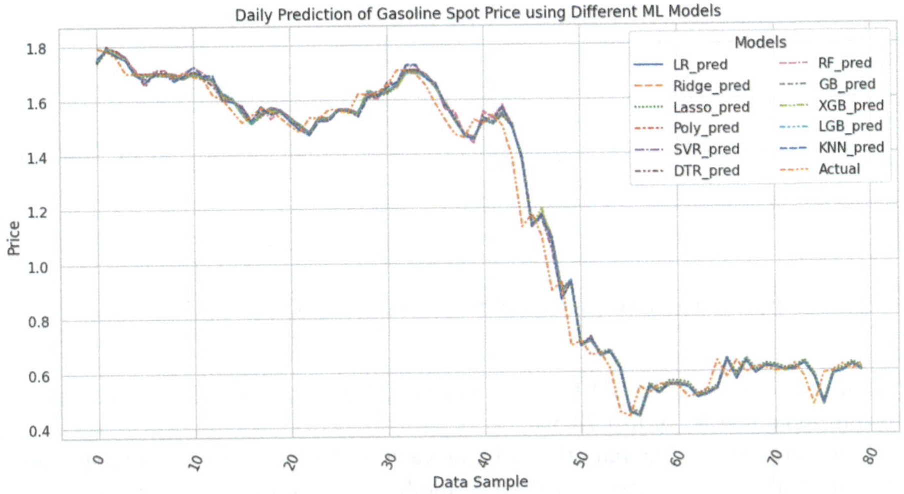

Fig. 7 Daily prediction of New York gasoline spot price using different stacking models

contrast, the other models were less accurate, often significantly overestimating or underestimating the predicted prices.

4.3.2 Performance of the DL Models

Similar to the ML models, Table 8 presents the performance of DL models in predicting the daily New York gasoline spot price. LSTM and GRU models excelled, with lower absolute and relative errors, achieving R^2 values of 99.1% and 99%, respectively, followed closely by the BiLSTM model with an R^2 of 98.8%. LSTM also proved to be more efficient in terms of elapsed time, making it the preferred

Table 8 Performance measures of the DL models for forecasting daily New York gasoline spot prices

Models	Elapsed time (s)	MAE	MSE	RMSE	MAPE	sMAPE	R^2
MLP	31.02	0.322	0.1245	0.353	13.797	14.863	0.785
LSTM	56.64	0.049	0.0051	0.071	2.349	2.328	0.991
BiLSTM	70.19	0.058	0.0067	0.082	2.658	2.683	0.988
GRU	63.73	0.054	0.0058	0.076	2.570	2.532	0.990

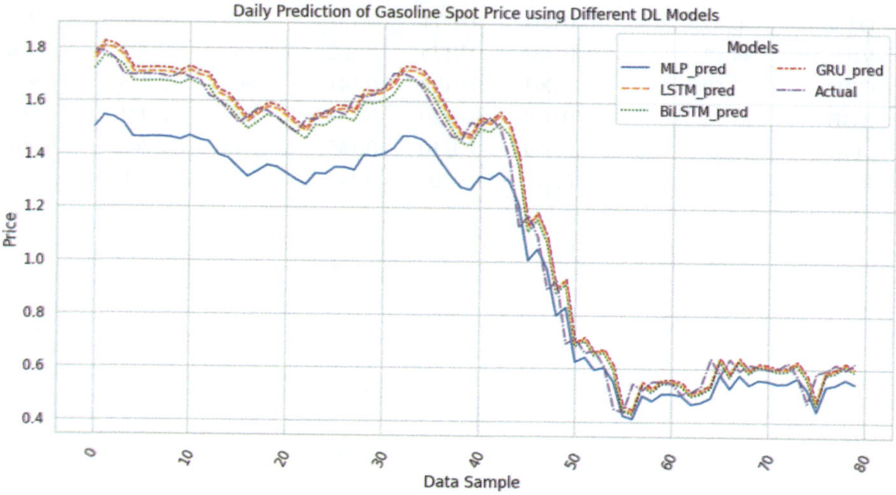

Fig. 8 Daily prediction of New York gasoline spot price using different stacking models

choice. Conversely, the MLP model performed poorly, with significantly higher errors and a low R^2 of just 78.5%.

Similarly, Fig. 8 illustrates the daily forecasts of the New York gasoline spot price based on the predictions from the DL models. As with the ML models, LSTM and GRU closely followed the actual gasoline prices, delivering nearly accurate forecasts, with BiLSTM also performing well. However, MLP struggled to provide accurate predictions.

4.3.3 Performance of the Hybrid Models

Finally, Table 9 displays the forecasting performance of four stacking ensemble models for predicting the daily New York gasoline spot price. Among these, the LRSDR and EAGHL models excelled, outperforming all other single ML, DL, and stacking models with the lowest absolute and relative errors and achieving an impressive R^2 of 99.2%. Additionally, the LRSDR model had the shortest elapsed time, making it the most efficient. It was followed closely by the DAESR and RGSKl models, which achieved R^2 values of 92.4% and 98.1%, respectively.

Table 9 Performance measures of the stacking models for forecasting daily New York gasoline spot prices

Models	Elapsed time (s)	MAE	MSE	RMSE	MAPE	sMAPE	R^2
RGSKL	9.40	0.063	0.0108	0.104	2.763	2.763	0.981
DAESR	1.31	0.082	0.0439	0.209	3.198	3.321	0.924
LRSDR	0.66	0.047	0.0046	0.068	2.259	2.248	0.992
EAGHL	2.91	0.047	0.0046	0.068	2.267	2.259	0.992

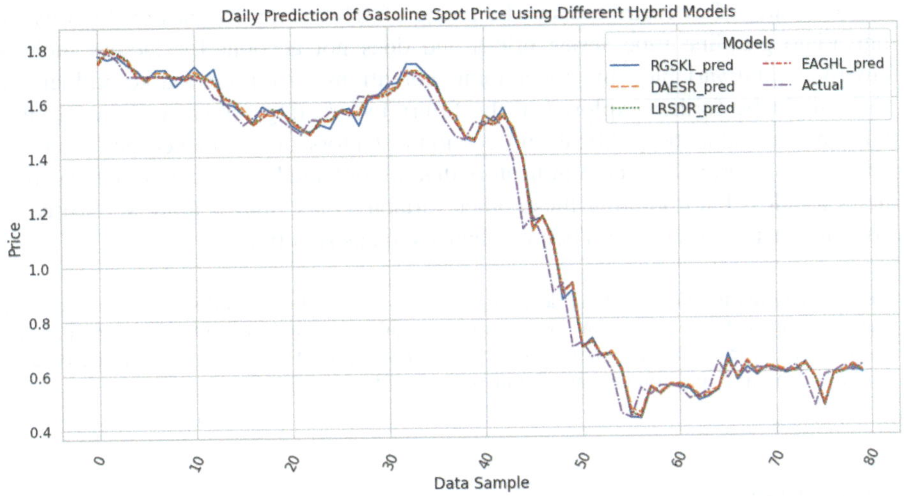

Fig. 9 Daily prediction of New York gasoline spot price using different stacking models

Figure 9 showcases the predicted versus actual curves for various stacking models. Notably, the stacking LRSDR model closely tracks the actual curve, reflecting near-accurate predictions. The EAGHL and RGSKL models also align closely with the actual data.

Similarly, based on all the performance metrics discussed in this study, the optimal model for forecasting the daily New York gasoline spot price is the stacking LRSDR model.

5 Conclusion and Future Work

In this research, we examined several stacking ensemble learning models that integrate various ML regression techniques to improve time series prediction. Our findings indicate that the stacking LRSDR model outperforms benchmark methods in terms of elapsed time, prediction accuracy, and various error metrics. For the purpose of making production decisions for the industry, the suggested LRSDR model exhibits robustness in forecasting across various granularities of gasoline time series. Furthermore, the findings can assist interested parties in creating profitable investment plans to optimize returns in times of market turmoil.

This research also highlights areas requiring further exploration. While our ML approach has proven effective for modeling diverse fluctuation patterns in energy markets, it would be valuable to assess its performance in other commodity and financial markets, like foreign exchange rates, stock prices, and precious metal prices. For instance, comparable models have had difficulty adjusting to the current decoupling of oil and gas prices, despite advances in gasoline forecasting.

There are a few restrictions to be aware of. This study focuses exclusively on univariate gasoline time series prices and does not account for factors such as environmental variables, macroeconomic conditions, or foreign market influences. This might be further explored in the future by creating forecasting models that take into account the political and economic factors that influence gas prices in addition to market sentiment indicators that are obtained from textual data such as news articles. Incorporating these other variables may offer a more sophisticated comprehension of the forces behind changes in gas spot prices.

Acknowledgments We extend our deepest gratitude to the Center for Multidisciplinary Research and Development (CeMRD) for their resources, guidance, and support. The expertise and encouragement from CeMRD members have been invaluable. This research was made possible by the collaborative environment and cutting-edge facilities at CeMRD.

References

Abdollahi, H., & Ebrahimi, S. B. (2020). A new hybrid model for forecasting Brent crude oil price. *Energy, 200*, 117520.

Agbaji, A. L., Morrison, R., & Lakshmanan, S. (2023). *ESG, sustainability and decarbonization: An analysis of strategies and solutions for the energy industry*. OnePetro.

Allcott, H., & Wozny, N. (2014). Gasoline prices, fuel economy, and the energy paradox. *Review of Economics and Statistics, 96*, 779–795.

Altman, N. S. (1992). An introduction to kernel and nearest-neighbor nonparametric regression. *The American Statistician, 46*, 175–185.

Borenstein, S., & Shepard, A. (1996). Sticky prices, inventories, and market power in wholesale gasoline markets.

Breiman, L. (2001). Random forests. *Machine Learning, 45*, 5–32.

Busse, M. R., Knittel, C. R., & Zettelmeyer, F. (2013). Are consumers myopic? Evidence from new and used car purchases. *American Economic Review, 103*, 220–256.

Cho, K., Van Merriënboer, B., Bahdanau, D., & Bengio, Y. (2014). On the properties of neural machine translation: Encoder-decoder approaches. arXiv preprint arXiv:1409.1259.

Dong, Y., Jiang, H., Guo, Y., & Wang, J. (2024). A novel crude oil price forecasting model using decomposition and deep learning networks. *Engineering Applications of Artificial Intelligence, 133*, 108111.

Eliwa, E. H. I., El Koshiry, A. M., Abd El-Hafeez, T., & Omar, A. (2024). Optimal gasoline price predictions: Leveraging the ANFIS regression model. *International Journal of Intelligent Systems, 2024*, 8462056.

Fathi, S., Srinivasan, R., Fenner, A., & Fathi, S. (2020). Machine learning applications in urban building energy performance forecasting: A systematic review. *Renewable and Sustainable Energy Reviews, 133*, 110287.

Freund, Y., & Schapire, R. E. (1997). A decision-theoretic generalization of on-line learning and an application to boosting. *Journal of Computer and System Sciences, 55*, 119–139.

Friedman, J. H. (2001). Greedy function approximation: A gradient boosting machine. *Annals of Statistics* 1189–1232 (2001).

Hamilton, J. D. (2009). Causes and consequences of the oil shock of 2007-08. Tech. Rep., National Bureau of Economic Research.

Hasan, M., Das, U., Datta, R. K., & Abedin, M. Z. (2023a). *Model development for predicting the crude oil price: Comparative evaluation of ensemble and machine learning methods* (pp. 167–179). Springer.

Hasan, M., Hassan, M. M., Faisal-E-Alam, M., & Akter, N. (2023b). *Empirical analysis of regression techniques to predict the cybersecurity salary* (pp. 65–84) Routledge.

Hasan, M., et al. (2023c). *Ensemble based machine learning model for early detection of mother's delivery mode* (pp. 1–6). IEEE.

Hasan, M., et al. (2024a). A blending ensemble learning model for crude oil price forecasting. *Annals of Operations Research* 1–31.

Hasan, M., Ahmed, T., Islam, M. R., & Uddin, M. P. (2024b). Leveraging textual information for social media news categorization and sentiment analysis. *PLOS One, 19*, e0307027.

Hasan, M., et al. (2024c). Performance discrepancy mitigation in heart disease prediction for multisensory inter-datasets. *PeerJ Computer Science, 10*, e1917.

Hastie, T., Tibshirani, R., Friedman, J., & Franklin, J. (2005). The elements of statistical learning: Data mining, inference and prediction. *The Mathematical Intelligencer, 27*, 83–85.

Hyndman, R. J., & Athanasopoulos, G. (2018). *Forecasting: Principles and practice*. OTexts.

James, G. (2013). An introduction to statistical learning.

Jin, B., & Xu, X. (2024). Price forecasting through neural networks for crude oil, heating oil, and natural gas. *Measurement: Energy, 1*, 100001.

Ke, G., et al. (2017). LightGBM: A highly efficient gradient boosting decision tree. In *Advances in neural information processing systems* (Vol. 30).

Khan, S. M., et al. (2023). A systematic review of disaster management systems: approaches, challenges, and future directions. *Land, 12*, 1514.

Knittel, C. R., & Sandler, R. (2011). *Carbon prices and automobile greenhouse gas emissions: The extensive and intensive margins* (pp. 287–299). University of Chicago Press.

Lewis, M. S., & Marvel, H. P. (2011). When do consumers search? *The Journal of Industrial Economics, 59*, 457–483.

Li, J., Tang, L., & Wang, S. (2020). Forecasting crude oil price with multilingual search engine data. *Physica A: Statistical Mechanics and Its Applications, 551*, 124178.

Li, J., Hong, Z., Zhang, C., Wu, J., & Yu, C. (2024). A novel hybrid model for crude oil price forecasting based on MEEMD and Mix-KELM. *Expert Systems with Applications, 246*, 123104.

Liu, L., et al. (2024). A robust time-varying weight combined model for crude oil price forecasting. *Energy, 299*, 131352.

MultiMedia LLC. (2024). Gasoline explained—U.S. Energy Information Administration (EIA). https://www.eia.gov/energyexplained/gasoline/

Lu, H., Ma, X., Ma, M., & Zhu, S. (2021). Energy price prediction using data-driven models: A decade review. *Computer Science Review, 39*, 100356.

Montgomery, D. C., Peck, E. A., & Vining, G. G. (2021). *Introduction to linear regression analysis*. John Wiley & Sons.

Qin, Q., Huang, Z., Zhou, Z., Chen, C., & Liu, R. (2023). Crude oil price forecasting with machine learning and google search data: An accuracy comparison of single-model versus multiple-model. *Engineering Applications of Artificial Intelligence, 123*, 106266.

Rabbi, M. F., Moon, M. H., Dhonno, F. T., Sultana, A., & Abedin, M. Z. (2022). *Foreign currency exchange rate prediction using long short-term memory, support vector regression and random forest regression* (pp. 251–267). Springer.
Rumelhart, D. E., Hinton, G. E., & Williams, R. J. (1986). Learning representations by back-propagating errors. *Nature, 323*, 533–536.
Sajid, S. W., Anjum, K. R., Al-Shaharia, M., & Hasan, M. (2023). *Investigating machine learning algorithms with model explainability for network intrusion detection* (pp. 121–136). Routledge.
Salamai, A. A. (2023). Deep learning framework for predictive modeling of crude oil price for sustainable management in oil markets. *Expert Systems with Applications, 211*, 118658.
Smola, A. J., & Schölkopf, B. (2004). A tutorial on support vector regression. *Statistics and Computing, 14*, 199–222.
Tibshirani, R. (1996). Regression shrinkage and selection via the lasso. *Journal of the Royal Statistical Society Series B: Statistical Methodology, 58*, 267–288.
Wolpert, D. H. (1992). Stacked generalization. *Neural Networks, 5*, 241–259.
Yuan, J., Li, J., & Hao, J. (2023). A dynamic clustering ensemble learning approach for crude oil price forecasting. *Engineering Applications of Artificial Intelligence, 123*, 106408.
Zhang, T., Tang, Z., Wu, J., Du, X., & Chen, K. (2021). Multi-step-ahead crude oil price forecasting based on two-layer decomposition technique and extreme learning machine optimized by the particle swarm optimization algorithm. *Energy, 229*, 120797.
Zhang, X., Cheng, S., Zhang, Y., Wang, J., & Wang, S. (2024). An attention-PCA based forecast combination approach to crude oil price. *Expert Systems with Applications, 240*, 122463.
Zhao, Z., Sun, S., Sun, J., & Wang, S. (2024). A novel hybrid model with two-layer multivariate decomposition for crude oil price forecasting. *Energy, 288*, 129740.
Zheng, L., Sun, Y., & Wang, S. (2024). A novel interval-based hybrid framework for crude oil price forecasting and trading. *Energy Economics, 130*, 107266.

Md. Amir Hamja is currently pursuing his MSc in Statistics from the Department of Statistics at Hajee Mohammad Danesh Science and Technology University, Dinajpur, Bangladesh, and he completed BSc (Hons.) in Statistics from the same university in 2023. Currently, he is a Research Assistant in the Center for Multidisciplinary Research and Development (CeMRD). His research interests include Federated Learning, Machine Learning, Deep Learning, Cyber Security, Health Informatics, Business Intelligence, Time Series, Public Health, and Biostatistics.

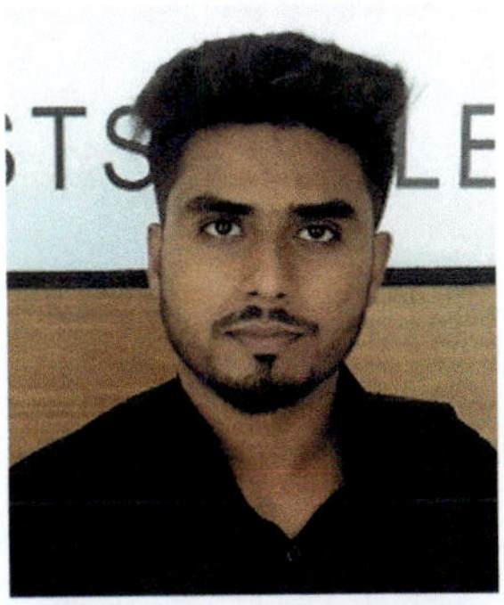

Md Rakinus Sakib is currently pursuing an M.Sc. in Statistics at the Department of Statistics, Hajee Mohammad Danesh Science and Technology University (HSTU), Dinajpur, Bangladesh. Where he also completed his B.Sc. (Hons.) in Statistics in 2024. His research interests encompass a wide range of areas, including Machine Learning, Deep Learning, Cybersecurity, Health Informatics, Business Intelligence, Time Series Analysis, Public Health, and Biostatistics.

Mahmudul Hasan is currently pursuing a PhD in Information Technology (IT) at Deakin University, Melbourne, Australia. He earned his BSc (Eng.) and MSc (Eng.) degrees in Computer Science and Engineering (CSE) from Hajee Mohammad Danesh Science and Technology University, Dinajpur, Bangladesh, in 2021 and 2023, respectively. He previously served as a Lecturer in the Department of CSE at the University of Creative Technology, Chittagong (UCTC), Bangladesh. He is the Founder and Director of the Center for Multidisciplinary Research and Development (CeMRD) and a moderator of "Be Researcher BD," the largest online research forum in Bangladesh. Additionally, he has taught online as a Data Science instructor to students in the USA, Italy, Denmark, South Korea, and Australia. He is also the founder of the online educational platform "MHM Academy." His research interests include federated learning, machine learning, deep learning, cybersecurity, health informatics, renewable energy, computational sociology, and business intelligence.

Md Sabir Hossain (a member, IEEE) received the bachelor's and master's degrees in computer science and engineering from the Chittagong University of Engineering and Technology, with an outstanding result. He is currently pursuing his PhD at Information and Computer Science Department of King Fahd University of Petroleum and Minerals (KFUPM), Saudi Arabia. Previously, he served as a faculty member (assistant professor) at the Chittagong University of Engineering and Technology. His research interests are algorithmic complexity analysis, data mining, machine learning, big data, and information visualization. He is the initiator and lead visionary of a research leveraging platform named "Be Researcher World Forum."

The Implications of Energy Transition and Development of Renewable Energy on Sustainable Development Goals of Two Asian Tigers

Rajib Bhattacharyya

1 Introduction

Of late, the transition of global energy resource base and shift towards de-carbonization has been one of the most critical and sensitive issues in the debates and discussions on sustainable development goals index (SDGI), climate change and geopolitical policies to establish economic power and supremacy over other nation. The impact of energy transition not only limited within the energy sector of an economy, but it involves a whole gamut of changes over the entire economy as it includes issues like substitution of fossil fuels, electrification, de-carbonization, technological upgradation, which may have far-reaching implications on agriculture, industry, services, infrastructure and others. It also involves changes in trade, fiscal and labour market policies. The issue of energy transition from fossil fuel to wind, solar, hydro and zero-carbon energy has serious implications in the context of energy crisis, energy affordability, security and sustainability in the long run. The Energy Transition Index (ETI) attempts to measure the emerging landscape of the performance of energy systems and readiness for energy transition across countries. China and India, the two populous giants, are highly vulnerable to climate change, but have shown significant improvements in ETI performances.

The consumption of cleaner energy per capita is not only considered as an important indicator of good and healthy life but also a significant component of a nation's green GNP measure of growth. Innovations in technology, mitigation policies of climate change, achievement of Sustainable Development Goals (SDGs) and geopolitical changes have brought into focus the issue of transition in the global energy system. The two recent macroeconomic shocks (Covid-19 and the Russia-Ukraine War) had resulted in serious disruptions to movement of goods and energy

R. Bhattacharyya (✉)
Goenka College of Commerce and Business Administration, Kolkata, India

M. Z. Abedin, W. Yong (eds.), *Machine Learning Technologies on Energy Economics and Finance*, International Series in Operations Research & Management Science 367, https://doi.org/10.1007/978-3-031-94862-6_13

and this has led to severe inflation in the global market. It has forced countries to rethink about energy crisis, security and sustainability in the long run. It has also pushed economies to reallocate its resource base to address the issue of its energy affordability and energy sustainability in the long run. This has been supplemented by the quest for renewable energy instead of the traditional fossil fuels and the strive towards net-zero-carbon emission. But this urge for energy transition has resulted in a geopolitical divide between developed and developing countries on the basis of priority in the process of transition. Energy prices shoot up when the global market recovery began after the end of the Covid-19 lockdown and the Russia–Ukraine war broke out. This excess demand was combined with insufficient supply and underinvestment in the energy market, mainly due to uncertainties and poor returns during shocks and also due to environmental, social and governance (ESG) factors on the part of the investors. The other challenge in this shifting process is the 'speed of transition'. Pressure has been imposed to bring down the carbon emission target from 2050 to 2030 to reduce the global warming by 1.5°c per year. Hence, the issues linked to energy transition are multi-dimensional: socio-economic (reduction of inequality and poverty along with economic growth), ecological (mitigating climate change) and geopolitical (concerned with national security and energy resource constraints). It is a dynamic adjustment towards an equilibrium where changing technology necessitates more demand for energy which has to be balanced by better and cleaner energy supply.

Hence, the conflict of interest between fossil fuel-rich countries and supporters of green energy nations. Countries like USA, Russia, Saudi Arabia, Canada, Australia, Venezuela, Brazil, Mexico, Iran, Iraq are fossil fuel-rich nations and if zero-carbon energy transition is achieved, these nations may lose in three ways: (i) capital loss due to large stock of fossil fuel will remain unexplored; (ii) additional economic loss due to the fact that fossil fuel rents will no longer be available to finance public sector; (iii) suffer positional loss due to geopolitical relative advantage and will be challenged by nations having solar, wind, hydro, geothermal, biomass, nuclear power.

1.1 Importance of India and China in Global Energy Transition Scenario

To make it compatible with the changing demands, a new formulation of Energy Transition Index (ETI) was released by the World Economic Forum (WEF), June 2023 edition, which takes a broader view of the energy triad: equity, security and sustainability. In the last decade there had been an improvement in global ETI score by around 10 percent with Nordic countries like Sweden, Denmark, Norway and Finland maintaining the top positions. This was supported by an enhancement in global scores of readiness transition by 19 percent. China is a soul exception in the global energy landscape to have shown an improvement in readiness transition scores by 43 percent (which is double the global average). This remarkable achievement made it possible for China as the only Asian country to enter into the top 10 performing nations in energy transition. It is also being pointed out in that report that the two countries—Singapore and India are making improvements in all aspects of energy system performance.

1.2 Major Issues Involved in Energy Transition

First, unlike the earlier energy transition driven by inter-fuel competition, the present energy transition is based on implementation of government policies and regulations with the prime aim of combatting global climate change. At the initial stage, the transition from hydrocarbon to renewable and carbon-free alternative may lead to failure of markets to price environmental externalities due to its high cost. The government support and investment policies may act as a catalyst through application of fiscal policy for the energy sector.

Second, technological transformation is required to achieve zero-carbon energy breakthrough. Solar power technologies, wind potential, nuclear reactors can radically reshape the global energy landscape.

Third, financial investment is one of the prime factors that drive technological innovations and government policy. A proper balance between public and private investment is the key to green energy transition as per the International Energy Agency (IEA). The Central banks initiative in the form of Network for Greening the Financial System (NGFS) is a step in this regard.

Fourth, development of energy networking and infrastructure is very crucial for this transition. This implies developing new infrastructure to supply de-carbonized energy and replacing the older pipeline network shipping fleets and distribution outlets. Energy integration system and digitalization process need to be given more focus along with developing new infrastructural base.

Fifth, energy access equity and justice are really an important matter of concern. With about 2.6 billion people still deprived access to clean cooking fuels (World Energy Outlook, 2020), energy transition may exacerbate energy inequality and poverty. Hence energy justice was a major issue in COP26.

2 Literature Survey

A large body of literature has developed on the energy transition and climate change issues. Here we mention some of the notable ones. Hafner and Tagliapietra (2020)s edited book on 'The Geopolitics of the Global Energy Transition' has been a pioneer in discussing the geopolitical impacts between developed and developing nations with regard to transition from fossil fuel to zero-carbon state. The shifting of power from coal, oil, natural gas to solar, wind, hydro, geothermal, biomass, nuclear requires not only a change in innovation, infrastructure, education and human capital, finance and investment, but political will, commitment and implantation. Singh et al. (2019) explored how ETI serves as composite and comprehensive world index which tracks the country wise performance of the energy system and has implications on macroeconomic, institutional, social and geopolitical levels. It also shows the direction and potentiality of a country to make transition. Yergin (2022) demonstrated the probable obstacles in the path of reducing net carbon emission to zero. Henderson and Sen (2021) discussed the major challenges of nations in the path of transition to a new system. They stress on the Intergovernmental Panel on Climate Change (IPCC) report on climate Change and IEA analysis. IRENA (2018) report talks about transition on global energy system with a road map 2050. It emphasizes energy efficiency and renewable energy as the two fundamental pillars of energy transition and stresses the need for scaling up renewable energy needs at a six times faster rate to meet the aims of the Paris Agreement. Energy Statistics India (2023), the report of the Government of India provides a theoretical and empirical comprehensive overview of India's step forward towards energy transition. It highlights the importance of deploying renewable and energy efficient technologies in line with the UN Summit 2015. World Economic Forum (2023) report June 2023 analyses the details of the ETI scores and rankings, performance of sub-indices and country performance profiles. World Energy Outlook (2023) describes the various dimensions of the energy transition and also the key challenges ensuring a just and secure clean energy transition. World Energy Trilemma (2024) published a full report on 'Evolving with Resilience and Justice' to focus on the World Energy Trilemma Index depending on three core indicators: energy security, energy equity and environmental sustainability of Energy Systems. It discusses about the multiple paths followed by different countries to ensure cleaner, affordable and reliable energy framework, pointing out the deficiencies in supply infrastructure and investment. Janardhanan (2022) in his paper tried to examine three dimensions of the role of China in India's energy transition: (a) identification of factors responsible for China's comparative advantage and dominance in the overseas market, (b) China's dual role (catalytic and inhibiting role) in the process of energy transition in India and (c) Scope and opportunities of China–India bilateral ventures in the development of clean energy. Mori (2022) in his book, through the various chapters, has extensively discussed China's carbon policy (leakage, relocation) and its role in the energy transition in different countries of Asia like Japan, Vietnam, Indonesia and India. The book tries to assess how the policy intensifies pressure and motivates the Chinese companies. Isoaho et al. (2016) in their paper attempted to focus on transformation of electric power system in both India and China to decouple economic growth from unsustainable resource consumption. It tries to analyse the whole issue from the political economy angle. Odhiambo (2009) in his paper tried to examine the intertemporal causal relationship between energy consumption and economic growth in Tanzania during the period of 1971–2006 using ARDL bound testing model and found the relationship to be a stable one. Bhattacharyya (2019) examines the pattern and composition of energy use in two most populous countries, China and India. It also uses ARDL bound test to establish the short and long run relationship between energy use, per capita GDP, energy intensity, electric power and extent of urbanization. The study found long run association between the variables in case of India but not in case of China.

3 Objective of the Study and Methodology

The present paper attempts to provide various macro dimensions of indicator wise comparison and explores the potentials for renewable energy between the two fastest growing emerging nations—India and China. Empirically it tries to estimate the impact of some macroeconomic variables, viz. economic growth (ECOGR), access to clean fuels and technologies for cooking (% of population) (CFT), CO_2 emissions from fossil fuel combustion and cement production or manufacturing (CM) (tCO2/capita, 2021), CO_2 emissions from fuel combustion per total electricity output (CME) (Mt CO_2/TWh, 2019) and renewable energy share in total final energy consumption (%, 2019) (RE) on the Sustainable Development Goal Index (SDGI) using the ARDL model. The study also seeks to examine the uni-directional and bi-directional short run causality between the dependent and independent variable in terms of pairwise Granger Causality test with the help of time series data available from WDI, WEF, WEO, GSIR.

4 The Energy Transition Index Framework

As per the methodologies developed by the World Economic Forum (WEF, 2023), the basic aim of the ETI tool is to assess two fundamental issues: (a) a nation's present state of energy system performance and (b) preparedness for energy transition. ETI is based on two main pillars: (i) Energy System Performance (weight = 60 percent) and (ii) Readiness for Energy Transition (weight = 40 percent). Again, system performance depends on three sub-pillars (weight = 33 percent each): (a) Equitability (including energy access, affordability economic development) (b) Security (supply security, resilience, reliability) and (c) Sustainability (energy efficiency, greenhouse gas (GHG) mitigation, clean energy). On the other hand, Readiness for Energy Transition depends on two sub-pillars (weight = 50 percent each): (a) regulatory framework and investment and (b) enabling factors (like education and human capital, innovation, infrastructure).

5 China and India's Standings and Role in ETI

Figure 1 shows the ETI scores and rankings across various regions of the world and also across nations. In the last decade, the region of Emerging and Developing Asia,

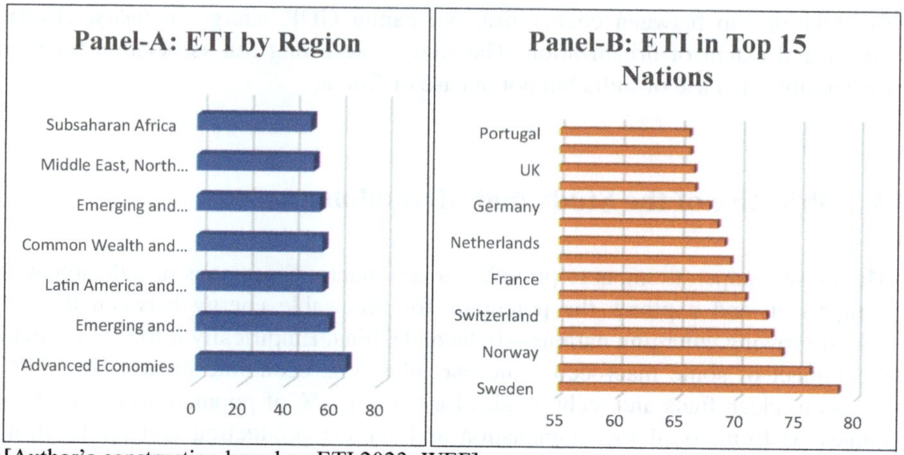

[Author's construction based on ETI 2023, WEF]

Fig. 1 Present scores of ETI and their ranking across regions and nations. [Author's construction based on ETI 2023, WEF]

which includes the two most populous countries of the world, China and India, has improved the ETI scores by 12 percent. This region has shown more than 10 percent improvement in equitability dimension, but the achievement is poor in respect of sustainability and security aspects. China's ETI score is 64.9 and its rank is 17th among 120 nations in 2023. China is the largest producer and consumer of energy in the world energy landscape and so its role is of vital importance for shaping the future trajectory. China has also been identified as the largest emitter of GHG. But it has improved greatly in both system performance and readiness transition aspects in the last decade. Though China had to face tremendous energy security challenge in the process of transition from fossil fuel to green energy, but in recent years its industry has been following the path of green development. It has moved a long way to introduce green finance to increase the supply of renewables and its investment was about 380 billion dollars in 2021. Moreover, it is one of the first nations in the world to launch the green bond project. One of the successful exhibitions is reflected in its industrial clusters where the powering is done by green and renewable electricity.

As compared to China, India is way behind in terms of ETI performance scores. India's score is 54.3 and its rank is 67th among 120 nations in 2023. But India has been highlighted as the country which has made significant improvement in all the three aspects: equitability, security and sustainability in the last decade. The main contributors to this success have been substitution of liquefied petroleum gas (LPG) in place of traditional wood, charcoal and others and also enhancement in the development of renewable energy. India has rapidly increased the share of renewable energy in power generation—more than 30 percent with solar and 92 percent of increased capacity by onshore wind. It has also set a target to install 500 GW of non-fossil power generation capacity by 2030. Indian government has introduced the Energy Conservation (Amendment) Bill 2022, which imposes the mandate to use renewable energy for big energy-intensive consumers and also initiated the carbon credit scheme. India also aims to develop a competitive 'Green Hydrogen' ecosystem and promote the production and distribution as well as consumption of green hydrogen through the policy of National Green Hydrogen Mission.

5.1 *A Comparative Performance of ETI and Its Components in China and India (2012–2023)*

Based on the time series data available from the WEF the present study has tried to analyse why China's performance is relative much better than India in terms of ETI scores and rankings.

For this we look separately into the progress with respect to the two pillars of ETI, i.e. energy system performance and preparedness for energy transition (Fig. 2). Throughout the entire period of our analysis, India's system performance was ahead of China, exception being years after 2021. But in case of readiness transition from the beginning China was ahead of India and the gap between them enlarged particularly after 2021 (post pandemic situation). This combined effect has helped China to improve its rank much faster than that of India.

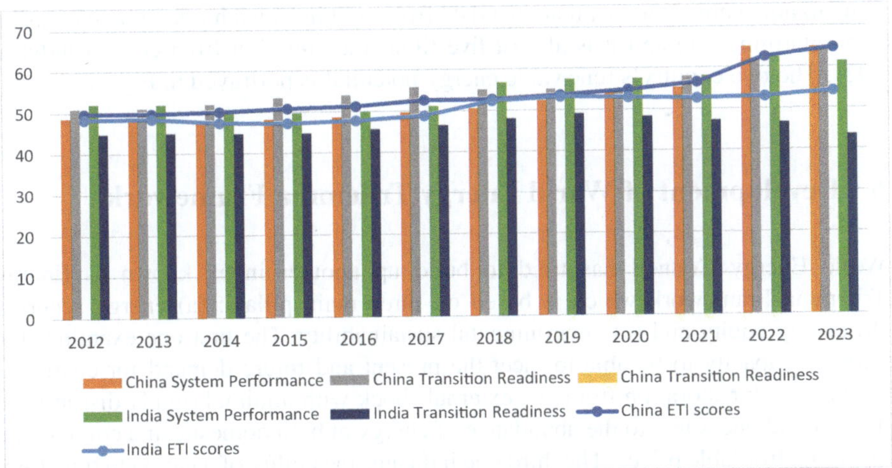

[Source: Author's construction based on ETI scores available from WEF dataset.]
[https://www.weforum.org/publications/fostering-effective-energy-transition-2021/in-full/rankings/]

Fig. 2 ETI and its components in China and India (2012–2023). [Source: Author's construction based on ETI scores available from WEF dataset.]. [https://www.weforum.org/publications/fostering-effective-energy-transition-2021/in-full/rankings/]

Figure 3 shows the comparison between China and India with respect to some indicators commonly used in the ETI calculations. Panel A in Fig. 3 shows carbon intensity in industrial consumption and panel B shows carbon intensity in road consumption. In case of industrial consumption, the carbon intensity of China was much more than that of India in the 1990s but gradually the gap has been narrowed down. It is measured in units kgCO2/2015USD [IEA data series (1990–2021)]. But the opposite is seen in case of carbon intensity road consumption. Here, India lies ahead of China. Though the gap narrowed between them in 2004, but after that it increased till 2021. This is because China's industrial activity is manyfold more than that of India, while China's transition to green energy transport has paid off dividends as compared to India. Panel C in Fig. 3 shows the comparison with regard to carbon emission per unit of GDP. In this case China was much ahead in 1990, but it went down sharply and finally surpassed India after that global financial crisis (2008). Panel D shows greenhouse gas (GHG) emission from various sources (coal, oil, natural gas and others).

The bottom panel E exhibits share of renewables, low carbon sources and fossil fuels (coal, oil and gas) in power of the two nations. In case of India the share of renewables and low carbon components in power has improved from 16 percent (1990) to 23.4 percent (2021, share of renewables) and from 16 percent (1990) to 27.2 percent (2021, share of low carbon). But in case of China the share of both has in fact fallen. Among all the sources the share of coal in power contributes to almost one-third. In case of India its share has been reduced between 1990 and 2021, but for China the share of coal has increased. The share of oil has decreased in both the nations but it has increased in case of gas for both the countries. One very important point to note here is the fact that the GHG from coal in China has kept on increasing at an alarming rate and it is almost five times the emission from coal in India in 2021. The two country's renewable energy potential is portrayed in Fig. 4.

6 Development of World Energy Trilemma Framework

World Energy Council has tried to build up another index known as Energy Trilemma Framework which is based on three main pillars: (a) energy security, (b) energy equity and (c) environmental sustainability. The first one examines the nation's capacity to be able to meet the present and future demand for energy as well as quick restoration from any external shock with minimal supply disruptions. The second one refers to the abundance of energy at both domestic and commercial levels at affordable prices. The third one indicates the ability of a nation to transform to a green energy system to mitigate environmental damage and climate change (Table 1).

Initially, when about 15 years ago this framework was developed it included variables like shocks on the supply side, ability to access scarce resource, energy efficiency, strategic reserves and exposure to commodity prices, but now it includes newer variables like demand-driven energy shocks (as faced by Europe after the

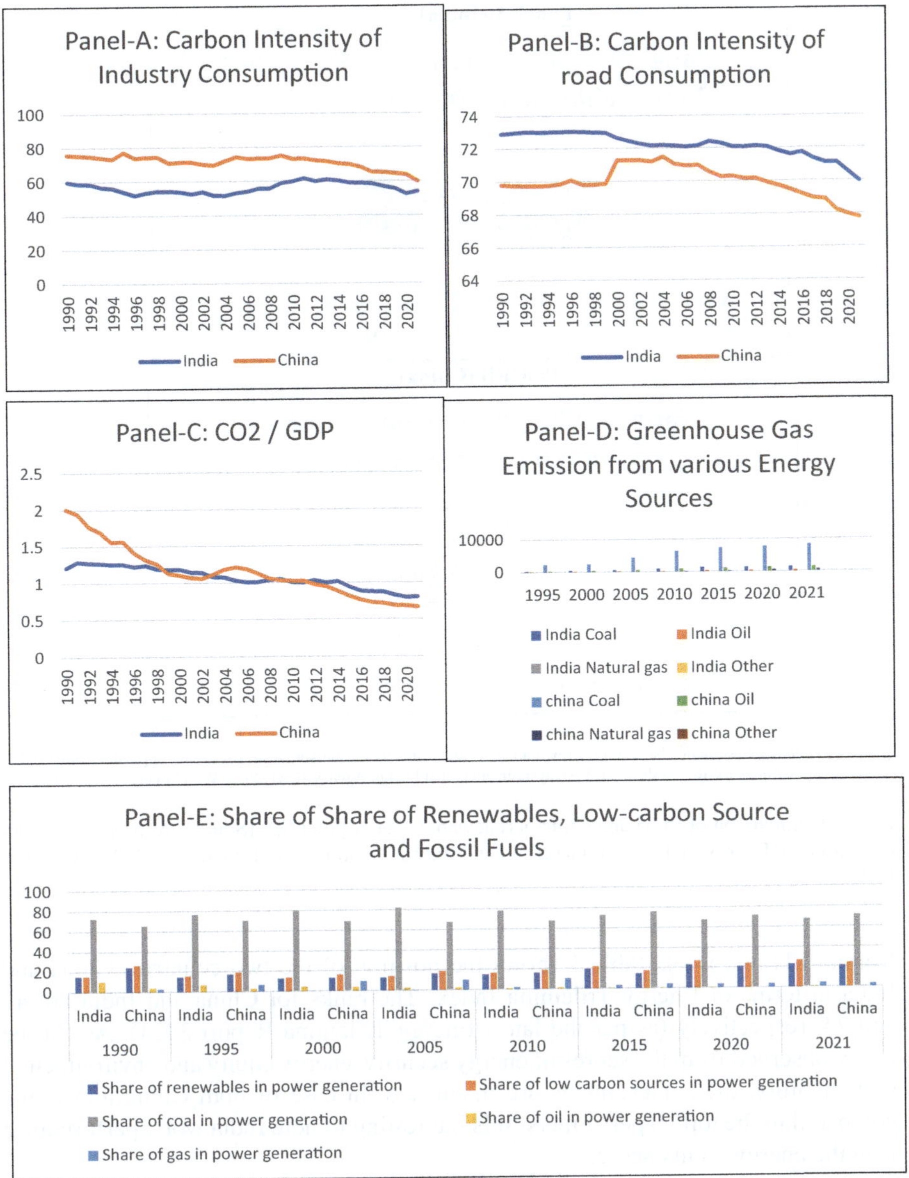

[Source: Author's construction based on IEA Data Services https://www.iea.org/data-and-statistics/data-product/world-energy-statistics-and-balances]

Fig. 3 Comparison between China and India (indicator wise). [Source: Author's construction based on IEA Data Services. https://www.iea.org/data-and-statistics/data-product/world-energy-statistics-and-balances]

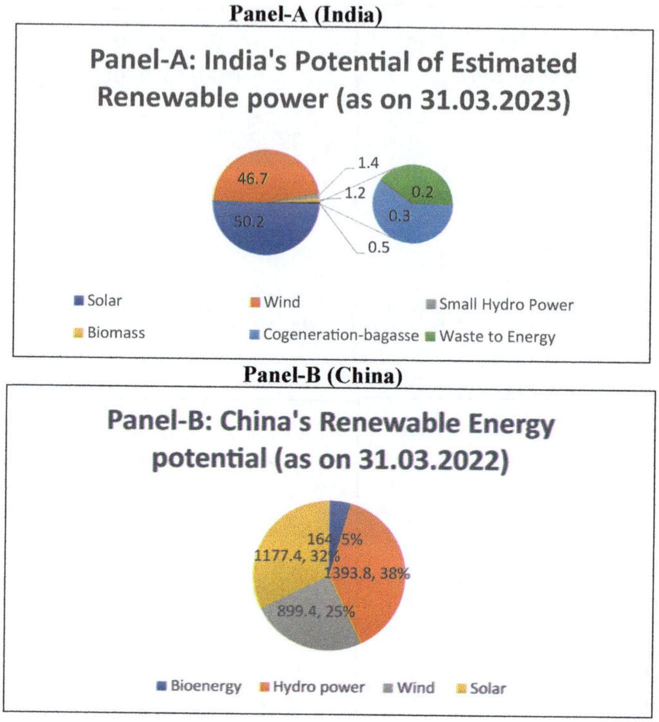

[**Source:** Author's construction based on Energy Statistics India. Ministry of Statistics and Programme Implementation (2023) and Executive Summary-Renewables 2023-Analysis-International Energy Agency(IEA) (2023) World Energy Outlook]

Fig. 4 Comparison of India and China's renewable energy potential. [Source: Author's construction based on Energy Statistics India (2023) and Executive Summary- Renewables 2023- Analysis-IEA]

Russia–Ukraine war). Table 1 shows the position of the two countries China and India in terms of Energy Trilemma Index. The ranks for China and India are 47 and 74, respectively (as per the latest Energy Trilemma Report 2024). As can be easily observed from the scores of energy security, energy equity and environmental sustainability, the achievements are much less, in case of both China and India, compared to the top 10 performers. It is interesting to note India worst performance is in the energy equity score.

Table 1 China and India's position in Energy Trilemma Index

Energy Trilemma index rank	Country name	Trilemma score	Energy security score	Energy equity score	Environmental sustainability score
1.	Denmark	83.2	72.2	95.8	83.5
1.	Sweden	83.1	73.4	93.4	85
2.	Finland	82.7	75.9	92.3	80.8
3.	Switzerland	82.1	64.5	98.1	85.7
4.	Canada	81	76.6	96.2	72.8
5.	Austria	80.9	71.8	95.3	78.6
6.	France	80.6	69.4	93.7	83.2
7.	Estonia	80.2	69.9	94.8	78.5
7.	Germany	80.2	72.9	94.4	76.6
8.	UK	80	67.7	95.7	79.2
8.	Norway	79.9	62.7	94.4	84.3
9.	New Zealand	79.6	68.2	95.4	76.4
10	US	78.9	72.7	97.3	69
47.	China	64.4	66.3	73	56.4
74.	India	55.6	61.7	49.5	56.5

Source: https://trilemma.worldenergy.org/#!/energy-index, World Energy Council

7 Empirical Analysis

Here we try to estimate the impact of economic growth (ECOGR), access to clean fuels and technologies for cooking (% of population) (CFT), CO_2 emissions from fossil fuel combustion and cement production or manufacturing (CM) (tCO2/capita, 2021), CO_2 emissions from fuel combustion per total electricity output (CME) (Mt CO_2/TWh, 2019) and renewable energy share in total final energy consumption (%, 2019) (RE) on the sustainable development goal index (SDGI) using the ARDL-ECM framework.

7.1 Autoregressive Distributed Lag (ARDL) Approach

Now after checking for unit root, we proceed for the testing of co-integration between the variables, based on ARDL framework. Pesaran et al. (2001) suggested the autoregressive distributed lag (ARDL) approach to test for co-integration as an alternative to co-integration model for Engle-Granger (1989). The ARDL-ECM model has been developed to check both long run and short run relationship between dependent variables, i.e. sustainable development goal index (SDGI) denoted by SDGI and the five explanatory variables are economic growth (ECOGR), access to clean fuels and technologies for cooking (% of population) (CFT), CO_2 emissions from fossil fuel combustion and cement production or manufacturing

Table 2 Stability analysis (results of LM test)

Country	Test result (Breusch-Godfrey serial correlation LM test)				CUSUM test
India	F-statistic	44.13297	Prob. F (2,4)	0.0951	Stable
	Obs*R-squared	17.60118	Prob. Chi-Square(2)	0.0876	
China	F-statistic	38.37184	Prob. F (2,4)	0.0697	Stable
	Obs*R-squared	20.91013	Prob. Chi-Square(2)	0.0671	

(CM) (tCO2/capita, 2021), CO_2 emissions from fuel combustion per total electricity output (CME) (Mt CO_2/TWh, 2019) and renewable energy share in total final energy consumption (%, 2019) (RE). In general, the ARDL restricted error correction model (RECM) is shown below. We have taken an unrestricted ARDL model with no trends with 2 lags and estimate the following equation:

For India

Estimated equation:

$$d\,(SDGI\,(-2)) = \mathrm{c} + \alpha 0\, d\,(ECOGR\,(-1)) + \alpha 1\, d\,(ECOGR\,(-2))$$
$$+\beta 0\, d\,(CFT\,(-1)) + \beta 1\, d\,(CFT\,(-2)) + \delta 0\, d\,(CM\,(-1))$$
$$+\delta 1\, d\,(CM\,(-2))\;\; \gamma 0\, d\,(CME\,(-1)) + \gamma 1\, d\,(CME\,(-2)) + \mu 0\, d\,(RE\,(-1))$$
$$+\mu 1\, d\,(RE\,(-2)) + \theta 0\;\,(ECOGR\,(-1)) + \theta 1\;\,(CFT\,(-1))$$
$$+\theta 2\,(CM\,(-1)) + \theta 3\,(CME\,(-1)) + \theta 4\,(RE\,(-1))$$

For China

Estimated equation:

$$d\,(SDGI\,(-2)) = \mathrm{c} + \alpha 0\, d\,(ECOGR\,(-1)) + \beta 0\, d\,(CFT\,(-1))$$
$$+\beta 1\, d\,(CFT\,(-2)) + \delta 0\, d\,(CM\,(-1)) + \delta 1\, d\,(CM\,(-2))\;\; \gamma 0\, d\,(CME\,(-1))$$
$$+\mu 0\, d\,(RE\,(-1)) + \mu 1\, d\,(RE\,(-2)) + \theta 0\;\,(ECOGR\,(-1))$$
$$+\theta 1\;\,(CFT\,(-1)) + \theta 2\,(CM\,(-1)) + \theta 3\,(CME\,(-1)) + \theta 4\,(RE\,(-1))$$

The details of the results of these estimations are added in the appendix section (Section-A).

Then we check Residual Diagnostics (Serial correlation) using LM Test. Then stability analysis check is being performed using CUSUM Test (Table 2), and finally, in order to find whether there exists a long run association between the variables.

Wald test is performed. For checking short run relationship, we have incorporated the error term [ECT (−1)] from our basic long run model and again estimated our model with 2 lags. The error term indicates the speed of adjustment towards long run equilibrium. Again, serial correlation is tested for short run model using LM test. Then long run causality is checked for each of the five independent variables—

ECOGR, CFT, CM, CME and RE using Wald test. This has been done separately for the two countries, India and China.

7.2 *Checking Long Run Association Between Variables*

Here our objective is to determine whether there exists any long run association between SDGI, ECOGR, CFT, CM, CME and RE. This has been tested through Wald test (Table 3). *F* test is used to determine whether the long run relationship exists between the variables through testing the significance of the lagged levels of the variables. When the long run relationship exists, the *F* test will show which variable should be normalized. The null hypothesis of no co-integration amongst the variables in estimated equation, i.e., coefficients of SDGI (-1), ECOGR (-1), CFT (-1), CM (-1), CME (-1) and RE (-1) are all zeros.

We discover a very interesting difference in the long run association between the variables in the two countries. In case of India there is no long run association between the independent variables, i.e. ECOGR, CFT, CM, CME, RE and the dependent variable SDGI, but in case of China the long run association exists. So, the ETI does have a significant long run impact on SDGI in China, but not in case of India and this probably explains why China has so rapidly improved its SDGI rank to move to the 66th position while India is still lagging behind in the 112th

Table 3 Long run association between variable (Wald test)

Country	Test result for Wald test				Interpretation
India	Test statistic	Value	df	Probability	*F-Stat* 0.304482 < *lower Bound 3.79.* SDGI, ECOGR, CFT, CM, CME and RE *has no long run association*
	F-statistic	0.304482	(5, 2)	0.8772	
	Chi-square	1.522412	5	0.9105	
	Null Hypothesis: C(12) = C(13) = C(14) = C(15) = C(16) = 0				
China	Test statistic	Value	df	Probability	*F-Stat* 14.58679 > *upper Bound 4.85.* SDGI, ECOGR, CFT, CM, CME and RE *has long run association*
	F-statistic	14.58679	(4, 4)	0.0118	
	Chi-square	58.34715	4	0.0000	
	Null Hypothesis: C(11) = C(12) = C(13) = C(14) = 0				

position. So, India is unable to convert the short run association to a long run one which actually China was able to do.

7.3 Modified Model after Incorporating the Error Term

For India

$$d\,(SDGI\,(-2)) = \mathrm{c} + \alpha 0\, d\,(ECOGR\,(-1)) + \alpha 1\, d\,(ECOGR\,(-2)) \\ +\beta 0\, d\,(CFT\,(-1)) + \beta 1\, d\,(CFT\,(-2)) + \delta 0\, d\,(CM\,(-1)) \\ +\delta 1\, d\,(CM\,(-2))\;\; \gamma 0\, d\,(CME\,(-1)) + \gamma 1\, d\,(CME\,(-2)) \\ +\mu 0\, d\,(RE\,(-1)) + \mu 1\, d\,(RE\,(-2)) + \theta 0\;\,(ECT\,(-1))$$

For China

$$d\,(SDGI\,(-2)) = \mathrm{c} + \alpha 0\, d\,(ECOGR\,(-1)) + \beta 0\, d\,(CFT\,(-1)) \\ +\beta 1\, d\,(CFT\,(-2)) + \delta 0\, d\,(CM\,(-1)) + \delta 1\, d\,(CM\,(-2))\;\; \gamma 0\, d\,(CME\,(-1)) \\ +\mu 0\, d\,(RE\,(-1)) + \mu 1\, d\,(RE\,(-2)) + \theta 0\;\,(ECT\,(-1))$$

The details of the regression results after incorporating the error term are shown in appendix, Section B. This shows that ECT (-1) is negative and statistically significant and the speed of adjustment towards long run equilibrium is 72.64 percent and 70.42 percent for India and China, respectively.

7.4 Short Run Association Between Variables

In the short run we have incorporated the error term ECT (-1) as one independent variable and again checked for serial correlation using LM test. Here, from LM test we have found that there is no serial correlation and the CUSUM test is also stable.

Finally, Wald test has been applied to check whether there exists short run causality. The test finds that in case of India there exists short run causality from CFT and CM to SDGI, but in case of China the short run causality only exists from CFT to SDGI and not for other independent variables.

Table 4 Results of pairwise Granger causality test

India	Null hypothesis:	F-statistic	Probability.	Granger causality test result
	ECOGR does not Granger Cause SDGI	2.68441	0.1031 > 0.05	Accept null; no causality
	SDGI does not Granger Cause ECOGR	4.58768	0.0294 < 0.05	Reject null; causality exists
	CFT does not Granger Cause SDGI	6.46662	0.0103 < 0.05	Reject null; causality exists
	SDGI does not Granger Cause CFT	0.18536	0.8328 > 0.05	Accept null; no causality
	CM does not Granger Cause SDGI	3.83632	0.0469 < 0.05	Reject null; causality exists
	SDGI does not Granger Cause CM	11.7302	0.0010 < 0.05	Reject null; causality exists
	CME does not Granger Cause SDGI	2.72547	0.1027 > 0.05	Accept null; no causality
	SDGI does not Granger Cause CME	6.00564	0.0142 < 0.05	Reject null; causality exists
	RE does not Granger Cause SDGI	0.90833	0.4273 > 0.05	Accept null; no causality
	SDGI does not Granger Cause RE	3.30217	0.0692 > 0.05	Accept null; no causality
China	**Null hypothesis:**	**F-statistic**	**Probability**	**Granger causality test result**
	ECOGR does not Granger Cause SDGI	3.04115	0.0800 > 0.05	Accept null; no causality
	SDGI does not Granger Cause ECOGR	3.78708	0.0485 < 0.05	Reject null; causality exists
	CFT does not Granger Cause SDGI	4.33524	0.0343 < 0.05	Reject null; causality exists
	SDGI does not Granger Cause CFT	3.04940	0.0796 > 0.05	Accept null; no causality
	CM does not Granger Cause SDGI	2.95392	0.0851 > 0.05	Accept null; no causality
	SDGI does not Granger Cause CM	0.91583	0.4229 > 0.05	Accept null; no causality
	CME does not Granger Cause SDGI	0.31494	0.7352 > 0.05	Accept null; no causality
	SDGI does not Granger Cause CME	3.58234	0.0576 > 0.05	Accept null; no causality
	RE does not Granger Cause SDGI	2.12346	0.1592 > 0.05	Accept null; no causality
	SDGI does not Granger Cause RE	4.84394	0.0268 > 0.05	Reject null; causality exists

7.5 *Pairwise Granger Causality Test*

Now we want to see the short run causality between the dependent and independent variables in terms of pairwise Granger causality test. The results of this test are summarized below in Table 4.

The results of the pairwise Granger causality (Table 4) show that for India there are uni-directional causality running from SDGI to ECOGR, CFT to SDGI and SDGI to CME and bi-directional causality between CMI and SDGI. But for China there is only uni-directional causality running from SDGI to ECOGR, CFT to SDGI and SDGI to REC.

8 Policy Implications and Conclusion

The whole world is passing through a critical juncture of time. In the last two decades it has been hit hard by the two crises—the global financial crisis and the Covid 19 crisis. After that the revival process again received a setback due to the Ukraine war. It caused severe disruption in movements of goods and energy. This caused energy prices to shoot up steadily and pushed the inflation rate to rise faster.

But the transition of the globe from the use of fossil fuel to de-carbonized cleaner energy requires not only a shift in technology, infrastructure, skill development, but also involves a transformation of the occupational labour force and society. It may lead to financial loss of existing fossil fuel-rich nations as their resource base will not be further utilized for energy consumption as well as financing the government expenditure. So, transition to a zero-carbon state is a long-term affair involving decisions on various segments of the economy as well as geopolitical events shaping the global scenario.

References

Bhattacharyya, R. (2019). Changing patterns of energy use and its linkage with some macroeconomic variables in India and China. In R. C. Das (Ed.), *Handbook of research on economic and political implications of Green trading and energy use* (AFAE book series) (pp. 163–180). IGI Global.

Hafner, M., & Tagliapietra, S. (2020). The geopolitics of the global energy transition, lecture notes in energy. Springer Open (73). Available at: http://www.springer.com/series/8874.

Henderson, J., Sen, A. (2021). The energy transition: Key challenges for incumbent and new players in the global energy system, The Oxford Institute for Energy Studies (OEIS) paper, ET:01, September.

International Energy Agency (IEA). (2023) World Energy Outlook.

IRENA. (2018). Global energy transformation: A roadmap to 2050, International Renewable Energy Agency, Abu Dhabi. Available at www.irena.org/publications.

Isoaho, K., Goritz, A., & Schulz, N. (2016). Governing clean energy transitions in China and India: A comparative political economy analysis, WIDER Working Paper: United Nations University World Institute for Development Economics Research, (Vo. 28, pp. 1–40).

Janardhanan, N. (2022). India's energy transition: Is China an inhibitor or a catalyst? In China's carbon-energy policy and Asia's energy transition: Carbon leakage, relocation and Halos by Akihisa Mori. (Vol. 9, pp. 195–209). Routledge.

Ministry of Statistics and Programme Implementation. (2023). National Statistical Office, Government of India, Energy Statistics India. Available at https://www.mospi.gov.in/.

Mori, A. (2022). *China's carbon-energy policy and Asia's energy transition:carbon leakage, relocation and Halos*. Routledge.

Odhiambo, N. M. (2009). Energy consumption and economic growth nexus in Tanzania: An ARDL bounds testing approach. *Energy Policy, 37*, 617–622.

Pesaran, M. H., Shin, Y., & Smith, R. J. (2001). Bounds Testing Approaches to the Analysis of Level Relationships. *Journal of Applied Econometrics, 16*(3), 289–326. http://www.jstor.org/stable/2678547.

Singh, H. V., Bocca, R., Gomez, P., Dhalke, S., & Bazilian, M. (2019). The energy transitions index: An analytic framework for understanding the evolving global energy system. *Energy Strategy Reviews, 26*, 1–10.

World Economic Forum. (2023). Fostering effective energy transition, Insight Report, Staff Report, June.

World Energy Council World Energy Trilemma. (2024). Evolving with Resilience and Justice. Available at: https://trilemma.worldenergy.org.

Yergin, D. (2022). Bumps in the energy transition, the scramble for energy, finance and development, December: 8–13.

Rajib Bhattacharyya is an Associate Professor (W.B.E.S.-A) in the Department of Economics, Goenka College of Commerce and Business Administration, Kolkata, India. He obtained his bachelor's degree in Economics from St. Xavier's College, Kolkata and his master's degree in Economics, from the University of Calcutta. He did his M.Phil. and Ph.D. in Economics, both from the University of Calcutta. He has 22 years of teaching experience at the UG and PG levels. His fields of interest are international trade, finance, Indian economic development, and women empowerment. He has contributed a good number of articles in reputed national and international journals/books, viz. IGI Global (USA), Emerald (UK), Taylor & Francis and Springer. He has delivered invited lectures in many UGC sponsored seminars/webinars/conferences. He has also worked as a member of the Editorial Board of various national and international level journals. He had been conferred the best paper award in an international seminar on 'Recent Trends and Perspectives in Economics' on January 2016. He has served as an invited reviewer in Indian Economic Journal (Sage), Foreign Trade Review (Sage), Journal of Asian and African Studies (Sage), International Journal of ICT Research and Development in Africa (IGI Global). He also has the experience of guiding Ph.D. students.

Recently Edited Books

1. The Indian Economy – Issues, Policies and performance (1st Edition, 2024)—Authored by Dr. Rajib Bhattacharyya and Dr. Ananya Ghosh Dastidar (Routledge, London) (imprint-Routledge, India), Taylor and Francis Group.
2. COVID-19 Pandemic and Global Inequality- Reflections in Labour Market, Business and Social Sectors—Edited by Rajib Bhattacharyya · Ramesh Chandra Das Achintya Ray (Springer Nature, Singapore, 2023).
3. The Covid-19 Pandemic, India and the World- Economic and Social Policy Perspective—Edited by Rajib Bhattacharyya, Ananya Ghosh Dastidar and Soumyen Sikdar. Routledge London and New York (Taylor and Francis (Sept. 2021).
4. Comparative Advantage in the Knowledge Economy: A National and Organizational Resource—Edited by Dr. Rajib Bhattacharyya [Emerald Publishing Limited, UK] [8th June 2021].
5. Role of IT-ITES in Economic Development of Asia-published by Springer Nature, Singapore, 2020—Edited by Prof, Soumyen Sikdar and Prof. Ramesh Das and Rajib Bhattacharyya.
6. 'Gains and Pains of Financial Integration and Trade Liberalization – Lessons from Emerging Economies'—Edited by Dr. Rajib Bhattacharyya (Emerald Publishing Limited, UK).

Member of the editorial advisory boards of the following journals:

- *Handbook of Research on Globalization, Investment, and Growth-Implications of Confidence and Governance* (pp. 1-526). Hershey, PA: IGI Global. doi:https://doi.org/10.4018/978-1-4666-8274-0.
- *Handbook of Research on Global Indicators of Economic and Political Convergence* (pp. 1-614). Hershey, PA: IGI Global. doi:https://doi.org/10.4018/978-1-5225-0215-9.

- *International Journal of Sustainable Economics Management* (Special Issue), IGI Global, Hershey, PA, USA, Volume-6, Issue-1, Jan-Mar, 2017, ISSN: 2160-9659 • e -ISSN: 2160-9667.
- *Handbook of Research on Economic, Financial, and Industrial Impacts on Infrastructure Development*, IGIGlobal, April, 2017; Hershey, PA, USA.
- *Handbook of Research on Military Expenditure on Economic and Political Resources*; IGIGlobal Hershey, PA, USA. [Projected Release Date: May, 2018].
- *Handbook of Research on Micro financial Impacts on Women Empowerment, Poverty, and Inequality*; IGI Global Hershey, PA, USA. [Projected Release Date: August, 2018].
- *International Journal of Research on Social and Natural Sciences* (IJRSNS) Volume 1, Issue 1, December 2015.
- *Asian Journal of Research in Business Economics and Management*, Asian Research Consortium (ISSN: 2249-7307 (Online), 2250-1673(Print).
- *International Journal of ICT Research and Development in Africa*, Hershey, PA: IGI Global.
- Invited to serve as a Reviewer in 'Journal of Asian and African Studies' of Sage Publishing House in 2019(PUBLONS).
- Invited Reviewer of Indian Economic Journal (SAGE).
- Invited Reviewer of Foreign Trade Review (SAGE).

9783031948619